Industrial Change in the United Kingdom

Industrial Change in the United Kingdom

Edited by William F. Lever

Longman Scientific & Technical
Longman Group UK Limited
Longman House, Burnt Mill, Harlow
Essex CM20 2JE, England
and Associated companies throughout the world.

© *Industrial Activity and Area Development Study
Group of the Institute of British Geographers 1987*

All rights reserved; no part of this publication may be reproduced, stored in a retrieval system, or transmitted in any form or by any means, electronic, mechanical, photocopying, recording, or otherwise, without the prior written permission of the Publishers.

First published 1987

British Library Cataloguing in Publication Data

Industrial change in the United Kingdom.
  1. Great Britain – Industries – Location
I. Lever, W.F.
338.6′042′0941   HC260.D5

ISBN 0-582-30141-6

Library of Congress Cataloging-in-Publication Data

Industrial change in the United Kingdom.

  Bibliography: p.
  Includes index.
  1. Great Britain – Industries – Location.   2. Industry and state – Great Britain.   3. Technological innovations – Economic aspects – Great Britain.   I. Lever, William F.
HC260.D5153   1987   338.0941   82-2777
ISBN 0-582-30141-6

Set in 9/11 pt Linotron 202 Times Roman
Produced by Longman Singapore Publishers (Pte) Ltd.
Printed in Singapore.

CONTENTS

	List of Figures	vii
	List of Tables	ix
	Preface	xi
	Introduction	xiii
	Acknowledgements	xv
Chapter 1	Industrial change in the United Kingdom *David Keeble*	1
	Introduction to Industrial Location Theory	21
Chapter 2	Neoclassical location theory *David M. Smith*	23
Chapter 3	Behavioural approaches to industrial location studies *Peter A. Wood*	38
Chapter 4	Structural approaches to industrial location *Martin J. Boddy*	56
	Introduction to the Factors of Production	67
Chapter 5	Labour and capital *William F. Lever*	69
Chapter 6	Industrial buildings and economic development *Steve Fothergill, Michael Kitson and Sarah Monk*	86
Chapter 7	Technological change *John B. Goddard and Alfred Thwaites*	96
Chapter 8	Industrial change, linkages and regional development *J. Neill Marshall*	108

Introduction to Industrial Enterprise 123

Chapter 9   The small firm sector
Colin M. Mason   125

Chapter 10   The multiplant enterprise
Michael J. Healey and H. Douglas Watts   149

Chapter 11   Multinational enterprises
F. E. Ian Hamilton   167

Chapter 12   Public sector industries
Graham Humphrys   196

Introduction to Policies for Industry 209

Chapter 13   National policy
William F. Lever   211

Chapter 14   Regional policy
Alan R. Townsend   223

Chapter 15   Urban policy
William F. Lever   240

Index   258

LIST OF FIGURES

1.1	UK manufacturing employment and output trends 1964–83	1
1.2	Small manufacturing establishments in the UK 1930–80	8
1.3	Manufacturing employment change 1971–78	11
1.4	Regional percentage shares of UK manufacturing employment, 1965–85	13
1.5	(a) High-technology manufacturing industry, 1981; (b) Research and development services, 1981	16
2.1	Weber's analysis of the effect of a cheap labour location	24
2.2	Lösch's transformation of the demand curve into market areas	27
2.3	The derivation of market areas, showing the impact of competition	27
2.4	Regional differences in market areas under different conditions	28
2.5	Derivation of the spatial margin to profitability from space cost and revenue curves	30
2.6	A surface of female labour costs in the components section of the British electronics industry	31
2.7	Optimum location and isodapanes for a fireworks factory	32
2.8	A simplified interpretation of the emergence of regions of industrial specialization in nineteenth-century England	33
2.9	Hypothetical space cost and revenue profiles for two industries in a nation with a metropolitan core and depressed periphery	34
2.10	A model of the spatial impact of industrial plants based on an analogy with market area analysis	34
2.11	A welfare formulation of a simple industrial development problem	35
3.1	Production policy selection, input needs and location	39
3.2	Corporate environments of the firm	41
3.3	Major directions and forms of transition within and between segments of the economy	51
5.1	Unemployment and vacancies: UK 1965–84	74
5.2	The UV curve in a local labour market	78
6.1	The stock of industrial floorspace in England and Wales 1964–82	87
9.1	Small firm competition winners	142
9.2	Companies on the Unlisted Securities Market (April 1985)	143
10.1	The multiplant enterprise: Cadbury Schweppes, 1984	152
10.2	The locational hierarchy	153
10.3	Corporate components of change in an accounting framework	155
10.4	Locational adjustment	157
10.5	Activity locations in a hypothetical multiplant enterprise	159
11.1	The locational pattern of employment change in GKN, 1978–82	181
14.1	Regional policy: assisted areas 1934–84	225
14.2	Regional policy: assisted areas 1982–84	232
14.3	Regional policy: assisted areas 1984	235
15.1	Enterprise Zones 1985	243
15.2	Economic policy status of local authorities in London, 1983	247

List of Figures

15.3	District unemployment rates 1971 and 1981: Clydeside	251
15.4	Area-based employment initiatives 1985: Clydeside	254

LIST OF TABLES

1.1	Manufacturing output growth and decline: the ten most extreme industries, 1975–81	4
1.2	Individual industry trends in output and employment: Great Britain, 1971–83	5
1.3	The urban–rural shift in UK manufacturing employment, 1971–81	10
1.4	Regional trends in manufacturing GDP, 1976–81	12
1.5	Regional trends in manufacturing employment, 1971–84	12
3.1	The segmented economy	49
5.1	Working population, UK	73
5.2	Sex composition of the workforce, 1974–84	73
5.3	Unemployment ratios, standard regions	76
5.4	UK capital investment	79
5.5	Average profitability and type of area	80
5.6	Net capital investment per head: Clydeside	81
5.7	Sources of capital for investment	81
5.8	Inward and outward capital flows, UK, 1972–83	81
6.1	Industrial floorspace by type of area: England and Wales 1967–82	88
6.2	Components of change in industrial floorspace by type of area: England 1974–82	88
6.3	The quality of the industrial building stock by type of area	90
6.4	Multiple regression analysis of establishment growth	92
6.5	Available industrial land by type of area, 1982	93
7.1	Trends in shares of substantial innovations by area	99
7.2	Percentage of innovations of foreign origin by receiving area	99
7.3	Establishment location by status: incidence of product innovation	100
7.4	Percentage of establishments with external sources of product innovation by region	101
7.5	Independent enterprise by on-site development of technology and location	101
7.6	Adoption of new technology by assisted area status	102
7.7	Adoption rates, by industrial sector and country	102
7.8	Characteristics of R and D employment by establishment location	103
7.9	Logit analysis of product innovation	104
7.10	On-site R and D facilities and urban status	105
7.11	The mean employment effect of product and process innovation	105
7.12	Estimated total annual regional impact of product and process innovation	106
8.1	Impact of organizational variables on linkages in provincial regions	111
8.2	The location of major motor vehicle component manufacturers, 1977–79	115
8.3	The spatial distribution of business service offices	117
8.4	Type of service organization used by manufacturing establishments	118
8.5	Location of service suppliers to manufacturing and business service industries in the Northern region by ownership status	119

List of Tables

9.1	The Bolton Committee's definition of small firms and revisions by the Wilson Committee	126
9.2	The distribution of small firms in the UK by sector, 1976	127
9.3	Small firms in the manufacturing sector, 1935–79	128
9.4	Proportion of manufacturing employment in small enterprises: international comparisons	128
9.5	New company registrations in Great Britain, 1971–84	129
9.6	New registrations on the VAT register 1974–82	130
9.7	Manufacturing employment change by size of firm	135
9.8	Regional distribution of new business starts; 1980–83	138
9.9	Regional comparisons of new manufacturing firm formation	139
9.10	Factors associated with low and high levels of new firm formation	140
10.1	Extent of multiplant operation in enterprises in the UK, 1958–79	149
10.2	Multiplant operation in the UK by industrial class, 1980	150
10.3	Distribution of unskilled workers within a corporate system: GEC Telecommunication, 1976	153
10.4	Net employment loss in selected large multiplant enterprises: 1978–79 to 1982–83	154
10.5	Urban–rural contrasts in the East Midlands, 1968–75	161
11.1	Estimated stock of direct investment abroad by area or country of origin 1914–78	169
11.2	World rankings of UK large industrial organizations within sectors 1982	170
11.3	The largest UK industrial organizations 1982	171
11.4	Changes in employment in the UK and overseas 1979–82 by sectors and leading UK-owned multinationals	175
11.5	Indices of the shift overseas of UK engineering firms 1973–82	179
11.6	Distribution of foreign-owned multinationals in manufacturing in the UK by country of origin 1963, 1975 and 1979	182
11.7	UK regional pattern of foreign-owned manufacturing 1979	185
11.8	Changing regional shares of foreign-owned manufacturing employment in the UK 1971–79	186
11.9	Foreign ownership in manufacturing in selected UK regions by country of origin, 1977	187
12.1	The major public sector enterprises in the UK	199
12.2	Employment in selected industries in the standard regions of Great Britain 1981	200
12.3	The regional distribution of public sector industry employment in Great Britain in 1981 and changes since 1971	201
12.4	UK coal mining by area 1982/83	204
13.1	Demand and output, 1978–82	212
13.2	Demand, output and prices, 1982–84	213
13.3	Composition of GDP, 1972–83	214
13.4	Visible trade of the UK	219
13.5	Manufacturing trade balance of the UK	219
13.6	Balance of imports and exports, by industrial sector	220
13.7	Export markets	220
13.8	Import sources	221
13.9	Balance of trade with Europe and the Commonwealth	221
14.1	Unemployment peaks and regional policy	224
14.2	Employment in interregional moves, 1945–75	227
14.3	The timing of major plant closures	231
14.4	Changes in regional policy, 1984	234
14.5	Estimates of change in employees in employment, 1981–84	238
15.1	Best and worst local unemployment rates, 1971 and 1981	241
15.2	Employment change in London	245
15.3	UDG allocations, London, 1983–84	246
15.4	Urban Programme funding and RSG loss, 1979/80–1983/84	248
15.5	Employment change in Clydeside	250
15.6	Small firm financial support in Clydeside/Scotland	252

PREFACE

Since the mid-1970s, the United Kingdom, in common with other industrialized European countries, has been experiencing one of the most traumatic periods of industrial change in its twentieth-century history. Drastic corporate restructuring and massive job losses by large firms in the face of deepening global recession have been accompanied by a resurgence in the number of small manufacturing businesses, while rapid technological change has nonetheless not been sufficient to prevent an alarming widening of the UK trade gap in high-technology products. This in turn is only one manifestation of growing internationalization of investment, production and control, via the increasing dominance of giant multinational corporations. UK government regional policies have been substantially weakened, whereas urban policy has been strengthened, within the context of new European Community policies on competition, regional and social development, and industrial innovation. The geographical impact of all these changes on manufacturing industry has been substantial, at both regional and urban scales, affecting different localities and communities in different ways.

These changes provide the context for this book. Its theme is the changing location of industrial activity in the United Kingdom in the 1980s, and the major theories, factors of production, types of enterprise and government policies, consideration of which is essential to any informed understanding of current trends in the country's industrial geography. The book thus systematically reviews such contemporary influences on UK industrial location as the interaction between capital and labour, technological change, the growth of small firms, multinational enterprise, and regional and urban policies, as well as succinctly summarizing three of the main approaches to industrial location – neoclassical, behavioural and structural – which are currently available for interpreting locational change.

This theme, and the book's origin, however both require some comment. After all, as Chapter 1 documents, the United Kingdom appears to be experiencing a drastic process of 'de-industrialization' or manufacturing decline. Why then does the book concentrate chiefly, though not exclusively, on the geography of manufacturing activity, with only occasional reference to sectors such as services?

One of several reasons for this concentration is that despite de-industrialization, the UK's manufacturing base remains of vital importance for national and regional economic prosperity, through its significance for overseas trade and domestic wealth-creation on which much service industry in turn depends. Trends in manufacturing investment, productivity, competitiveness and technological innovation are thus, and will remain, of crucial importance to UK national and regional economic development. Equally, the most dramatic structural and spatial changes in UK economic activity since 1970 have been in *manufacturing* employment, organization and investment, as traced in the chapters of this book, rather than in services. Contemporary processes of manufacturing change are indeed of such complexity and interest in themselves, involving both the largest and smallest of enterprises, and operating at all geographical scales from global to local, as to justify separate and in-depth analysis. The originality and vitality of recent theoretical and empirical research by British industrial geographers is also almost certainly greater than that in any other country, a claim supported by the novel insights and findings presented in this book.

Lastly, despite the growing importance in the UK of service industry, especially for jobs, relatively little research has so far actually been carried out on the nature and determinants or service industry growth and location. More work is urgently needed here, a judgement which explains the setting-up in 1984 of an Institute of British Geographers Working Party on Producer Services as a joint initiative by members of the Industrial Study Group (including authors of chapters of this book) and of other study groups of the institute. The Working Party's activities, in investigating *inter alia* the role of services as part of an increasingly complex process of production in industrialized economies, and changes in the pattern of control and supply of producer services in response to technological and corporate restructuring, should help considerably in this respect.

The origin of this book also deserves comment. It is the specific result of an initiative by the **Industrial Activity and Area Development Study Group** of the Institute of British Geographers, of which most its authors are members. The Study Group exists to promote theoretical, empirical and policy-related research into the changing geography of manufacturing and service industry and urban and regional economic development. But to be of value, academic research also needs to be disseminated. This book represents a deliberate attempt by the Study Group to make available to the widest possible audience – students, teachers, planners, researchers in other countries, and interested readers from whatever background – the fruits of very recent British geographical research in this field. The result, planned and produced under the able editorship of Professor William Lever, is an exceptionally informed and up-to-date appraisal of the key determinants and forces of industrial location change at work in the United Kingdom of the 1980s.

David Keeble
Chairman, Industrial Activity
and Area Development Study Group,
Institute of British Geographers

INTRODUCTION

As David Keeble has stressed in his Preface, the theme of this volume is industrial change. Contemporary commentators on all aspects of life have drawn attention to the observation not merely that the world is changing but that it is changing at an increasing rate. Some have suggested that change is occurring so rapidly that many people and many institutions have great difficulty in adjusting. No sooner have they absorbed the impact of one change than it is overtaken by another. This book is therefore a record of two elements of change – one of the change in the British industrial system, the other of the capacity of a group of academics to understand and to interpret those changes.

Accounts of the changes in British industry highlight several key dates in the post-war period. The 1945 Distribution of Industry Act initiated the post-war era of regional policy; the 1960 Local Employment Act completely recast regional policy; in 1963 and 1966 the four-year business cycle changed radically, so that a rise in unemployment could no longer be expected to be followed by a subsequent decline of equal magnitude; the 1973 rise in the price of crude oil threw the world into recession (although advocates of 'long wave theories' were able to point to precedents in the 1920s and in the 1870s which had nothing to so with crude oil prices), and a second price rise in 1979 had a similar effect.

Since 1979, however, the rate of change has accelerated; manufacturing industry in Britain and in much of the developed world has declined in output for two major reasons: the rise in the rate of the world's consumption of manufacturing products halted, and the proportion of the world's goods produced in the developed economies declined. In terms of employment, almost more important than the decline in the level of output was the increase in productivity per capita so that more goods could be produced by fewer workers. The consequence has been a major rise in unemployment. As geographers, from the 1960s onwards, have become increasingly concerned about the relevance of their work to contemporary problems, it was inevitable that industrial geographers in Britain would be drawn into the debate and be asked to advise the growing range of policy-making bodies addressing the issue of maintaining and renewing Britain's industrial base and assisting those sections of the workforce currently without employment.

This book is a part of that process. It begins with a general overview of industrial change, highlighting the key areas – the small firm sector, the urban – rural shift, the changing balance between manufacturing and services, the changing balance between the regions, and Britain's position in the global economy. There follows a group of three chapters which address the process of industrial location and change from three differing but complementary perspectives. A second major section, with four chapters, examines the major factors which influence manufacturing industry – labour and capital, land and premises, technology, and linkages. The 'geography of enterprise' is a phrase which has gained increasing currency in the last decade as geographers have turned their attention more to the individual firm, be it a major multinational, multiproduct corporation or a one-man operation, as the most instructive focus of study, rather than, say, an individual sector, product market or region. The third major section of this volume therefore turns to various types of enterprise and looks in turn at small firms, multiplant companies, multinationals and government-owned enterprises. The final section of the book examines policy for industry – first national, then regional and finally urban.

The book comprises chapters written by experts in their several respective fields, bringing together the most recent theoretical, empirical and policy evaluative work in those fields. However, by tightly structuring the book, all the chapters knit together to form a cohesive whole, so that it may be read, like a story, from cover to cover or on an individual chapter basis. It is at this point that I should like to thank the authors, seventeen in all, for the enthusiasm with which they tackled this project and their willingness to not only produce their own chapters but to offer and receive comments on other contributions, thereby making the volume a joint product.

The book covers the whole of the contemporary state of British industry from the economic geographer's perspective. Yet in reviewing the book's contents, one becomes aware of naggingly unanswered questions about the future of British industry – what is the future of the public and private sectors' shares of industry, and should public control be reduced? How will the world economy move into the 1990s, as China becomes a major world industrial power? How will the centrally-planned economies maintain their isolation, or is it inevitable that they will become more integrated into the global economy? How large can the informal sector, and the black economy, become in countries like Britain? And is the small firm sector really the basis of future national and multinational corporatism? Already, one begins to sense the genesis of the sucessor volume to *Industrial Change in the United Kingdom*.

William F. Lever
University of Glasgow

ACKNOWLEDGEMENTS

We are indebted to John Wiley & Sons Inc for permission to reproduce figs 2.6, 2.7, 2.8, 2.9, 2.10, 2.11 from figs 11.6, 12.4, 9.7, 15.1, 14.3, 14.4 (D. M. Smith 1981).

CHAPTER 1

Industrial change in the United Kingdom

David Keeble

Since the mid-1960s, manufacturing industry in the United Kingdom has undergone major changes in its scale, nature and location. Some of these changes, such as the decline of older nineteenth-century industries (steel, textiles) or the growth of large multiplant companies, reflect longstanding trends operating throughout the twentieth century. But others, such as the impact of EEC integration, the development of high-technology industries such as electronic computers, or the onset of the urban–rural manufacturing shift, are new and arguably different from previous trends. This initial, introductory, chapter therefore seeks to provide a brief empirical overview of the main changes which have been affecting UK manufacturing industry during the 1970s and earlier 1980s, as a context for the detailed appraisals of particular influences, theories and components of change which follow in subsequent chapters.

The most logical starting point for any evaluation of recent changes, and especially of urban and regional shifts in manufacturing location, is consideration of the macro-economic environment of UK industry during the last twenty years. Specifically, five major groups of influences on manufacturing industry can be identified, which have provided the context for investment and disinvestment decisions, and hence for spatial shifts of production and employment. These are global recession and national de-industrialization, the impact of technological change, perhaps operating partly in terms of Kondratieff-type long waves of innovation, trends in sectoral performance and employment characteristics, changes in corporate organization and the balance of large and small firms, and institutional changes, notably EEC enlargement and increasing state intervention through sectoral and spatial policies.

1.1 The changing environment, structure and organization of UK manufacturing

De-industrialization

As Fig. 1.1 shows, the 1970s and 1980s have witnessed a catastrophic decline in the level of UK manufacturing employment, together with a

Fig. 1.1 UK Manufacturing employment and output trends 1964–83

substantial fall in output. Between 1966, the peak year for manufacturing jobs in the United Kingdom, and June 1983, the manufacturing sector shed no less than 3.14 million workers, or 37 per cent of the 1966 total. Job losses have been particularly severe since 1979, with the onset of the worst global capitalist recession since the 1930s. After rising to 1973, the volume of UK manufacturing output has also fallen, by 18 per cent to 1983. In aggregate, the geography of industrial change in the UK in the 1970s and 1980s is thus the geography of decline, not the geography of growth.

This manufacturing decline is clearly largely due to the worsening global economic recession which has afflicted the industrialized capitalist economies ever since the oil price crisis of 1973. However, the rate and nature of the UK's manufacturing decline appear to be more extreme than elsewhere, leading many commentators to characterize it as 'de-industrialization' (Blackaby 1981; Thirlwall 1982). Cairncross (1981) notes that de-industrialization has been used to refer to four separate trends, namely absolute decline in manufacturing employment, a decline in manufacturing's share of total employment, a progressive failure to achieve a sufficient surplus of manufactured exports over imports to keep the economy in external balance, and industrial contraction so severe as ultimately to jeopardize the country's ability to pay for essential imports (Chisholm 1985). Certainly the UK's manufacturing decline warrants characterization as de-industrialization on at least the first three of these criteria. As a consequence of massive job loss, the share of manufacturing in total UK employment has slumped from 36.9 per cent in 1966 to only 25.8 per cent in 1985, while a progressive failure during the 1970s and early 1980s for UK manufactured exports to grow (+44% by volume, 1970–82) as fast as manufactured imports (+154%, 1970–82) led eventually in 1983 to the UK's first balance of payments *deficit* in trade in manufactured goods since the Industrial Revolution, of £2,100 million. While the continuing fall of the country's share of world trade in manufactures over the last twenty years (e.g. from 12.5% to 9.% by volume, 1971–83) is probably historically inevitable, the relatively poor export performance by British manufacturers and rising import penetration across a wide range of industries do suggest fundamental problems of declining international competitiveness, 'which is both a consequence and a cause of a deteriorating economic environment, characterized by low profits, the holding back of new investment and insufficient job opportunities' (Martin 1982a: 375).

Recession and de-industrialization, then, are fundamental to the changing geography of manufacturing in the United Kingdom in the 1980s. Intense competition and declining home demand have forced many manufacturing firms and factories to close, especially in older industries such as steel and shipbuilding which are concentrated in particular regions. Moreover, those which have survived have experienced a drastic decline in profitability, from average rates of 17 per cent in the 1950s to only 8 per cent by 1972–75 and 2 per cent by 1982–83 (The Treasury 1983: 8). Inevitably, investment levels have also plummeted, from £8.2 bn in 1979 to only £5.5 bn in 1982. The locational implications of these macro-economic trends are clearly likely to be important; but their precise nature is debatable. Over and above structural/sectoral impacts on particular regions, falling sales and substantially reduced investment have undoubtedly worked sharply to diminish the volume of industrial migration by existing firms to new UK locations, such as the assisted regions, given the long-established and close relationship between firm growth and migration (Keeble 1976: 127–32). However, it could also be argued that sharply declining profits may have forced firms, especially multiplant firms, to intensify restructuring of their activities in order to concentrate production in their most profitable plants; and some workers have argued (Massey 1979; Cooke and Pires 1984) that these are to be found in assisted regions as a response to low wages and a non-militant, often female, workforce. Equally, it may plausibly be argued that firms which are forced by competitive pressures to invest in new technology or machinery are likely to become increasingly, not decreasingly, sensitive to the availability of regional policy financial grants during a profits squeeze, when internal resources for such new investment are at a minimum. So the likely impact of recession on the industrial geography of Britain is debatable.

Technological change

The concept of technological change is surprisingly difficult to define, but may broadly be taken as involving an increase in the accumulated body of technical knowledge and/or the number of firms or individuals who possess and use this knowledge (Goddard and Thwaites 1983). In the United Kingdom, the impetus for technological change comes primarily from product competition in the market place between firms, especially larger firms, and increasingly at a global scale, although government policies (e.g. for nuclear power, aircraft and defence industries, space programmes) have played an important secondary role. While technological progress in manufacturing is fundamentally dependent on research, invention and development of new processes and products, perhaps the most dramatic examples involve *innovation*, that is the first commercial application of technological development, and the diffusion of significant manufacturing innovations within an economy. The last thirty years have witnessed a rapid rate of innovation in UK manufacturing. Sussex University Science Policy Research Unit workers have identified no less than 2,300 significant innovations in thirty manufacturing sectors since 1945 (Townsend *et al.* 1981). There is no doubt that the impact of many of these upon particular industries, their products, employment and location has been profound. Examples range from the introduction of basic oxygen processes in the steel industry in the 1960s to the microprocessor and silicon chip revolution in the computer electronics industry in the 1970s.

Recently, a number of studies of innovation and technological change have adopted frameworks based on, and aroused renewed interest in, Kondratieff-type theories of long waves of economic change (Mensch 1979; Mandel 1980; Freeman, Clark and Soete 1982; Freeman 1983; Van Duijn 1983). Mensch's work in particular has argued not only that Kondratieff was right in 1925 to identify historically-specific long waves – of the order of 50 to 60 years – of economic growth and decline in capitalist economies, but that these are associated à la Schumpeter with bursts of innovation adoption concentrated in short periods. Mensch points out that the long-run cycle of economic recession (1930s and 1980s) and intervening growth *since* Kondratieff's paper was published bears out the latter's ideas quite strikingly. This fourth Kondratieff cycle, like its predecessors, has possessed three phases, of recovery and boom, of stagflation characterized by rising unemployment and inflation, and finally of severe depression. And Mensch claims to document empirically for each Kondratieff cycle a peak of innovation during depression (e.g. in 1935, associated with drugs, television, radar and synthetic materials) which triggers subsequent recovery. He thus predicts that deepening recession since the early 1970s will eventually lead to a burst of technological innovation beginning in the 1980s and peaking in 1989, a fifth Kondratieff upswing.

While the existence of long waves of economic change in capitalist economies is supported by various economic indicators, Mensch's analysis and predictions of technological innovation are more controversial (Freeman, Clark and Soete 1982). Three more general points concerning recent and continuing technological change in British manufacturing industry should nonetheless be noted here. First, the last twenty years have clearly witnessed a powerful process of technological innovation aimed at increasing automation and mechanization of manufacturing production, and reducing the labour input. Its extreme form is robotization, with Britain by 1982 ranking fifth in the world for numbers of robots installed in manufacturing plants (713; c.f. 14,000 in Japan and 2,300 in West Germany), mainly in large companies such as BL, Ford and British Aerospace (Johnstone 1982). But most companies, large and small, have increasingly been forced to install more automated machinery, given the benefits this yields in terms of cost savings, and continuity and reliability of production. As a result, average capital intensity in UK manufacturing has increased very considerably, possibly doubling since 1965 (Department of Trade and Industry 1983: 2), a process affecting all manufacturing sectors. For this reason alone, future growth in UK manufacturing *employment*, though not production, seems extremely unlikely. The possible locational implications of this increasing capital intensity will be considered later.

Secondly, rates of technological change clearly vary markedly between different industries. While very difficult actually to quantify, there is almost certainly a broad relationship between rapid technological change and high levels of research and development expenditure in particular industries. Sectors characterized by these two attributes may be termed 'high-technology' industries (Kelly 1986). They usually employ an above-average proportion of administrators, technical and clerical staff, and record relatively high levels of recent output growth. Thus the three leading industry groups by proportion of non-production staff in 1983 were instrument engineering (36.9%), electrical engineering (37.7%) and chemicals (39.2%): while four of the five fastest-growing individual industries in the UK since 1975 (Table 1.1, excluding plastics) are high-technology industries by Kelly's criteria. In contrast, none of the five fastest-declining

Table 1.1 Manufacturing output growth and decline: the ten most extreme industries, 1975–81

	Output index 1981 (1975 = 100)		Output index 1981 (1975 = 100)
Electronic computers	253.5	Iron castings	55.5
Radio and electronic components	144.7	Leather, leather goods and fur	65.5
Radio, radar and electronic capital goods	124.3	Shipbuilding and marine engineering	68.5
Plastic products	122.4	Textiles	71.3
Pharmaceutical chemicals and preparations	115.9	Other metal goods	73.3

Source: Annual Abstract of Statistics, 1983

industries (Table 1.1) may be so described, while these also included two of the three lowest-ranking industry groups, leather and textiles, by proportion of non-production staff (the third was clothing). Clearly, rapid innovation in the high-technology industries is allied with above-average growth in demand for their products, whether from the private sector (e.g. personal computers) or the public sector (e.g. information technology), as an explanation for their exceptional expansion under conditions of deepening recession. However, rapid market growth may itself be facilitated by technological change, and resultant massive cheapening of products or extensions of their use into new markets, as has been happening in the micro-electronics field. In short, industry differences in rates of technological change represent one if not the major determinant of recent inter-industry variations in output and employment growth or decline, with all that means for differences in manufacturing performance between different regions and areas of the UK.

The third point worth emphasizing is that, whether couched in terms of the debate over innovation bursts as a key mechanism in Kondratieff upswings or not, much research does reveal that the recent geography of manufacturing innovation in Britain is spatially biased in favour of particular areas and regions. This issue, and the allied question of the possible locational determinants of high-technology industry, are taken up briefly later in this chapter, and more fully in Chapter 7.

Sectoral and employment trends

The simple facts of sectoral variations in output and employment change in British manufacturing industry over the period 1971–83 are set out, in broad Standard Industrial Classification (1968 SIC) Order categories, in Table 1.2. The differences which this reveals are considerable. The most severely declining industries on both measures were metal manufacture, textiles and the leather products category. The first two had both lost by 1983 over *half* of their 1971 employed workforce, compared with an average manufacturing industry loss of (only) 32 per cent. This appalling rate – and volume – of employment decline is almost certainly unparalleled in Britain's twentieth-century history, notwithstanding the 1930s. Other industries recording above-average, but somewhat less rapid, employment decline include the country's largest single manufacturing industry, mechanical engineering, together with vehicles, clothing and footwear, bricks/cement, and the tiny coal and petroleum products category. In each case, however, output decline was less severe, particularly in the clothing/footwear and bricks/cement sectors.

The most striking contrast to these declining industries is afforded by two industries, electrical engineering and chemicals, which recorded increases in their volume of output over the period of 45 and 27 per cent respectively. Not surprisingly these were amongst the three sectors experiencing the least severe employment losses (the third being paper, printing and publishing). The food and drink, and instrument engineering industries also recorded some degree of output growth, while the 'other' manufacturing category more or less maintained its production volume, notwithstanding the recession.

Recent sectoral variations in manufacturing performance are then quite marked, and indeed appear to have intensified appreciably with recession since 1977. The reasons for these differences almost certainly centre on demand variations for different kinds of goods, closely linked to different rates of technological change and innovation, and on Britain's particular inability to compete under conditions of intense global recession in the production of more traditional,

Table 1.2 Individual industry trends in output and employment: Great Britain, 1971–83

SIC order	1971–83 Output change (%)	1971–83 Employment change (%)	SIC order	1971–83 Output change (%)	1971–83 Employment change (%)
3 Food, drink and tobacco	+12.0	−22.8	12 Other metal goods	−32.3	−30.6
4 Coal and petroleum products	−22.3	−47.6	13 Textiles	−38.3	−50.8
5 Chemicals and allied industries	+26.6	−16.2	14 Leather, leather goods and fur	−42.7	−40.9
6 Metal manufacture	−33.6	−53.4	15 Clothing and footwear	−12.9	−41.2
7 Mechanical engineering	−23.6	−35.9	16 Bricks, pottery, glass, cement	−16.3	−36.1
8 Instrument engineering	+7.3	−24.7	17 Timber, furniture	−11.5	−22.8
9 Electrical engineering	+45.3	−21.4	18 Paper, printing and publishing	−4.1	−19.5
10 Shipbuilding and marine engineering	−25.6	−28.8	19 Other manufacturing	−0.2	−31.5
11 Vehicles	−21.0	−36.0	All manufacturing	−8.3	−32.2

Sources: Department of Employment Gazette; Employment Gazette; Annual Abstract of Statistics; British Business

'low-technology' and sometimes labour-intensive manufactures. A classic example of the latter is textiles and clothing, where foreign imports, primarily from EEC and Third World countries, jumped from 47 to 64 per cent of the UK market, 1978–82 (Searjeant 1983). Rising import penetration has characterized many other industries, such as footwear (58% import penetration by 1982), electrical consumer goods and motor vehicles. Demand variations underpin both the collapse of metal manufacture and the remarkable output growth of the electronics industry since 1978. In the latter case, relative consumer affluence (average real household disposable income per head in the UK *rose* by 28 per cent, 1971–83, notwithstanding recession), allied of course with rapid technological change, has helped fuel a consumer demand boom for such products as micro-computers, hi-fi equipment and video recorders: while demand for electronic capital goods (business computers, word processors) from both private and public sector organizations has also rocketed. Recent variations in individual industry performance thus represent an important structural influence on the geography of manufacturing change in the UK.

National trends in the type of jobs, and of workers, in manufacturing have been less striking. Thus the proportion of women in the GB manufacturing workforce has changed only slightly, falling from 29.7 per cent in 1971 to 29.4 per cent in 1977 and 28.9 per cent in 1985. Part-time employees (mainly women) are not important in manufacturing, their proportion fluctuating from 6.8 per cent in 1971 to 7.8 per cent in 1977 and 6.7 per cent in 1985. These are clearly not dramatic or substantial changes, and probably mainly reflect sectoral shifts within manufacturing. The changing quality of manufacturing jobs is much more difficult to measure. Some workers (e.g. Massey and Meegan 1978: 287) have drawn attention to 'deskilling' processes of rationalization and technological change, involving a shift from craft and technically skilled jobs to semiskilled assembly work. Certainly Cameron, Kirwan and McGregor (1981: 21) found that while manual female employment in manufacturing declined appreciably between 1973 and 1979, the proportion of these engaged in assembly and repetitive activities increased, from 22.6 to 25.2 per cent. However, they also found that the dominant trend in job type within manufacturing was the relative growth of non-manual, non-production employment, especially for men. This growth has concentrated on professional occupations in management and administration, and in highly-qualified scientific, engineering and technical categories. This seems to imply a 'skilling' process in terms of job quality within manufacturing, reflecting almost certainly technological change,

R and D growth, and the development of high-technology industries such as electronics (Morgan and Sayer 1983: 51–2: Kelly, 1986).

Changes in corporate organization

Trends in the corporate organization of manufacturing industry, and their spatial implications, have been the object of much recent industrial location research in Britain (Keeble and McDermott 1978; McDermott and Taylor 1982; Hayter and Watts 1983). This research has focused on the centralization and internationalization of capital in large multiplant and multinational organizations, and the impact of this on the geography of production and employment in different areas (Watts 1980, 1981; Seneschall 1984). Some work has, however, also investigated the process, extent and significance of new, small firm formation, at the opposite end of the size and organization spectrum (Storey 1982).

The increasing concentration of production, employment and manufacturing capacity in large multiplant firms has been documented by Prais (1976) for the period up to 1970. His work shows, for example, that the share of British manufacturing net output accounted for by the country's top 100 manufacturing organizations increased from only 16 per cent in 1909 to 41 per cent in 1970, with the most rapid period of growth coming after 1950, and that this substantial increase in manufacturing concentration primarily reflected a marked increase in the average number of plants operated by these giant companies, from only 27 in 1958 to no less than 72 in 1972. This increase has involved both new branch plants on 'greenfield' sites and plant acquisition through takeover or merger. The latter has almost certainly dominated, especially during and since the 1970s, when rates of new branch plant establishment have been far lower than previously. In the 1960s, however, quite large numbers of new branches were set up by large firms, often in the assisted regions of the UK (Keeble 1976: 139). The distinction between new and acquired branch plants appears to be important for the regional implications of the branch plant formation process (Smith 1979). It should also be noted, as Prais stresses, that most giant firm branch plants are medium, not large in size (average 1972 employment in plants owned by the top 100 enterprises was only 430); and that over the last 25 years, average giant firm plant size has been falling steadily, at least in terms of employment.

The growth in dominance of giant multiplant firms in Britain has varied considerably between industries. In 1970 (Prais 1976), dominance was greatest in the vehicle industry, food, drink and tobacco (brewing is a classic example), chemicals, and metal manufacture (after steel nationalization in 1968). It was quite appreciable in mechanical and electrical engineering. Generally, these are industries where barriers to entry, for reasons of production technology, scale economies and hence exceptional investment requirements, are high and large firms have a powerful advantage over small ones. An excellent specific example of this is the high-technology UK pharmaceutical industry, where very high cost of essential R and D activity can only be recouped, if at all, after long (twenty-year) periods of drug research, trials and testing (Howells 1984a). Not surprisingly, the industry is dominated by a relatively few giant companies (the top nine accounted for 47 per cent of UK pharmaceutical net output in 1981), most of them in fact foreign-owned and/or operating globally, not just nationally. At the other extreme, furniture, clothing and other metal goods are basically small-firm industries, because of very low barriers to entry in terms of technology, investment and skill requirements, and in the first two cases a relatively unstandardized, fashion-conscious market offering unusual scope for small companies.

Discussion of inter-industry variations thus suggests one possible reason for the striking growth of large corporations up to 1970, namely increasing scale economies in production in particular industries as a result of technological change. This does seem to be important in the motor vehicle case (Seneschall 1984; Ward 1982). However, Prais argues that much more significant considerations are financial advantages via the workings of the capital market, and 'spontaneous drift', or concentration as an inevitable result of random, or stochastic, firm growth processes over time. Marketing advantages and falling transport and communication costs may also play some part. Certainly other workers have stressed the vital significance for *multinational* companies of the growth of global aviation and telecommunication systems as an essential prerequisite for the post-war expansion of such giant organizations.

The spatial implications of multiplant firm growth are very important, and can be considered at two broad scales, international and intranational. Most large British manufacturing companies now serve global markets, and operate production facilities and subsidiary companies in other countries. Equally, foreign multinationals have been accounting for an ever-increasing share of UK manufacturing capacity over the last twenty years, Britain's leading enterprises including such giant American organizations as Esso, Ford, General Motors and IBM. By 1979, foreign-owned manufacturing establishments were responsible for no less than 21 per cent of UK net output and 15 per cent of UK manufacturing employment, reflecting a marked growth in importance and foreign direct investment since 1963, when only 7 per cent of employment was in such companies. Three-quarters of this output was by US-owned firms (Dicken 1982).

This growth in importance of foreign firms clearly owes much to increased acquisition activity, particularly by American companies, in Britain since it became a member of the EEC (Smith 1982). Certainly this has been the dominant factor in the rapid recent growth of foreign ownership in the UK paper and packaging industry, for example (Gould 1984). But foreign-owned companies have also frequently set up new 'greenfield' manufacturing operations in Britain, a recent example being a spurt of Japanese manufacturing investment in the late 1970s and 1980s. Though modest (28 plants, employing about 6,000 workers by 1984) in scale, this investment is almost wholly in new plants, most of them in consumer electronics assembly, such as colour television sets and video recorders (Dicken 1983). An important development here is the Nissan car assembly plant which began production in 1986 in Washington, County Durham. An initial £50 m. investment will provide 600 jobs and assemble 24,000 cars a year; but a second phase could involve a £300 m. investment, 2,500 jobs and 300,000 vehicles per annum. Clearly, this Japanese investment can be understood only in the context of Britain's membership of the EEC, and the increasingly perceived need by Japanese companies to develop a visible manufacturing presence within the Community, for market share and trading policy reasons.

Growth in the importance of foreign-owned manufacturing in Britain is not only due to acquisition activity and new 'greenfield' developments, but also to the better performance of foreign companies once established here compared with indigenous firms. This is partly because foreign companies are concentrated in relatively newer industries, with more buoyant markets and products. There is also evidence, however, that foreign-owned firms are more efficient and profitable, allowing for structural differences, benefiting from the economies of scale made possible by the global operations of the parent company (Brech and Sharp 1984). Thus Ford of Britain, which has increasingly been operating as part of a European-wide integrated production system (Seneschall 1984), has been consistently more profitable, has invested much larger sums, and has maintained employment levels since the 1960s appreciably better than BL, Britain's only significant indigenous car assembly company. Equally, many significant manufacturing innovations have reached Britain via diffusion to local foreign (especially American) subsidiaries from overseas parent R and D facilities. This is almost certainly one important process underlying the recent surge of high-technology computer and software development in the so-called 'M4 Western Corridor' between London and Bristol (Breheny *et al*. 1983).

At the same time, however, the growth of foreign manufacturing investment in the UK, of course, also renders British manufacturing activity, output and jobs potentially more vulnerable to shifts in production policies decided on in New York, Detroit or Tokyo, rather than in London, Birmingham or Newcastle, and determined by the changing global requirements and strategies, rather than the local British circumstances, of the multinational organization. Thus union opposition in 1984 to the closure of Ford's Dagenham foundry operation apparently partly reflected the fear that this heralded a switch of Sierra engine production to Cologne, on grounds of efficiency and cheapness, while export sales from both the British and German Ford operations have been affected since 1970 by a global policy of Ford investment in local production facilities for locally-growing markets, as in Spain and Commonwealth countries (Seneschall 1984). So the growth of multinational production systems makes rapid adjustment to international variations in market demand (Spain versus Britain) and changing production costs (Britain versus Germany) increasingly feasible (see Ch. 11).

This is, of course, equally true for British-owned multinationals operating plants overseas. Direct investment abroad by UK firms has increased in recent years (£7.5 to £11.1 thousand million, 1977–79 to 1980–82), following relaxation of government restrictions on capital movements (see Ch. 5). Critics of this relaxation argue that this represents investment diverted from Britain, and hence a cause of lost output and jobs. Thus it has been suggested (Goddard 1983a: 20) that increased foreign investment reflects a shift of production by multinationals to Third World locations in search of low labour costs, as part of an inexorable development of a world capitalist production system. In fact, however, while some investment has indeed been inspired by this motive, empirical research (Paine 1979; Seneschall 1984) shows that the prime reason for British and European manufacturing investment abroad in the 1970s and 1980s has been overseas market growth and the need for local production to serve this. Much the most important destination for such investment has thus been the USA (Grant 1983: 96), not the Third World. So the labour-cost perspective is probably misleading as a general characterization of recent trends in global investment behaviour by UK companies. Equally, the extent of actual diversion of manufacturing investment and jobs from Britain, while very difficult to measure, may not be as great as some have feared.

At the intranational scale, the growth of multiplant companies also – of course – increases the potential vulnerability of local and regional economies to investment and disinvestment decisions taken elsewhere for wider company objectives. However, equally important is that at this scale, the development of multiplant organizations permits, uniquely, the development of *functional specialization* of different establishments, and their location in different areas which offer the most favourable environments for each different function. This very important point relates directly to Haig's classic conceptualization of the manufacturing firm as 'a packet of functions', functions which in the multiplant firm can be 'separated and located at different places' (Haig 1926: 416). Thus whereas in the single-plant firm all activities must, by definition, be carried on at one location, in the multiplant firm component production may be hived off to areas offering readily available and cheap female labour, or government regional policy incentives, R and D may be located in areas attractive to high-level scientific and technical staff, and company headquarters may be sited at the centre of decision-making in the national space economy (Keeble 1971: 35–7). There is considerable evidence that the growth of multiplant firms in the UK has in turn, through this mechanism, increasingly replaced the assisted regions' historic sectoral specialization with a new functional specialization centred on branch plant production (Massey 1979). Conversely, the South East has

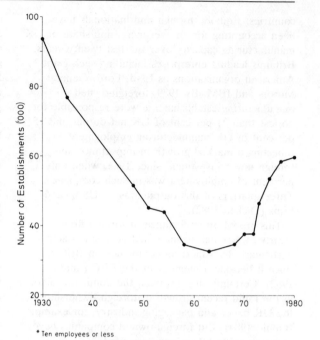

* Ten employees or less

Fig. 1.2 Small manufacturing establishments in the UK 1930–80

experienced a relative concentration of high-level (R and D, head office) manufacturing functions, notwithstanding exceptionally rapid manufacturing decline up to 1977. Massey suggests that one result of this is widening regional disparities in occupational structure and the quality of job opportunities, and the creation of a new 'spatial division of labour' within the UK.

At the opposite end of the corporate spectrum is the new, small, independent firm. Between 1930 and 1960, the number of small manufacturing firms in the UK declined very steeply (Prais 1976). However, this decline ceased in the 1960s (Ganguly 1982), while Census of Production returns suggest a striking reversal in the 1970s, with a growth in numbers of small manufacturing establishments (one to ten employees) between 1970 and 1980 of 22,000 or 58 per cent (Fig. 1.2). Total employment in this small firm category also rose, by 52,000 or 23 per cent, compared with a *decline* in the total UK manufacturing workforce of 20 per cent, and in employment in giant manufacturing enterprises (employing 5,000 or more workers) of no less than 26 per cent. Moreover, this growth in number of small enterprises almost exactly fits independent findings, based on VAT registrations rather than Census of Production data, of a growth in the total population of UK manufacturing firms of 18,000 (from 111,000 to 129,000) between 1974 and 1981

(Gudgin, 1984). The latter suggests that substantial growth in numbers of small firms chiefly reflects high new firm formation rates, rather than a process of shrinkage by previously larger firms.

This UK growth in small firm numbers must of course be kept in proportion. Despite growth, firms of ten or less employees still provided only 4.3 per cent of UK manufacturing employment in 1980, compared with 2.8 per cent in 1970. But the trend is extremely interesting, both because of the exceptionally rapid decline of giant firms noted above, and because it has also been noted in other industrialized countries, such as France, Italy and Belgium (Aydalot 1983; 1984). Indeed, the significance of this small firm reversal in France even leads Aydalot (1983: 24) to claim that at the regional scale, 'large companies that contract their employment needs can no longer control regional development. This responsibility has passed into the hands of the regions themselves' via the creation of new, small enterprises. This means, he claims, 'that we have to make a transition from a theory of localizing large business to a theory of local dynamics, a theory involving the role of the local environment in creating new activities'. Certainly new firm research in Britain (see Ch. 9) does suggest that a high formation rate may be an important long-run influence on regional manufacturing performance (Gudgin 1978) while the rise of new high-technology industries in southern England since the mid-1970s in part reflects a surge of new companies, and especially favourable local environments for high-technology enterprise formation (Gould and Keeble 1984: Keeble and Kelly, 1986).

Institutional changes

Perhaps the most important institutional changes which have affected British manufacturing industry since 1970 have been the entry of the UK to the European Economic Community in 1973, and the changing policies, sectoral and spatial, of central government. The effects of British entry to the European Economic Community on manufacturing industry are very difficult to isolate, because of the coincidence of entry with the onset and deepening of global recession since 1983. However, several points should be noted. First, Community membership and the abolition of tariffs between Britain and the other EEC countries has led to a rapid and substantial shift of trade in manufactured goods away from the rest of the world to EEC partners. Mayes (cited in Grant 1983: 91–2) has demonstrated this EEC-inspired shift for virtually all manufacturing industry groups. Overall, the share of Britain's total exports going to EEC countries rose from 31 per cent in 1972 to 45 per cent in 1984, while the EEC's import share rose from 32 to 42 per cent (see Ch. 13). Secondly, this shift has resulted in a substantial and worsening trade *deficit* in manufactured goods with the rest of the EEC. This serious trend is clearly a product of the intensified competition facing British producers after entry, and their apparent inability to compete 'in the sophisticated, consumer-orientated markets of the EEC' (Grant 1983: 94). Indeed, Grant (1983: 102) shows that between 1970 and 1978, the only industries 'where the UK increased its share of European output were those with a stagnant or declining output', such as clothing and footwear, rubber, printing, textiles and leather: that is, 'low technology products whose output is increasingly shifting towards developing countries'. Thirdly, a significant proportion of the increasing trade in manufactured goods with the rest of the EEC seems to have taken the form of flows of components, parts and semi-finished products being moved by EEC multinationals between plants in different EEC countries, particularly in such industries as motor vehicles (Seneschall 1984). As a result, 'a good deal of the country's imports and exports (now) consist of intra-firm trade' (Dicken 1983: 194), dependent upon the global policy decisions of multinational companies. This has been facilitated by EEC membership.

Perhaps more positive points concern the impact of membership on foreign direct investment in British manufacturing and of EEC regional policy. Has membership encouraged non-European companies to locate production facilities in Britain, as a springboard for the EEC market? The answer to this question seems to be a qualified yes. Certainly this is the logic behind Japanese investment in Britain, while the later 1970s did witness an increase in the share of US manufacturing investment going to the UK, from 12.6 per cent in 1976 to 15.8 per cent in 1980 and 15.7 per cent in 1981 (Grant 1983). Much of this is, moreover, in high-technology industries, as with Wang Laboratories' £38 m electronics plant at Stirling, opened in 1984, with two-thirds of its output scheduled for Europe, and Digital Equipment's first-ever non-US £7 m R and D software centre at Reading, established in 1984. However, it is also true that some other EEC countries, such as Ireland, have done better than Britain in capturing an increasing share of foreign manufacturing investment since 1973; while some observers (e.g. Dicken 1982) adopt a pessimistic view of Britain's long-run prospects as a foreign manufacturing base for Europe. Lastly, it should be noted that Britain has received the second largest share (after Italy) of payments from the Community's Regional Development Fund, for industrial and infrastructure projects in its assisted regions, since 1975. While still tiny in comparison with the EEC's agricultural budget, the ERDF has increased quite considerably in real terms, as has the Community's Social Fund, from which the UK also receives help towards manpower training and job creation in areas of high unemployment.

Central government economic, sectoral and spatial policies are of the greatest importance in providing the context for national industrial trends. These are, of course, considered in detail in Chapters 13–15, and only three brief points will therefore be emphasized here. First, many observers argue that monetarist government economic policies since 1979, operating through high interest rates and an inflated pound, have markedly intensified the impact of global recession upon British manufacturing industry in the 1980s. Resultant low levels of capital investment, and high and rising import penetration, bode ill for any significant industrial recovery in the foreseeable future. Secondly, a wide range of evidence supports the view that until the later 1970s, government regional policy did exert a considerable influence on the location of manufacturing investment and resultant jobs in the UK. Moore, Rhodes and Tyler's most recent (1986) estimates here suggest a policy impact on the assisted areas of 604,000 (gross) and 450,000 (net) manufacturing jobs, 1961–81.

Lastly, however, their work and more general considerations suggest a major weakening of the effect of regional policy in the 1970s, and specifically since 1979. One aspect of this has been the shift in government industrial spending away from regional, and towards sectoral, policies. Since 1979, the present Conservative government has cut back regional policy incentives and designated areas (Keeble 1980; Regional Studies Association 1983), while boosting financial allocations to innovation-oriented and small firm programmes. Thus by the financial year 1982/83, the government's Support for Innovation Scheme was involving grant allocations of over £160 m a year, while a further £130 m was to be allocated under the Small Engineering Firms Investment Scheme by the end of 1983. Provisions in the 1983 budget alone brought the number of special measures aimed at helping small firms and enterprise creation since 1979 to 108, according to government claims, and involved an anticipated revenue cost of £275 m in a full year, chiefly with regard to the Business Expansion and Loan Guarantee Schemes (Harris 1984). The locational significance of this shift from regional to sectoral policies could be considerable, given the already much higher small firm creation rates of southern England (section 1.2), and the concentration here of electronics and other high-technology industries. Not surprisingly, no less than 56 per cent of SFI allocations, 1982/83, went to firms in the South East plus East Anglia, with a further 9 per cent to the South West. Sectoral and innovation-oriented policies, arguably desirable for national industrial regeneration, may thus be having significant spatial effects, widening rather than diminishing regional disparities in industrial structure and performance in the United Kingdom.

1.2 The changing location of manufacturing industry in the UK

Major locational trends

The preceding discussion of key trends in the industrial environment, structure and organization of UK manufacturing leads naturally to consideration of the main recent shifts in the location of manufacturing activity. The most up-to-date and spatially detailed measure of these shifts

Table 1.3 The urban–rural shift in UK manufacturing employment, 1971–81

	Manuf. employment 1971 ('000)	Manuf. employment change, 1971–81 ('000)	(%)
Conurbations (8)	3,758	−1,288	−34.3
More-urbanized counties (14)	1,743	−318	−18.2
Less-urbanized counties (22)	1,869	−287	−15.4
Rural counties (20)	686	−78	−11.3
United Kingdom	8,056	−1,971	−24.5

Source: Unpublished Department of Employment statistics

The changing location of manufacturing industry in the UK

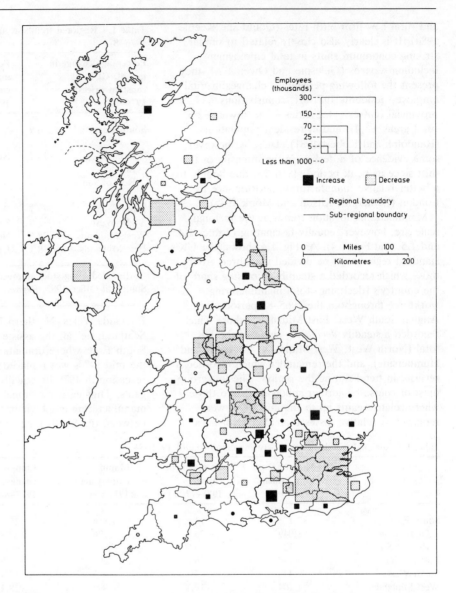

Fig. 1.3 Manufacturing employment change 1971–78

is, almost inevitably, employment, rather than output or investment. Nonetheless, some regional statistics on manufacturing GDP trends are also presented below, while it should be noted that spatial shifts in employment often correspond surprisingly closely with those in manufacturing output (Keeble 1976: 14; Keeble, Owens and Thompson 1983).

The single most powerful trend in manufacturing location in the United Kingdom since the 1960s is arguably the marked relative shift of manufacturing capacity, employment and output from the conurbations and cities to small towns and rural areas (Table 1.3 and Fig. 1.3). This urban–rural manufacturing shift involves a striking *continuum*

of rates of manufacturing job change with density and scale of urbanization (Fothergill and Gudgin 1982: 22), is closely paralleled in most other industrialized countries (Keeble, Owens and Thompson 1983), and cannot be explained by structural/sectoral mix variations in manufacturing composition between cities and smaller settlements (Fothergill and Gudgin 1979; Keeble 1980). While involving a not inconsiderable degree of manufacturing migration by existing firms (Keeble 1978; Fothergill and Gudgin 1983: 40), the shift is even more powerfully a reflection of the differing performance of existing, non-mobile, urban and rural firms (Fothergill and Gudgin 1982; 1983), together with the minor factor of differing urban and rural new firm birth rates (Gould and Keeble 1984). It is clearly also closely related to similar striking continuum shifts in total employment, including services (Gillespie and Owen 1984: they present the following percentage changes in total employed residents 1971–81, conurbations −14,2, provincial centres −4.2, cities −2.0, towns +2.6, rural areas +5.4), and in resident population (Randolph and Robert 1981). There is, however, some evidence of a decline in the intensity of the shift since its peak period, 1966–73, due largely to a better relative manufacturing performance by London (Fothergill, Kitson and Monk 1985).

Manufacturing location trends at the regional scale are, however, equally fascinating (Tables 1.4 and 1.5, and Fig. 1.4). As Fig. 1.4 shows, the UK's standard regions can be divided into three groups: those which recorded a steadily *increasing* share of the country's (declining) total manufacturing workforce throughout the 1965–83 period (East Anglia, South West, East Midlands), those which recorded a steadily *decreasing* share of the UK total (North West, West Midlands, Yorkshire and Humberside), and the remainder which recorded a *reversal* in trend during the period. The last group must of course be further subdivided, into those where relative gains up to the mid-1970s were replaced by relative losses thereafter (Northern

Table 1.4 Regional trends in manufacturing GDP, 1976–81

	Percentage change in manuf. GDP, at constant prices* 1976–81		Percentage change in manuf. GDP, at constant prices,* 1976–81
South East	−6.9	Yorkshire and Humberside	−18.8
Greater London	−16.2	North West	−16.0
Rest of South East	−1.4	North	−19.4
East Anglia	−0.7	Wales	−17.9
South West	−3.0	Scotland	−16.0
West Midlands	−24.3	Northern Ireland	−26.5
East Midlands	−8.1	United Kingdom	−13.6

* After deducting stock appreciation. *Source*: Central Statistical Office 1983, *Regional Trends*, Table 9.4

England, Wales, Northern Ireland, and perhaps Scotland; i.e. all the assisted regions), and the South East, where dramatic relative decline up to the mid-1970s was replaced with the onset of recession in 1979 by equally dramatic relative gains. Thus earlier regional convergence in manufacturing employment levels, particularly as between the South East and the much smaller

Table 1.5 Regional trends in manufacturing employment, 1971–84

	Manuf. Employment 1971 ('000)	Employment change 1971–77 (%)	Manuf. employment 1977 ('000)	Employment change, 1977–84 (%)	Manuf. Employment 1984 ('000)
South East	2,173	−14.6	1,856	−15.2	1,574
Gt London	1,049	−26.0	776	−22.8	599
Rest of S.E.	1,124	−3.9	1,080	−9.7	975
East Anglia	190	+6.8	203	−11.8	179
South West	439	−3.2	425	−13.4	368
West Midlands	1,104	−10.1	992	−28.1	713
East Midlands	618	−3.6	596	−17.3	493
Yorkshire and Humberside	777	−8.0	715	−28.1	514
North West	1,131	−11.1	1,005	−30.9	694
North	461	−5.9	434	−33.2	290
Wales	324	−4.6	309	−31.4	212
Scotland	669	−8.1	615	−28.1	442
Northern Ireland	170	−16.5	142	−29.6	100
United Kingdom	8,056	−9.5	7,292	−23.5	5,579

Sources: Department of Employment Gazette, Employment Gazette

Note: 1984 manufacturing employment is defined according to the 1980 Standard Industrial Classification (SIC). This adopts a wider definition of manufacturing than the previous 1968 SIC, and 1984 figures are to that extent 'inflated' relative to those for previous years (by an average of +3.6 per cent, comparing GB 1983 estimates for the two SICs).

The changing location of manufacturing industry in the UK

Fig. 1.4 Regional percentage shares of UK manufacturing employment, 1965–85

manufacturing economies of the assisted regions, has given way to subsequent divergence between these regions. Put another way, manufacturing recession since 1979 has clearly been regionally differentiated in its impact, hitting the assisted regions and the three big urban–industrial regions of the West Midlands, North West and Yorkshire and Humberside, more severely than the four 'favoured' regions of the South East, East Anglia, East Midlands and the South West (Martin 1982a; 1982b; Townsend 1983). The striking South East reversal is epitomized by Greater London, by far the most rapidly declining manufacturing centre of the UK up to 1977, but declining more slowly than the national average – and seven other regions – after 1977 (Table 1.5). This is, of course, in line with recent suggestions that the post-1979 recession has witnessed both a diminution of the urban–rural shift and a resumption of intensification of southern Britain–northern Britain regional manufacturing disparities (Townsend 1983; Owen, Coombes and Gillespie 1983).

Explaining the urban–rural shift?

The consistency and ubiquity of the urban–rural manufacturing shift, notwithstanding the recent regional-scale impact of recession, strongly suggests that it is the product of very powerful forces associated with basic environmental differences between Britain's big cities and rural areas. The nature of these forces is, however, a matter of considerable debate. *Constrained location theory*, developed by Fothergill and Gudgin (1982); Fothergill, Kitson and Monk (1985), argues that the primary environmental cause is factory floorspace supply constraints in urban relative to rural areas. The driving force is the continuing displacement of machinery by labour as a result of the continuing capital investment in new production technology documented in section 1. Such investment generates steadily-increasing space needs per worker employed, which cannot be accommodated in physically-congested cities. Manufacturing employment therefore inevitably falls, whereas in small towns and rural areas, factories can readily be enlarged to accommodate new machinery and layout requirements, and employment is maintained or even expanded. This relatively simple theory is supported by a remarkable volume of empirical evidence, including the significant finding that rate of *subsequent* employment loss in the total population of East Midlands factories 1968–82 was substantially and closely related to the *initial* degree of physical congestion and age of buildings on the factory site, irrespective of whether the factory was urban or rural (Fothergill, Kitson and Monk 1985).

An alternative approach, argued by Lever (1982) and recently investigated by Moore, Rhodes and Tyler, is the *production cost explanation*. This contends that one, if not the main, cause of the urban–rural shift in Britain is substantially higher operating costs in cities, with inevitable long-term consequences for firm profitability, investment, competitiveness and employment change. Moore, Rhodes and Tyler (1984), using 1982 Census of Production data, argue that production cost differences between London and the rural south west, are equivalent to over 30 per cent of gross profits for two thirds of all industries studied. This difference is seen as primarily reflecting wage and salary costs, rather than factory rents and rates. Lever (1982) reaches a similar conclusion for Clydeside manufacturing. However, Fothergill's more recent work (Fothergill, Gudgin, Kitson and Monk 1984) discounts the view that operating costs vary significantly between most urban and rural areas in Britain, and ascribes identified profitability differences to factory and space inadequacies in cities.

Lastly, the *capital restructuring* approach advocated by radical and Marxist geographers tends to focus attention on shifts in production from cities, especially by large multiplant enterprises, to exploit less skilled, less unionized and less costly labour in rural branch plants (e.g. Massey and Meegan 1978: 286). However Fothergill and Gudgin (1982: 94) have shown that the urban–rural shift cannot be ascribed solely or largely to the activities of multiplant firms, while much rural growth is in small, independent local companies. They also argue that urban–rural differences in labour costs, militancy and availability are too small and variable to explain either the remarkable consistency or extent of the observed shift. So while research continues into the production cost explanation, the current front runner in the debate over the causes of the urban–rural manufacturing shift is probably constrained location theory.

Explaining regional reversal?

As Fig. 1.4 graphically illustrates, regional manufacturing trends between 1960 and the mid-1970s were characterized by a striking convergence in employment levels between the big industrial – and generally non-assisted – regions (South East, West Midlands, North West) and smaller assisted and rural regions (Wales, Northern England, East Anglia, the South West). A great deal of research suggests that two factors were of key importance in this convergence. One was the impact of regional policy incentives and controls, *both* upon the assisted regions, especially Wales and the North, *and* upon the South East and West Midlands, the main traditional targets of control policy (Moore, Rhodes and Tyler 1986). Thus most of the 450,000 net assisted area jobs estimated by these researchers to have been created by regional policy between 1960 and 1981 (see section 1) were already in existence by the mid-1970s. Policy therefore had a substantial impact, including a negative effect upon the South East and Midlands, directly via policy-induced firm migration and indirectly via more efficient assisted area firm competition.

The second dominant factor in regional convergence was the urban–rural manufacturing shift. This benefited disproportionately those regions (East Anglia, South West, Wales, and much of the North) which lack large conurbations and cities (Fothergill and Gudgin 1979). Indeed, the latter show that urban–rural composition accounted for no less than two-thirds of the variation in indigenous (non-migrant) manufacturing employment performance between British regions during the 1966–71 period. In contrast, inherited regional industrial structure played only a limited and declining part in influencing regional manufacturing change during the 1960s and early 1970s, largely because industry-specific change rates nationally converged appreciably compared with the 1950s (Fothergill and Gudgin 1983). Far more important was interregional industrial movement by existing firms, especially those previously located in the South East and Midlands. Considerable long-distance branch plant migration by large companies to the assisted regions was strongly influenced by regional policy controls and incentives together, in the 1960s, with greater labour availability. In contrast, the South West and East Anglia benefited from the complete transfer of small South East firms to larger modern premises on 'greenfield' sites, affording room for expansion and a pleasant residential environment for industrialists and workers (Keeble 1976: Ch. 6; Townroe 1983).

Explaining post-1977 regional reversal of this earlier convergence trend is more problematic. However, its timing certainly suggests some association with the onset of acute national recession after 1978. The latter seems likely to have influenced, to some degree at least, each of four specific factors which may explain the reversal. First, Fothergill, Kitson and Monk (1985) argue that recession has resulted in a diminution of the urban–rural shift since 1973, because of reduced investment in new plant and machinery, the motor force behind the shift according to constrained location theory. They note, however, that since 1978, this chiefly reflects a marked relative improvement in London's manufacturing employment performance (see also Table 1.5), with the other conurbations actually doing much worse, relative to the country as a whole. This South East/remaining industrial Britain contrast during a period of acute national decline is taken up again below. Regionally, however, the diminished incidence of the urban–rural shift since 1978 has probably played some part in regional reversal, notably with respect to London and the South East, and perhaps Wales and the North.

Secondly, there is no doubt that the 1970s witnessed a progressively diminished impact of regional policy on job creation in the assisted regions (Keeble 1980; Moore, Rhodes and Tyler 1986). Thus the latter's estimates suggest a 1976–81 rate of policy-induced net manufacturing job creation only 60 per cent that of 1971–76 (73,000 jobs compared with 121,000 in 1971–76: both figures incorporate a loss of jobs previously created – by 1971 – by immigrant firms as a result of policy). Three reasons may account for this much reduced impact. The chief one is the marked effect of recession on national manufacturing output, expansion and investment, which drastically reduced the number of firms contemplating installation of new capacity to cater for market growth, and therefore potentially available for steering to the assisted regions. In the past, no less than 85 per cent of all mobile firms in the UK have migrated because of pressures induced by growth in demand (Keeble 1976: 128), a causal relationship which has of course worked via regional policy steering mechanisms to the advantage of the assisted regions. With recession and decline, not growth, since the mid-1970s, regional policy has had very little to bite on.

A second reason is the 'plateau' effect which Moore, Rhodes and Tyler argue is inevitable as time passes, and old policy-induced plants reach the end of their life-cycle and close. Such closures were by the 1970s inevitably offsetting new gains to a much greater degree than in the 1960s, irrespective of recession. The result was a diminished *net* rate of policy induced job creation, with regard to immigrant plants. Lastly, the period since the mid-1970s has seen a substantial reduction in the intensity of operation of regional policy itself (see Ch. 14), a policy shift in turn partly related to recession, public expenditure cuts, and the obvious nation-wide rather than just regional incidence of plant closures, redundancies and steeply-rising unemployment. Industrial Development Certificate controls were almost totally in abeyance after 1977 and abolished in 1982, while average annual regional policy expenditure fell by 30 per cent in real terms during 1976–80 compared with 1972–75 (Moore, Rhodes and Tyler 1986).

Industrial change in the United Kingdom

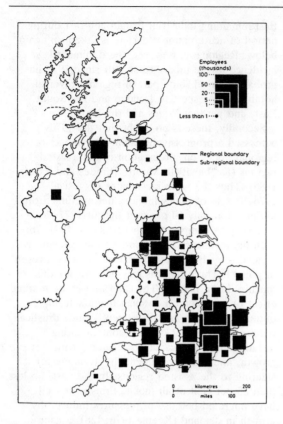 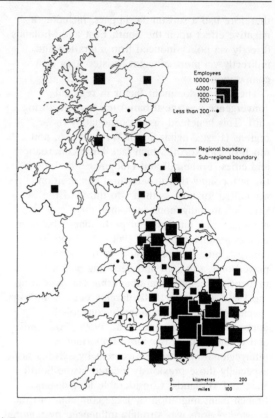

Fig. 1.5 (a) High-technology manufacturing industry, 1981; (b) Research and development services, 1981

A third major factor in regional reversal since 1977 is almost certainly the renewed impact of structural differences on regional manufacturing performance (Townsend 1983: 164). As noted in section 1, post-1979 recession has decimated several older industries which are historically concentrated in the assisted regions, such as steel, textiles and shipbuilding; while the above-average decline of vehicles and mechanical engineering has intensified still further the previous regional decline of the West Midlands (Fig. 1.4). Conversely, and unrelated to the recession, the striking recent growth of certain newer, and notably high-technology, industries appears to have been particularly biased towards the three southern regions of the South East, East Anglia and the South West. While the definition of high-technology industry is a matter of complexity and debate, seven unequivocally high-technology Minimum List Heading sectors can be identified on the basis of Kelly's criteria (1986), namely pharmaceuticals, scientific instruments and systems, broadcasting and sound reproducing equipment, aerospace, and electronic computers, components, and capital goods. In 1981, some 59 per cent of UK employment in these high-technology sectors was already concentrated in the three regions listed above, with 46 per cent in the South East alone (Figure 1.5a). By 1984, their share was almost certainly even greater. Thus employment in electronic computer manufacturing in the South East apparently increased by 15,000 jobs, or 71 per cent, between 1976 and 1981, compared with a growth of only 3000 jobs, or 14 per cent, in the rest of Great Britain (Kelly 1986). Again, Sayer and Morgan (1986) show that by far the leading county of Great Britain for electronics industry employment growth generally in the late 1970s was Berkshire. So a very recent high-technology industry boom in particular sectors and areas of the South East appears to be one important factor in that region's – and London's ? – improved relative manufacturing performance since 1977.

High-technology industry expansion, most strikingly exemplified by a national (UK) growth rate for computer electronics manufacturing employment of 31 per cent between 1977 and 1983, can of course only be understood in the context of the so-called microchip revolution and very rapid technological change (Freeman 1986). Southern England's high-technology industry concentration is accordingly almost certainly linked to the similar marked regional bias in production, though not process, innovation rates recently identified for UK manufacturing plants generally (Goddard 1983b; Goddard and Thwaites 1983; see Ch. 7). It must also reflect the even greater concentration of separate industrial research and development establishments in this area revealed by Fig. 1.5b. Thus no less than two-thirds of all 1981 UK employment in separate R and D units was located in these three regions, with over half (53 per cent) in the South East alone (see also Howells 1984b). These considerations perhaps lend support to Hall's (1981) thesis that the industrial geography of the fifth Kondratieff cycle will differ from that of the fourth. 'Tomorrow's industries are not going to be born in yesterday's regions . . . . Britain's future, if it has one, is in that broad belt that runs from Oxford and Winchester through the Thames Valley and Milton Keynes to Cambridge.'

Hall's vision of a possible new industrial geography is clearly provocative and debatable. But in a period of undeniably rapid technological change, and in the context of apparently radically new industrial location dynamics in other European countries such as France (Aydalot 1983), three interrelated locational determinants can be identified which arguably are exerting a considerable influence on the evolving industrial geography of Britain in the 1980s, via their impact on technologically-advanced industries. The first of these is the perceived outstanding residential attractiveness of southern England, with the exception of urban London, as an influence on the location and availability both of highly qualified researchers and of entrepreneurs (Keeble and Gould 1985; Keeble and Kelly 1986). The former are fundamental to successful technological change and hence firm performance and growth in many modern industries, yet are peculiarly free, because of their scarcity value and high incomes, to vote with their feet in favour of high-amenity locations. The existence of a marked perceptual preference for residence in the climatically-favoured coastal and rural areas of the South West, East Anglia and the South East outside London is well documented, applying equally to such varied categories as sixthform-school leavers (Gould and White 1968), industrial managers and executives (Gleave, reported in Keeble 1980: 952), and skilled electronics workers (Cambridge Recruitment Consultants 1980). And there is considerable evidence that this has directly influenced high-technology companies, such as Inmos (Bristol), Sinclair (Cambridge), Plessey (Christchurch) and IBM (Hursley, Hampshire) in their choice of R and D and associated production locations.

A second location determinant for newer technologically-oriented industry markedly favouring southern England may well be links with, and spin-off of entrepreneurs and highly-qualified personnel from, universities and public and private sector research institutions. While an area of some debate, workers such as Marsh (1983) and Segal (1985) have argued that certain major scientific universities, and notably Cambridge, have had a substantial recent catalytic effect on the development of a local technology-oriented complex of interrelated firms. Equally, Breheny, Cheshire and Langridge (1983) have posited an 'indigenous' explanatory hypothesis for high-technology industry growth in the M4 corridor which focuses on new firm spin-off from the government research institutions which were located there after the Second World War. Certainly the role of universities and research institutions in generating a local pool of highly-qualified manpower for technology-oriented industry, and in the former case 'as a cultural amenity to attract new engineers and scientists', has been stressed by recent North American research on this phenomenon (Steed and DeGenova 1983 and by Kelly 1986).

Of course, the concentration of R and D units in southern England which underpins Breheny's 'indigenous' spin-off hypothesis itself requires explanation. This directs attention to the third interrelated factor, namely the advantage the area enjoys of exceptional national and international communications accessibility, centred upon proximity to London. This proximity to the country's dominant focus of government, high-level decision-making, and information of all kinds is fundamental to the historic concentration of so many research establishments, public and private, within a hundred miles of the capital: while continuing growth of foreign-owned high-technology firms (IMB, Honeywell, Digital Corporation, Hewlett-Packard) in southern England has undoubtedly been influenced not only by research staff life-style preferences, but also by the perceived need for rapid access to and from parent organizations via Heathrow airport.

The final possible factor in post-1978 regional

reversal overlaps with the newer industries debate. It is that during recession, the southern regions, this time including the East Midlands, have gained relatively from the long-term impact of much higher new manufacturing firm formation rates generally than is the case with the assisted regions. Certainly recent work has identified a significant difference in birth rates during the 1970s as between Hampshire, East Anglia and the East Midlands, on the one hand, and the North, North West and Scotland, on the other (Gould and Keeble 1984; Keeble 1986). This may largely be explained by historic differences in plant size structure, given the overwhelming dominance of small firms and plants as incubators of new entrepreneurs (Lloyd and Mason 1983), exacerbated by corporate large branch colonization of assisted regions since 1960. Regional occupational differences and educational levels may also have played a part (Storey 1982), while within southern England as in other EEC countries (Keeble and Wever, 1986), the migration of potential entrepreneurs to high-amenity residential areas has boosted firm formation rates in such locations as Cambridgeshire and South Hampshire (Keeble and Gould 1986). Certainly the south's marked bias to small plants and firms, during a recession which appears to have savaged disproportionately large plant and large enterprise employment in the UK, would seem to provide one explanation for its relatively better manufacturing employment performance. Equally, the national growth of small new manufacturing companies since 1970 has undoubtedly been concentrated in these southern regions. These trends are thus likely to have intensified further regional industrial differences in Britain between the high-unemployment, externally-controlled, large branch plant economies of the assisted regions, with manufacturing occupational structures biased towards relatively less skilled jobs (Watts 1981), and the smaller firm, more indigenously-controlled, entrepreneurial economies of southern England, with their more buoyant job markets biased towards higher-level non-production occupations.

References

Aydalot, P. 1983 *New spatial dynamisms in western Europe: the French case*. Centre Economie-Espace-Environnement, Université de Paris 1 Panthéon-Sorbonne.

Aydalot, P. 1984 Questions for regional economy. *Tijdschrift voor Economische en Sociale Geografie* **75**(1): 4–13.

Blackaby F (ed.) 1981 *De-industrialisation*. Heinemann.

Brech, M., Sharp, M. 1984 Inward investment: policy options for the United Kingdom. *Royal Institute of International Affairs, Chatham House Papers* 21, Routledge and Kegan Paul.

Breheny, M., Cheshire, P., Langridge, R. 1983 The anatomy of job creation? Industrial change in Britain's M4 corridor. *Built Environment* **9**(1): 61–71.

Cairncross, A. 1981 What is de-industrialisation? In F. Blackaby (ed.) *De- industrialisation*. Heinemann, 5–17.

Cambridge Recruitment Consultants 1980 *Some recruitment trends in the electronics industry: a discussion paper*. CRC, Cambridge.

Cameron, G. C., Kirwan, R. M., McGregor, A. M. 1981 *Highly qualified manpower and less favoured regions*. Unpublished report, Department of Land Economy, University of Cambridge.

Chisholm, M. 1985 De-industrialization and British regional policy. *Regional Studies* **19**(4): 301–13.

Cooke, P., Pires, A. da R. 1984 *Productive decentralisation in three European regions*. Paper presented at Institute of British Geographers conference, University of Durham.

Department of Trade and Industry 1983 *Regional industrial policy: some economic issues*.

Dicken, P. 1982 The industrial structure and the geography of manufacturing. In R. J. Johnston, J. C. Doornkamp (eds) *The changing geography of the United Kingdom*. Methuen, 171–201.

Dicken, P 1983 Japanese manufacturing investment in the United Kingdom: a flood or a mere trickle? *Area* **15**(4): 273–84.

Fothergill, S., Gudgin, G. 1979 Regional employment change: a sub-regional explanation. *Progress in Planning* **12**(3): 155–219.

Fothergill, S., Gudgin, G. 1982 *Unequal growth; urban and regional employment change in the U.K.* Heinemann.

Fothergill, S., Gudgin, G. 1983 Trends in regional manufacturing employment: the main influences. In J. B. Goddard, A. G. Champion (eds) *The urban and regional transformation of Britain*. Methuen, 27–50.

Fothergill, S., Gudgin, G., Kitson, M., Monk, S. 1984 Differences in the profitability of the UK manufacturing sector between conurbations and other areas. *Scottish Journal of Political Economy* **31**(1): 72–91.

Fothergill, S., Kitson, M., Monk, S. 1985 *Urban industrial change: the causes of the urban–rural contrast in manufacturing employment trends*. HMSO for the Departments of Environment, and Trade and Industry.

Freeman, C. 1983 *Long waves in the world economy*. Butterworths.

Freeman, C. 1986 The role of technical change in national economic development.

Freeman, C., Clark, J., Soete, L. 1982 *Unemployment and technical innovation: a study of long waves and economic development*. Frances Pinter.

Ganguly, P. 1982 Small firms survey: the international scene. *British Business* 19 Nov: 486–91.

Gillespie, A. E., Owen, D. W. 1984 *The conurbations and the recession*. Paper presented at Institute of British Geographers conference, University of Durham.

Goddard, J. B. 1983a Structural change in the British economy. In J. B. Goddard, A. G. Champion (eds) *The urban and regional transformation of Britain*. Methuen, 1–26.

Goddard, J. B. 1983b The geographical impact of technological change. In J. Patten (ed.) *The expanding city*. Academic Press, 103–24.

Goddard, J. B., Thwaites, A. T. 1983 *Technological innovation in a regional context: empirical evidence and policy options*. Paper presented to OECD Workshop on Research, Technology and Regional Policy, Paris.

Gould, A. 1984 *The location, organisation and development of foreign-owned firms in the UK paper and packaging industry*. Unpublished Ph.D. dissertation, University of Cambridge.

Gould, A., Keeble, D. 1984 New firms and rural industrialisation in East Anglia. *Regional Studies* **18**(2):, 189–201.

Gould, P. R., White, R. R. 1968 The mental maps of British school-leavers. *Regional Studies* **2**(2): 161–82.

Grant, R. 1983 The impact of EEC membership upon UK industrial performance. In R. Jenkins (ed.) *Britain and the EEC*. Macmillan, 87–110.

Gudgin, G. 1978 *Industrial location processes and regional employment growth*. Saxon House.

Gudgin, G. 1984 Employment creation by small and medium-sized firms in the UK. Dept. of Applied Economics, University of Cambridge.

Haig, R. M. 1926 Towards an understanding of the metropolis. *Quarterly Journal of Economics* **40**(2): 179–208; (3): 402–34.

Hall, P. 1981 The geography of the fifth Kondratieff cycle. *New Society* 26 March: 535–7.

Harris, D. 1984 Small business wants a bigger break. *The Times* 9 Feb. 17.

Hayter, R., Watts, H. D. 1983 The geography of enterprise: a reappraisal. *Progress in Human Geography* **7**(2): 157–81.

Howells, J. R. L. 1984a *Location, technology and filter-down theory: an analysis of the United Kingdom pharmaceutical industry*. Unpublished research dissertation, University of Cambridge.

Howells, J. R. L. 1984b The location of research and development: some observations and evidence from Britain. *Regional Studies* **18**(1): 13–29.

Johnstone, B. 1982 March of the robots points the way to industrial survival, *The Times* 20 Dec.

Keeble, D. 1971 Employment mobility in Britain. In M. Chisholm, G. Manners (eds) *Spatial policy problems of the British economy*. Cambridge University Press, 24–68.

Keeble, D. 1976 *Industrial location and planning in the United Kingdom*. Methuen.

Keeble, D. 1978 Industrial decline in the inner city and conurbation. *Transactions of the Institute of British Geographers* **3**(1): 101–14.

Keeble, D. 1980 Industrial decline, regional policy and the urban–rural manufacturing shift in the United Kingdom. *Environment and Planning A* **12**(8): 945–62.

Keeble, D. 1986 The changing spatial structure of economic activity and metropolitan decline in the United Kingdom. In H. J. Ewers, H. Matzerath and J. B. Goddard (eds) *The future of the metropolis*. de Gruyter.

Keeble, D., Gould, A. 1986 Entrepreneurship and manufacturing firm formation in rural regions: the East Anglian case. In M. J. Healey, B. W. Ilbery (eds) *Industrialization of the Countryside*. Geobooks.

Keeble, D., Kelly, T. 1986 New firms and high technology industry in the United Kingdom: the case of computer electronics. In D. Keeble and E. Wever (eds.) *New firms and regional development in Europe*. Croom Helm.

Keeble, D., McDermott, P. 1978 Organisation and industrial location in the United Kingdom. *Regional Studies* **12**(2): 139–41.

Keeble, D., Owens, P. L., Thompson, C. 1983 The urban–rural manufacturing shift in the European Community. *Urban Studies* **20**(4): 405–18.

Keeble, D., Wever, E. 1986 Introduction. In Keeble, D., Wever, E. (eds.) *New firms and regional development in Europe*. Croom Helm.

Kelly, T. 1983 *The location of high-technology industry in Great Britain: computer electronics*. Unpublished Ph.D. dissertation, University of Cambridge.

Lever, W. F. 1982 Urban scale as a determinant of employment growth or decline. In L. Collins (ed.) *Industrial decline and regeneration*. Department of Geography, University of Edinburgh.

Lloyd, P., Mason, C. 1983 New firm formation in the UK. *SSRC Newsletter* **49**: 23–4.

Mandel, E. 1980 *Long waves of capitalist development: the marxist interpretation*. Cambridge University Press.

Marsh, P. 1983 Britain's high technology entrepreneurs. *New Scientist* 10 Nov: 427–32.

Martin, R. L. 1982a Job loss and the regional incidence of redundancies in the current recession. *Cambridge Journal of Economics* **6**: 375–95.

Martin, R. L. 1982b Britain's slump: the regional anatomy of job loss. *Area* **14**(4): 257–64.

Massey, D. 1979 In what sense a regional problem? *Regional Studies* **13**(2): 233–43.

Massey, D. B., Meegan, R. A. 1978 Industrial restructuring versus the cities. *Urban Studies* **15**(3): 273–88.

McDermott, P., Taylor, M. 1982 *Industrial organisation and location*. Cambridge University Press.

Mensch, G. 1979 *Stalemate in technology: innovations overcome the depression*. Ballinger, New York.

Moore, B. C., Rhodes, J., Tyler, P. 1984 Geographical variations in industrial costs. *Department of Land Economy, University of Cambridge, Discussion Paper 12*.

Moore, B., Rhodes, J., Tyler, P. 1986 *The effects of government regional economic policy*. HMSO for the Department of Trade and Industry.

Morgan, K., Sayer, A. 1983 The international electronics industry and regional development in Britain. *University of Sussex, Urban and Regional Studies Working Paper 34*.

Owen, D. H., Coombes, M. G., Gillespie, A. E. 1983 The differential performance of urban and rural areas in the recession. *Centre for Urban and Regional Development Studies, University of Newcastle upon Tyne, Discussion Paper 49*.

Paine, S. 1979 Replacement of the West European migrant labour system by investment in the European periphery. In D. Seers, B. Schaffer, M.-L. Kiljunen (eds) *Underdeveloped Europe: studies in core-periphery relations*. Harvester Press.

Prais, S J. 1976 *The evolution of giant firms in Britain*. Cambridge University Press.

Randolph, W., Robert, S. 1981 Population redistribution in Great Britain, 1971–1981. *Town and Country Planning* **50**: 227–31.

Regional Studies Association 1983 *Report of an inquiry into regional problems in the United Kingdom*. Geo Books, Norwich.

Sayer, A., Morgan, K. 1986 The electronics industry and regional development in Britain. In A. Amin, J. B. Goddard (eds.) *Technological change, industrial restructuring and regional development*. Allen and Unwin.

Searjeant, G. 1983 Textiles SOS to government. *The Times*, 3 Mar., 19.

Segal Quince and Partners 1985 *The Cambridge phenomenon: the growth of high-technology industry in a university town*. Segal Quince and Partners, Cambridge.

Seneschall, M. 1984 *The spatial evolution of the giant EEC-based manufacturing firm, 1970–81*. Unpublished Ph.D. dissertation, University of Cambridge.

Smith, I. J. 1979 The effect of external takeovers on manufacturing employment change in the Northern region between 1963 and 1973. *Regional Studies* **13**(5): 421–37.

Smith, I. J. 1982 The role of acquisition in the spatial distribution of the foreign manufacturing sector in the United Kingdom. In M. Taylor, N. Thrift (eds) *The geography of multinationals*. Croom Helm, 221–51.

Steed, G. P. F., DeGenova, D. 1983 Ottawa's technology-oriented complex. *Canadian Geographer* **27**(3): 263–78.

Storey, D. J. 1982 *Entrepreneurship and the new firm*. Croom Helm.

Thirlwall, A. P. 1982 Deindustrialisation in the United Kingdom. *Lloyds Bank Review* **144**: 22–37.

Townroe, P. 1983 United Kingdom. In L. H. Klaassen, W. T. M. Molle (eds) *Industrial mobility and migration in the European Community*. Gower, 352–86.

Townsend, A. R. 1983 *The impact of recession on industry, employment and the regions, 1976–81*. Croom Helm.

Townsend, J., Henwood, F., Thomas, G., Pavitt, K., Wyatt, S. 1981 Science and technology indicators for the UK: innovations since 1945. *Science Policy Research Unit, University of Sussex, Occasional Paper 16*.

The Treasury 1983 *Economic Progress Report* 160, September.

Van Duijn, J. J. 1983 *The long wave in economic life*. Allen and Unwin.

Ward, M. F. 1982 Political economy, industrial location and the European motor car industry in the postwar period. *Regional Studies* **16**(6): 443–53.

Watts, H. D. 1980 *The large industrial enterprise: some spatial perspectives*. Croom Helm.

Watts, H. D. 1981 *The branch plant economy: a study of external control*. Longman.

Introduction to Industrial Location Theory

Geographers have generally been concerned with the identification of regularities in the spatial distribution of phenomena on the earth's surface. Amongst these phenomena, the distribution of manufacturing activity has been one of the prime foci of geographers' attention. Early observations that some manufacturing enterprises tended to be located close to the sources of their more important inputs whilst others were located at some distance from them but close to their markets, whilst again others were located at sites which were neither close to inputs or markets, led geographers to delineate a theory of location based almost entirely upon the respective magnitude of transport costs. This transport cost-minimizing approach implicitly assumes that all the other costs of production – labour, land, capital, and so on – do not vary in space, whilst revenue (sales) is also unaffected by location. Such assumptions are clearly very restricting, and as David Smith describes in Chapter 2, the neoclassical location theory approach was extended to include changes in labour costs, in land costs and, perhaps most importantly, differences in revenue at different locations. He graphically conjures up the idea of a cost surface and a revenue surface which, when one is subtracted from the other, jointly generate a profit surface. This device enabled geographers to move away from a study of the one optimal, profit-maximizing location for an enterprise to the study of whole areas in which enterprises might operate profitably, and introduced the concept of the spatial margins of profitability. Since around 1960, geographers have tended to emphasize the weaknesses of neoclassical location theory on the grounds that it is too rigid in its assumptions and fails to take into account the changing nature of corporate structure. Whilst neoclassical theory is appropriate to single plant enterprises each independently seeking a profit-maximizing location, the emergence of multiplant, and indeed multinational, companies with a hierarchy of different functions allocated to different locations requires an approach which pays more attention to 'the geography of enterprise' and theories of corporate behaviour and decision-making. David Smith responds in his chapter to these criticisms in two ways. Firstly, he argues that neoclassical theory can be used in a normative way, indicating what *ought* to happen and setting up a datum against which actual decisions can be measured so that the resulting difference becomes a focus of further study. The second approach is to argue that the concept of the least-cost or profit-maximizing point is still a useful and relevant analytical tool which can be applied, for example, at the global scale. Thus the cheap labour locations of the Third World become as important in a Weberian sense as the markets of North America or Western Europe in the formulation of a locational triangle. David Smith further points out that there are wider welfare implications of the neoclassical approach which are often ignored but play on important part in structuralist theories of industrial location. Just as the market impact of a factory location can theoretically be thought of as a cone in which market access peaks at the factory location and declines regularly with increasing distance, so the welfare benefits (employment, income) and the disbenefits (pollution, hazard) similarly are cone-shaped.

In Chapter 3, Peter Wood compares the

development of behavioural approaches to the treatment of industrial location since the mid-1960s, much of it a studied response to perceived inadequacies of the neoclassical approach. Whilst a part of the behavioural approach focused on small firms with their restricted ability to gather and to utilize the information necessary to identify optimal locations, the major part of it is concerned with the behaviour of large and complex industrial organizations both in response to competitive pressures and in response to government policies. Behavioural theories, like more developed neoclassical theories, recognize that decisions about locational choice are only part of the set of all decisions and that many 'nonlocational' decisions concerning the choice and pricing of production factors, product mix, production technology, may all influence the locational choice of companies. Whereas neoclassical theory has been criticized for being over-rigid and deterministic, behavioural theory, as Peter Wood admits, has not developed under the influence of any strong theoretical orthodoxy, and pragmatism and policy relevance have been the two main characteristics of their approach. Much of their 'theory' has been borrowed and adapted from the literature of management science and theories of organizational behaviour. In consequence, the behavioural approach has been criticized for having concentrated too much on the managerial focus, disregarding or undervaluing the role of the workforce or the local community.

Martin Boddy, in Chapter 4, describes the structural approach to industrial location and economic change. This approach attempts to relate the changing geography of industry and employment to the underlying structure of capitalist society, economic and class relations, and the social and spatial organization of commodity production. The key issue in the structural approach is the interaction and potential for conflict between capital and labour in their local, national and international contexts. The major argument is that capital, in the face of continuing and deepening recession, is forced to defensively restructure in a number of ways, each of which have implications for the amount of labour demanded. The most common responses are likely to be the reduction of capacity, the more intensive use of existing capacity, and the replacement of existing plant by newer technologies. In most cases these responses are likely to reduce the demand for labour. Locationally, structural theories are relatively weak, but studies of individual establishments and sectors have produced explanations of the particular failure of the inner areas of the largest cities to retain industrial employment in terms of older capital vintage and the existence of a reserve of secondary labour in the largest cities. It is conventional to regard the three approaches to industrial location as sequential, with behavioural theories representing a critical response to perceived inadequacies in neoclassical theories, and with structural theories responding to criticisms of both neoclassical and behavioural theories. A more challenging view is to regard all three theories as acting simultaneously, each making some contribution to our understanding of industrial decision-making behaviour, but with differing theories having more explanatory power for different types of enterprise.

CHAPTER 2

Neoclassical location theory

David M. Smith

The term neoclassical location theory is used, somewhat loosely, to identify an approach to problems of industrial location which has its origin in the work of Alfred Weber. It is *neo*classical in the sense that it comprises a substantial elaboration of Weber's 'classical' theory directed towards determination of the least-cost location. However, the approach goes well beyond Weber's preoccupation with spatial cost variations to incorporate variations in revenue, so the term 'neo-Weberian', which is occasionally adopted, is inappropriate to capture the full breadth of neoclassical location theory.

It is important at the outset to distinguish the subject matter of this chapter from what is generally termed neoclassical economics. Neoclassical economics is what most people today accept as mainstream economic theory, at least in the capitalist world. It was codified during the latter part of the nineteenth century and subsequently given considerable mathematical sophistication as a means of specifying the conditions under which the allocation of resources and distribution of outputs could be held to be optimal. Neoclassical location theory shares some of the characteristics and assumptions of neoclassical economics. For example, both bodies of knowledge are normative, in the sense that their concern is with what *should* take place in given circumstances rather than with the actual conduct of economic life, and both require superhuman feats of knowledge, ability and rationality if optimality in decision-making is actually to be achieved. Furthermore, it is not difficult to integrate locational analysis in the neoclassical tradition with conventional production theory, as Isard (1956), Moses (1958) and Smith (1971) have shown. However, neoclassical location theory is not merely a branch of neoclassical economics, if for no other reason than that the two 'classical' origins are different. Neoclassical location theory is thus not subject by definition alone to the full critique (or condemnation) levelled at neoclassical economics in recent years by radical or Marxist economists.

This is not to say that neoclassical location theory is robust enough to withstand critique on other grounds; far from it. The rise of approaches to industrial location which stress business behaviour, industrial organization and economic structure can be attributed to some extent to the inadequacy of neoclassical location theory as a positive framework for understanding the industrial world as actually observed. The two chapters which follow will elaborate this argument. The present chapter seeks to identify the salient characteristics and abiding strength of neoclassical location theory, as a normative framework with an adaptability and contemporary relevance inadequately recognized both by those who see it as an indissoluble element of neoclassical economics and by those who have sought to build advocacy of new approaches on an exaggerated critique of the neoclassical position.

2.1 Weberian antecedents

The central features of Weber's theory of industrial location (Weber 1909) are too familiar to require extensive repetition here. The 'locational triangle' with a market at one corner and material sources at the other two is virtually obligatory content in geography text books from O-level onwards, as is the mechanical model of pulleys and weights (the Varignon frame) used to explain the determination of least-cost location. However, some consideration must be given to these simple conceptions, as from them arise both the fundamental strengths and the major limitations of neoclassical theory.

Weber was concerned with *optimality* in plant location, at least in that part of his work most familiar today. Optimality meant least-cost location, which was initially considered purely in terms of transportation. The problem was to find the single point within a locational triangle where the cost of shipping two materials from their respective sources and the finished product to its market would be minimized. The resolution was portrayed in mechanical terms: each corner of the triangle exerts a pull on plant location, proportional to the volume of material or finished product shipped from or to that point and inversely proportional to distance moved (or cost of shipping per unit of volume per unit of distance).

As an initial statement of the now familiar aggregate travel model, Weber's locational triangle was a conception of brilliant simplicity. His attempts to extend the framework to accommodate the complications of labour costs and agglomeration economies were clumsy if ingenious, and would hardly bear repetition today had it not been for one fortuitous by-product. This is best illustrated with respect to labour costs. Weber introduced this complication by assuming a point at which labour costs less than its (uniform) price elsewhere. He posed this question: in what circumstances would the optimal plant location be diverted from the point of least-transport cost to the point of low labour cost? His analysis is illustrated in Figure 2.1, where P is the least-transport-cost location within the triangle comprising material sources M_1 and M_2 and the market C. A source of cheap labour at L_1 would permit a reduction of £3.00 (per unit of output) in the cost of labour compared with point P. However, whether this would be sufficient to divert the plant from P to L_1 depends on the additional transport costs which would be incurred at L_1. Weber portrayed the rising transport costs away from P as a series of 'isodapanes', or lines joining points of equal additional transport costs compared with the least-transport-cost location. The one with the same value as the labour cost saving at L_1 was termed the 'critical isodapane'.

Thereby was derived the following rule: the

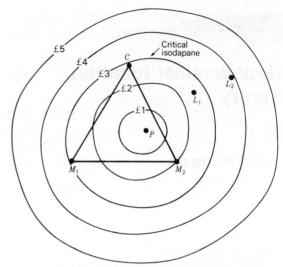

Fig. 2.1 Weber's analysis of the effect of a cheap labour location, using 'isodapanes' or lines joining points of equal additional transportation cost.
Source: Based on Smith (1981 Fig. 4.2)

optimum location would be diverted to the cheap labour location providing that it was within the critical isodapane. In Fig. 2.1 the critical isodapane has the value of £3.00, L_1 is within it, so L_1 becomes the optimum location; the saving of £3.00 in labour costs more than compensates for the increase in transport costs (less than £3.00). Beyond the critical isodapane, transport costs rise by more than £3.00, so a cheap labour location at L_2 with a saving of only £3.00 would be sub-optimal.

While the isodapane was, to Weber, an incidental device for solving a particular problem, it took on a much greater significance in the development of neoclassical location theory. It permitted a broadening of vision, from the single point of locational optimality to the spatially differentiated surface of transportation cost. And once the isodapane map is seen as a particular case of the cost surface in general, the way is open to conceive of the cost of single inputs, total production costs, distribution costs, revenue and even profit as three-dimensional surfaces analogous to topographical contour maps. It is difficult to imagine any abstract economic construct sitting more comfortably within the field of geography, yet it was half a century before the concept of the cost surface came into its own in the geographical literature. What continuity and development took place in the interim was largely the responsibility of Weber's successors in the field of spatial economics, most notably Palander (1935) and Hoover (1937; 1948).

Before leaving Weber, it is important to recognize the selectivity with which his contribution to location theory has been reproduced and incorporated into later approaches. It is for the narrowly technical analysis of transport costs and for the mechanics of the location triangle solution that Weber is most widely remembered. Incomplete and unperceptive readings of Weber's text are partly to blame, as are brief textbook summaries, but this selectivity also reflects the longstanding acceptance of industrial location as largely if not entirely a technical matter. The locational triangle, like Christaller's central-place hexagons, fits neatly into the a-social world of geometrical regularity promoted by the 'location analysis' school associated with geography's quantitative and model-building revolution. The elusive nature of the point of minimum aggregate travel or median centre in two-dimensional descriptive statistics, added to the attraction by posing a challenge to the narrow mathematical task of devising a suitable algorithm. The real world of social relationships, labour exploitation, unemployment and so on all but vanished.

If neoclassical location theory is a legitimate target for those advocating a structural approach, which sees industrial location as a social as well as a technical matter, Weber himself was hardly to blame. As Gregory (1980) has pointed out, Weber had been trained in political economy, and he placed explicit restrictions on his abstract formulation of location theory. Quoting Weber himself in English translation (1929: 12–13):

> We shall see that the kind of industrial location which we have today is not entirely explained by the 'pure' rules of location, and therefore is not purely 'economic'. It results to a large extent rather from very definite central aspects of modern capitalism and is a function of modern capitalism which might disappear with it. It results, we may say in hinting at the main point, from degrading labour to a commodity bought today and sold tomorrow, and from the ensuing laws determining the labour market and from the local 'agglomeration of workers' created thereby.

This problem would have to be solved by the 'realistic' theory, towards which his work was leading (Weber 1929: 225–6):

> For at this point it becomes necessary to consider particular economic systems. I do not wish to assert that the creation and development of labour locations can be explained by economic reasons; but if it can be so explained the reasons will be related to the position of which the particular economic system gives to labour. . . . The further realistic theory must therefore consider how labour is handled in the particular economic system which is studied.

Had these passages, and their implications, been given the subsequent attention afforded the locational triangle, the development of neoclassical location theory would doubtless have been broader and richer than has actually been the case.

2.2 The generalized variable-cost model

At a technical level, Weber's analysis foundered on his failure to incorporate convincingly considerations other than cost of transportation. The cul-de-sac explored for a solution of the problems of cheap labour and agglomeration was not entirely without interest, but the way out was to transcend the triangular starting point. To extend the framework from a three-cornered figure to one with four, five or n corners permits the incorporation of more material sources and markets and thus a more realistic transport situation. However, the crucial step is to include sources of non-material inputs from the outset. Thus, the source of cheap labour in Weber's analysis would be added to the original triangle *before* resolution of the optimum location. Similarly, sources of power, capital, cheap land and so on could be built into the initial locational figure (Smith 1981: 149–51).

While this generalization of the Weber model requires little mental agility, the problem of solving the least-total-cost location is formidable. Strict adherence to Weber's original formulation is virtually impossible as soon as the fairly regular cost–distance functions in transportation are recognized to be quite unsatisfactory representations of how the cost of most other inputs varies in geographical space. Both the original conception of corners exerting a pull on plant location and the family of operational research methods devised to solve the generalized Weber problem require relatively simple if not linear cost–distance functions with respect to specific origins of inputs. While the cost of land in a city might fall by some fairly regular distance–decay function from the peak value intersection, and while labour costs may sometimes be subject to systematic spatial variations within and between regions, these are exceptions to the generally much more haphazard pattern of spatial cost variation of non-transportation inputs. Even transport costs in reality are often far more complex in their spatial expression than a simple function of distance.

Weber himself anticipated a solution to the generalized model, with his concept of isodapanes. Each input can be regarded as having a spatial cost surface, which at any point represents the cost of acquiring the quantity necessary for a particular volume of output. The summation of all individual input cost surfaces gives a surface of total cost. At any possible plant location i:

$$TC_i = \sum_{j=1}^{n} Q_j U_{ij}$$

where TC_i is total cost at i
Q_j is required quantity of input j
U_{ij} is unit cost of j at i

and the summation is for n inputs. The optimum (least-total-cost) location is where TC is minimized. This formulation is dependent on two critical assumptions (to which the discussion will return later). First, the result holds for a particular volume of output, as input costs may vary with quantity required. Secondly, the production function or combination of inputs is the same in all locations, i.e. the input coefficients Q are spatial constants. These assumptions are connected, in that change in scale may bring substitution among inputs and a change in the production function.

However complex the general variable-cost model may be, it is true to its Weberian origins in one crucial respect. Just as Weber saw the pull or attraction of a corner of his triangle as proportional to volume to be shipped and cost of overcoming distance, so the significance of any input as an influence on the optimum location will depend on both its contribution to total cost and the extent of its spatial cost variations.

There are, of course, practical obstacles to the determination of the least-total-cost location of an industry. Derivation of an accurate total-cost surface requires individual input cost surfaces, for some (if not most) of which data may be incomplete or even unavailable. Computer graphic methods make interpolation from limited data fairly easy today, and surfaces constructed from different sets of control points can be amalgamated. For most practical purposes, however, (e.g. in actual plant site selection) something less than a full total-cost surface is generally sufficient. Inputs not regarded as significant can be omitted. More important, the problem can be framed in terms of selection among a limited number of discrete points rather than a continuous surface. Thus, the pragmatism of a comparative-cost analysis of selected locations replaces the elegance of the general variable-cost model.

2.3 Revenue as a spatial variable

A major shortcoming of variable-cost theory with its roots in the work of Weber is neglect of demand considerations. Implicitly or explicitly, revenue is assumed to be a spatial constant, leaving spatial variations in costs to determine the optimum (profit-maximizing) location. In part, this one-sided approach reflects the economist's predisposition towards stripping a situation down to essentials by making simplifying assumptions; only in this way was Weber able to achieve his initial penetration of the plant location problem in terms of transport costs. However, preoccupation with the cost side also arises from the greater conceptual and practical difficulty of dealing with demand and revenue.

The first steps in the direction of a variable-revenue approach came quite incidentally, in a context much closer to the concerns of mainstream economics than was plant location. During the 1920s and 1930s a debate was engaged as to the effect of space (and other things) on the theory of perfect competition. It was recognized that location in space may give a supplier some degree of monopolistic control over a market, and this led to consideration of possible competitive strategies. Most familiar to geographers is the contribution of Hotelling (1929), subsequently elaborated and popularized in textbooks as the case of ice-cream sellers competing for customers on a beach. A full exposition of what is alternatively referred to as the 'market area' or 'locational interdependence' approach will be found elsewhere (Smith 1981: 91–7); all that will be done here is to explain briefly how demand and spatial competition generate variations from place to place in the revenue which can be earned.

Total revenue obtainable by a plant in any location i can be defined as follows:

$$TR_i = \sum_{j=1}^{n} Q_j P_j$$

where TR_i is revenue earned at location i

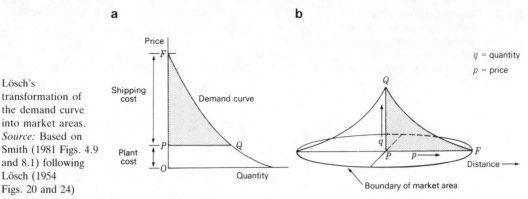

Fig. 2.2 Lösch's transformation of the demand curve into market areas. *Source:* Based on Smith (1981 Figs. 4.9 and 8.1) following Lösch (1954 Figs. 20 and 24)

Q_j is quantity sold in market j
P_j is price prevailing at j

and the summation is over n markets. Total revenue is by definition the product of quantity sold and price obtained, so the relationship between Q and P is important. Demand or consumption is usually held to be dependent on price such that as price increases, demand is reduced. The sensitivity of changes in demand to changes in price is the elasticity of demand. At the extreme of inelasticity, where demand is constant irrespective of price, spatial variations in price have no bearing on Q and the location problem becomes trivial. Thus, in location theory some elasticity of demand is generally assumed.

The conventional demand curve of elementary economics provides a starting point. Figure 2.2a displays demand horizontally and price vertically. The level of demand at the plant (P) is Q, but more important is F at which point demand ceases as the price is higher than the consumer is prepared to pay. It was Lösch (1954: 106) who achieved the spatial transformation of this diagram by the simple device of rotating it through 90 degrees and setting distance on the horizontal axis (Fig. 2.2b). Rises in price are thus portrayed as proportional to distance from the plant location (P). The *delivered* price reflects the addition of transportation and other distribution costs to the cost of production at the plant. The point F on the price/distance axis now represents the limit to the market area served from the plant. Lösch rotated this point about O to derive the market area in two-dimensional space. At any point in the distance dimension, demand can be read off the demand curve. Total demand becomes proportional to the area of the triangle PQF assuming a continuous series of market points from P to F, or to the volume of Lösch's demand cone in two-dimensional space (through which PQF is a section).

To follow Lösch further into his elegant derivation of 'economic landscapes' would be an indulgence; a more limited analysis is sufficient to take the argument from market areas through spatial competition to variations in revenue. Figure 2.3 represents the analysis of Fig. 2.2 in a manner

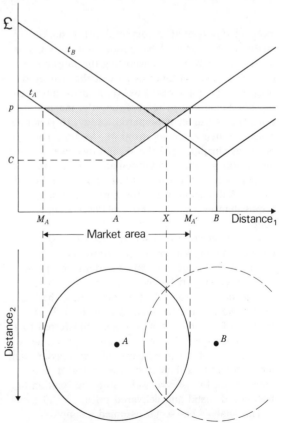

Fig. 2.3 The derivation of market areas, showing the impact of competition

Fig. 2.4 Regional differences in market areas under different conditions of: (a) transport costs; (b) demand; and (c) production costs

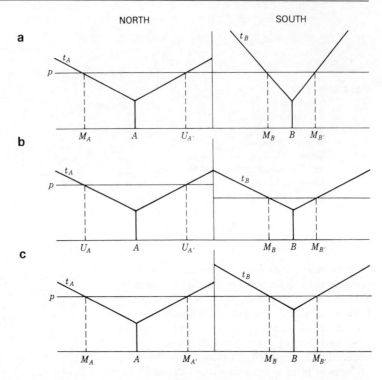

more readily related to competition for market territory. A plant at A has production costs C and passes on the cost of transporting the finished product to the consumer as part of the delivered price t_A. The gradients of delivered price intersect the horizontal line p, representing the maximum price the consumer is prepared to pay, to generate a market area extending from M_A to $M_{A'}$. This becomes circular when the second distance dimension is added. Total demand (or revenue) is proportional to the size of the market area and to the shaded portion in the top part of the diagram (equivalent to a section through Lösch's demand cone).

The introduction of a second plant (at B) enables the implications of competition to be considered. Production costs and delivered price (t_B) are assumed the same as for plant A. The intersection of t_A and t_B at a point below the line p makes the two market areas intersect, at X. A portion of A's market area is thus transferred to B, with a consequent loss of demand and revenue. The revenue that can be earned from a given plant location will thus depend on the extent of local competition for sales as well as on the relationship between demand and delivered price.

The analysis has so far assumed a simple demand function in which price is the only consideration. Other things that may be relevant to demand at any market point include the population (i.e number of consumers) and such characteristics as their incomes, taste and preferences. As these conditions vary spatially, so will demand and potential revenue.

Figure 2.4 shows how size of market area (and hence volume of sales and revenue) can vary in space with differences in three conditions represented in Fig. 2.3. Two regions are assumed ('North' and 'South') with different conditions with respect to: (a) delivered price; (b) consumer demand; and (c) production costs. In Fig. 2.4a, transport costs are higher in the South than the North, perhaps because of more rugged terrain, though other conditions are the same as in the North; the result is a smaller market area in the South, and lower revenue. In Fig. 2.4b, consumers have a different demand curve in the South than in the North, reflected in a lower maximum price that they are prepared to pay; the market area is smaller as a result. In Fig. 2.4c, production costs are higher in the South than in the North; again the size of the market and of total revenue is affected.

It is a simple step from a two-region case to recognition that the conditions upon which total revenue depends (including degree of spatial competition) can be continuous spatial variables. Revenue also becomes a continuous spatial variable, analogous to the total-cost surface, with one point at which total revenue will be maximized.

2.4 The spatial interaction of cost and revenue

It will be apparent from the two previous sections that the variable-cost and variable-revenue approaches to industrial location theory have developed almost entirely independently. Cost variations were introduced in Fig. 2.4c above, but not in a manner that involved full interaction with other variables. To go further requires coming to terms with the scale effect – something that was held constant even in the exposition of the variable-cost approach, as explained above. Attempts to fuse spatially-variable costs and demand (or revenue) together in a single framework founder on their interdependence.

To put the problem as simply as possible, we can identify the least-cost location with respect to a given level of output (or demand), but this need not be the level required to maximize total revenue; similarly, to find the revenue-maximizing location involves some assumption with respect to production costs, but these will depend on the scale required to maximize revenue at the location yet to be determined. Whether the profit-maximizing location will coincide with that of minimum costs or maximum revenue in any specific situation is an empirical rather than a theoretical question. Lösch (1954: 29) put it as follows:

> A geometrical solution becomes impossible as soon as price and quantity are added to the two spatial variables, for it can be applied to three variables at most. Yet algebraic treatment leads to equations of an insoluble degree. This complexity stems from the facts that, as already explained, there is more than one geographical point where the total demand of a surrounding district is at a maximum, and that from these points outward total demand does not decrease according to a simple function. We are thus reduced to determine separately for every one of a number of virtual factory locations the total attainable demand, and for similar reasons the best volume of production as a function of factory price (market and cost analysis). The greatest profit attainable at each of these points can be determined from the cost and demand curves, and from this place of greatest money profits, the optimum location can be found. Now the procedure is no longer theoretical, however, but simply empirical testing, since the result holds only for the locations actually examined and cannot be interpolated. As all points in an area can never be analyzed in this manner, we cannot exclude the possibility that among the locations not examined there may be one that would yield a higher return than the most advantageous of those investigated. There is no scientific and unequivocal solution for the location of the individual firm; but only a practical one: the test of trial and error. Hence Weber's and all the other attempts at a systematic and valid location theory for the individual firm were doomed to failure.

At the root of this problem is the mutual interdependence of the three basic production decisions of location, scale and technique. Different levels of demand (in different places) require different volumes of output. Different scales of production can require different techniques, and substitution among inputs can generate different patterns of cost variation in space and different least-cost locations. Thus, there can be a different optimum location for each scale of output. However, there can be only one optimum scale, and only one optimum combination of inputs, as is recognized in production theory, and there can be only one true optimum location. As each optimum is by definition associated with profit maximization, it is necessary to recognize a single global optimum, with respect to location, scale and technique simultaneously (Smith 1981: 190). Location theory and production theory are therefore indissoluble.

It will be clear from this argument that to conceive of a total-cost surface and a total-revenue surface interacting to generate a surface of profits requires some heroic assumptions. Major considerations on either the cost or the demand side must be held constant to stabilize the interdependencies, if meaning is given to the concept of optimum location as something other than a fleeting characteristic of an indefinable point. However, to bring cost and revenue surfaces together in some grand neoclassical synthesis is more than an exercise in intellectual obstinacy, for it leads to a further concept of significance barely diminished by its elusive nature.

The 'spatial margin to profitability' was first suggested by Rawstron (1958) in a paper the importance of which was barely recognized at the

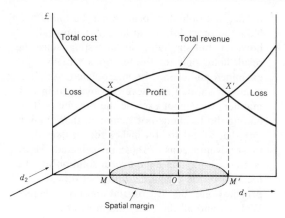

Fig. 2.5 Derivation of the spatial margin to profitability from space cost and revenue curves

time and which has been insufficiently appreciated ever since; Rawstron's exposition of spatial limits to the area in which profitable production is possible was a brilliant feat of intuitive insight. At a superficial level a statement of the obvious, when seen as an integral element in neoclassical location theory the spatial margin not only permits the incorporation of suboptimal decision-making but also contributes to the integration of location and production theory. It would be hard to imagine a more original geographical contribution to a field dominated, for the most part, by economists.

A simple diagram is required to explain the derivation of the spatial margin. In Fig. 2.5, sections through surfaces of total cost and total revenue are shown as a 'space cost curve' and 'space revenue curve' respectively. The interactions of these curves, where cost and revenue are the same, generate the points M and M' on the distance axis. These are the spatial margins to the area within which profitable operation is possible, total revenue exceeding total cost; beyond them, a loss would be sustained. Introducing the second distance dimension in the diagram (without turning the curves into surfaces) shows the spatial extent of the area within the margin. The optimal location is at O, where profit is maximized, but anywhere within the margin will provide some profit and thus accommodate firms unable or unwilling to seek and find the optimum.

The concept of the spatial margin enabled neoclassical theory to incorporate, if not explain, the economic fact of life of suboptimal decision-making. A hitherto normative body of theory, directed towards the determination of a single optimum point, could contribute at least one important behavioural principle: that freedom of choice is spatially constrained, if plant viability is to be achieved. How wide or narrow the margins will be in specific circumstances depends on the economic conditions generating the prevailing cost and revenue topography, over which the individual entrepreneur or corporation has at best only limited control, even in a world of multinational firms.

The spatial margin to profitability is, of course, subject to the same reservations as the optimum locations, in that its position will be the outcome of a set of intricate mutual interdependencies. Different spatial margins can apply to different volumes of output, via the effect of technique on scale, and they can shift in circumstances where the optimum location remains the same. Elusive though the margins may be, they do enable a further direct link to be made with production theory. In elementary economics there is a minimum and a maximum scale (where $TC = TR$) consistent with profitable operation, to which the term 'break-even points' is sometimes applied. Between these points in a plot of total cost against total revenue some profit can be made, though there is of course one only single optimum scale. The spatial margins are analogous to these break-even points. Furthermore, they are analytically related. Just as optimum scale creates an optimum location, as explained above, so the production limits to viability are at the same time limits to locational choice (Smith 1981: 190).

2.5 Practical applications

Whether normative or positive, a body of theory is of little utility unless its concepts can be given operational identity. Neoclassical location theory is capable of finding the optimum location only if spatial variations in cost and revenue can be identified, as surfaces or some equivalent adequate to the task at hand. The opportunities and limitations are considered fully elsewhere (Smith 1981); all that is offered here are a few illustrations and references to major applications in the literature.

Cost surfaces with respect to single inputs are fairly easily identified if transportation is the major element in spatial variability. In other cases, sample data for discrete points has to be found, and a surface interpolated. An example using computer graphics is provided in Fig. 2.6, which shows labour costs in a section of the British electronics industry. The information was compiled

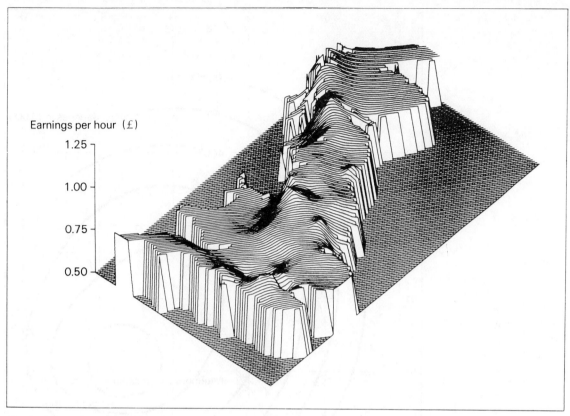

Fig. 2.6 A surface of female labour costs in the components section of the British electronics industry. *Source:* Robinson (1979), reproduced in Smith (1981 Fig. 11.6)

from enquiries made of a random sample of thirty-eight factories representing a broad range of the type of units operating in the industry. The surface reveals the cost advantages associated with cheap female labour in some of the peripheral regions.

Figure 2.7 shows the resolutions of an extended Weberian locational figure with respect to transportation costs incurred in the assembly of inputs and distribution of finished products on the part of a fireworks manufacturer. There are forty input sources and four markets. Input costs at source were assumed away in this case. The minimum cost location has been identified by one of the computer algorithms designed for the purpose, which has been extended to generate data required to specify the transport cost surface by way of isodapanes (Taylor 1975). The actual plant location is in the Thanet area of Kent, whereas the optimum is in London. The cost difference was estimated to be £2,100 per 500 tons of product.

While examples of single-input cost surfaces and solutions of generalized Weber-type transportation problems are not infrequent in the literature on industrial location, more comprehensive applications of neoclassical theory are still quite rare. The exceptions do indicate something of the strength and versatility of the neoclassical framework when applied with sufficient dedication, however. The study of the Swedish paper industry by Lindberg (1951) shows how the impact of technological change on plant location is amenable to analysis by cost surfaces, while a similar problem is addressed by Lewis (1970) for the paper industry in England and Wales. The work of Tornqvist (1962) on the Swedish light clothing industry is another example of the careful identification of cost surfaces. Kennelly (1954–55), on the Mexican steel industry, illustrates different approaches to the identification of the optimum location in the Weber tradition.

The spatial margin to profitability has proved predictably unyielding to those bent on giving it empirical identity. Of three successful attempts (Taylor 1970; McDermott 1973; Haddad and Schwartzman 1974), two express scepticism as to the utility of the concept. If the spatial margin is

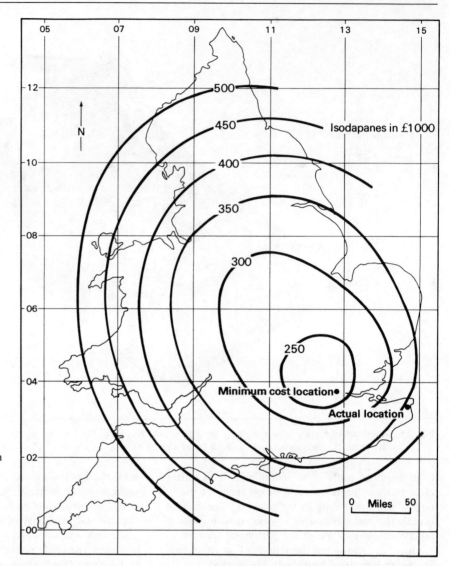

Fig. 2.7 Optimum location and isodapanes for a fireworks factory. *Source:* Taylor (1975), reproduced in Smith (1981 Fig. 12.4)

of any value in explaining reality, then it is probably more as an aid to informal speculative interpretations than as an element in rigorous analysis. For example, Fig. 2.8 employs hypothetical space cost and revenue curves in identifying tentative spatial margins to suggest the basis for regional specialization of manufacturing in nineteenth-century Britain. The concept of the spatial margin has been used in more detailed investigation of one of these regions – the knitwear or hosiery district of the East Midlands – as its spatial limits expanded and contracted during the second half of the nineteenth century (Smith 1970).

It is significant that practical applications of neoclassical location theory are almost exclusively based on the variable-cost version. Despite some not entirely convincing flirtations with the market potential concept as a surrogate for spatial variations in demand, the empirical identification of revenue surfaces has thus far defied human ingenuity and perseverance. The problem is more one of practicability than of inadequate conceptualization; demand is an intrinsically more elusive phenomenon than cost.

2.6 Neoclassical welfare formulations

As a normative body of theory, neoclassical location analysis is of greater direct relevance to welfare issues than are the behavioural, organizational and structural perspectives. This

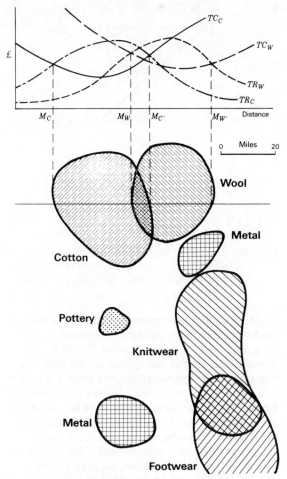

Fig. 2.8 A simplified interpretation of the emergence of regions of industrial specialization in nineteenth-century England. *Source:* Smith (1981 Fig. 9.7)

aspect of neoclassical theory has not been extensively developed, however, and is almost completely overlooked by ardent critics of the neoclassical framework. What follows is merely suggestive of some of the more obvious welfare applications.

The simple determination of least-cost location of an industrial plant can be considered an exercise in welfare maximization. The profit-seeking entrepreneur who maximizes profits by minimizing costs would also maximize personal satisfaction or utility (other things being equal) under the assumptions of *homo economicus*, by selecting the least-cost location. More generally, in a perfect market system with the welfare-maximizing properties bestowed on it in the purest version of neoclassical economics, profit maximization is a necessary element in the equilibrium solution and locational choice is inseparable from choice of scale and technique in achieving maximum profits.

More interesting, however, is to consider optimum location in a broader social sense, i.e. within a system where public involvement is regarded as a necessary and legitimate means of enhancing the general welfare. A cost-minimizing (or profit-maximizing) location is an efficiency solution to a resource-allocation problem. It can have the same status for a public facility as for a privately-owned plant, i.e. it minimizes the investment required for a given level of output or maximizes the positive difference between value of output and resources committed. Thus, in the public sector as in the private sector, the kind of optimum location generated by conventional neoclassical analysis is the best guide to where a new plant should be located, providing that there are no extraneous (e.g. non-pecuniary) social considerations – a matter to which we shall return below.

Weber-type models are thus applicable in the solution of public facility location problems. This even extends into the non-industrial sphere. For example, the classical Weber model has been used to identify the optimum location for a new maternity hospital (Smith 1979: 315–19). The problem was posed purely in terms of transportation: the solution sought was the location which would minimize the aggregate travel of mothers-to-be visiting the hospital. Women were thus the inputs to be shipped, and it was shown that their numbers could be weighted by indicators of need (i.e. infant mortality and car ownership) specific to their areas of origin. The optimum location was found by applying one of the algorithms designed to solve the generalized Weber transportation problem.

There are some fairly obvious applications of neoclassical location theory in the field of regional development policy. As measures aimed at subsidizing industry in what are assumed to be unattractive locations by normal business criteria are central to industrial dispersal policy in Britain and many other countries, it is surprising that little attention has been given to the identification of the actual cost (or profit) penalties involved. Figure 2.9 illustrates a hypothetical situation. There is a metropolitan core (e.g. London) and some depressed peripheral regions. Two industries are being considered as possible instruments in regional development strategy (itself a radical departure from traditional British practice of not being selective among industrial sectors). In industry 1,

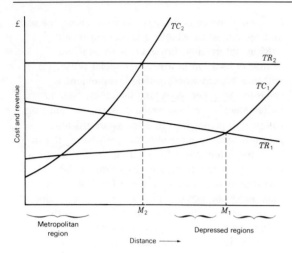

Fig. 2.9 Hypothetical space cost and revenue profiles for two industries in a nation with a metropolitan core and depressed periphery. *Source:* Smith (1981)

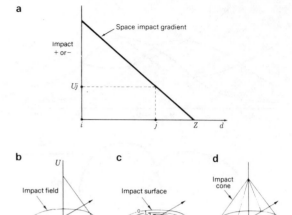

Fig. 2.10 A model of the spatial impact of industrial plants based on an analogy with market area analysis. *Source:* Based on Smith (1977 Fig. 4.7; 1981 Fig. 14.3)

costs rise gradually away from the metropolitan region and revenue falls; the spatial margin is somewhere within the depressed regions. Industry 2 has revenue as a spatial constant but a steep rise in costs with distance from the metropolis; the margin (M_2) is closer to the metropolis than for industry 1.

In these circumstances it is clear that industry 1 is potentially more mobile and offers more promising prospects of diversion from core to periphery by state subsidies. Policy could be designed accordingly. The variable-cost analysis required to generate this conclusion could also be used to calculate the level of subsidy which might encourage relocation into the depressed regions, and to identify the most significant input costs to be subsidized. Such a policy could well be more effective than the blanket subsidies which have characterized the unimaginative British industrial location policy for decades.

Industrial location planning can be conceived of as an attempt to induce a certain local 'impact' (Smith 1981: 359–85). If the impact is positive, e.g. employment or income generation, then the state would be sensible to seek the location from which this would be maximized. If negative, e.g. noise or air pollution, then the social or welfare optimum would be where impact is minimized. Such problems are in principle amenable to solution by an analogy of market area analysis in the neoclassical tradition. Figure 2.10 shows a 'space impact gradient' (a), falling away from a plant location i along the single distance dimension d; it might measure local income generated (positive) or smoke emission from the factory chimney (negative). Impact ceases at point Z, which identifies the limit of the spatial 'impact field' (b). The intensity of impact (U) at any point j, which is a function of distance from i, generates an 'impact surface' (c), an obvious example of which would be noise contours about an airport. The 'impact cone' (d) describes the total volume of impact and is analogous to Lösch's demand cone. The volume of impact will vary with location, according to the nature of the impact gradient and the density of population.

The application of such devices as illustrated in Fig. 2.10 is constrained by practical and conceptual problems. At a practical level, the variables involved and the technical production relations (i.e. how much impact will be generated) are often hard to identify. When physical properties amenable to instrumental measurement are involved, e.g. noise or smoke emission, the problem is fairly straightforward, but assessing the impact of a new plant on local employment or income still remains hazardous despite years of experimentation with local multipliers and input–output models.

The notion of a welfare optimum in industrial location inevitably involves some hard conceptual issues. To maximize (positive) impact, or some property like accessibility *in aggregate*, may conflict with distributional equity. For example, the locations which maximize the generation of new jobs or minimize the number of people within a critical smoke emission level still leave the

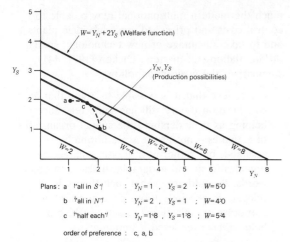

Fig. 2.11 A welfare formulation of a simple industrial development problem. *Source:* Smith (1981 Fig. 14.4)

conditions in question unequally distributed and there may be locations which are less satisfactory with respect to aggregate impact but which have benefits and penalties more equally distributed among localities. Scale becomes a relevant consideration: a larger number of smaller facilities more evenly distributes impact, but there is likely to be an efficiency cost.

The formidable intellectual apparatus of neoclassical welfare economics can be brought to bear on questions of social, as opposed to private, optimality in plant location. While conventional welfare economics has been discredited as an account of how the world actually works, it does provide an effective framework for structuring problems of public policy. In theory, the welfare-maximizing decision is determined by the interaction of technical conditions governing production possibilities from given resources and a social preference function which purports to identify those elements of consumption, in the broadest sense, on which welfare depends. In practice, the welfare function can comprise planning objectives.

Figure 2.11 provides a simple illustration. There are two regions: North (N) and South (S). A certain volume of new investment is planned, with the objective of increasing income by use of industrial development. The South is a depressed region and preference is given to development there such that an increase in income would be assigned twice the social value as the same increase in the North: hence the social welfare function of $W = Y_N + 2Y_S$. This function can be drawn as lines of equal increments of social welfare (welfare contours). Three alternative investment strategies are examined: to locate a single plant in the South, a single plant in the North, or to set up two plants involving half the investment in each region. Technical conditions, including inter-regional multiplier effects, are such that each strategy generates some increment of income in both regions, the specific outcomes being as shown in the figure. The three different combinations of income, plotted on the graph, can be thought of as points on a production possibilities frontier showing what can be achieved with given resources. The optimum strategy is read off the graph where the lowest welfare contour (maximum welfare attainable) is tangent to the production possibilities frontier: this is to locate plants in both regions. The solution is confirmed by putting actual values for Y in the welfare function.

More complex problems of public resource allocation can be structured in a similar manner. Programming methods provide means of solution. However, it requires a strong element of state control to make the most of these methods. Taken to their technical limits, they are capable of specifying the optimal spatial arrangement of productive capacity and service infrastructure to satisfy a standard of living defined in terms of specific quantities of goods and services supplied, against a set of resource constraints. The most interesting example of this type of industrial location planning is probably in Siberia, though there are questions as to how far practice follows what the models prescribe (Smith 1981: 410–18).

2.7 Conclusion

It will be clear from the content of this chapter that the application of neoclassical theory to industrial location problems is limited by operational difficulties. (In this respect, it is no different from behavioural and structural approaches, which have not been conspicuously embellished by convincing case studies.) The variables affecting spatial differences in cost and revenue are often hard to measure accurately, and their inter-relationships pose conceptual as well as practical obstacles. Furthermore, such common characteristics of the modern industrial world as intricate input–output linkages and associated agglomeration economies comprise additional if not unsurmountable complexities.

The contemporary critique of neoclassical theory goes further, however. Despite extensive developments on Weber's original conceptions, neoclassical theory has tended to reproduce an outdated view of the industrial world similar to that of neoclassical economics, in which single plant and single product firms with fairly simple production functions respond to exogenously-determined costs and prices and seek maximum profits in a well-informed and single-minded manner. Advocates of alternative perspectives stress the complexity of industrial organization and decision-making in a world dominated by multiplant, multiproduct, multinational (and, of course, multilocational) firms, the diversity of business behaviour where motives may not always be pecuniary and *homo economicus* is a figment of textbook fiction, and the importance of the wider economic and social structural conditions to which firms respond and over which they seek, often through influence on government, to exercise a degree of control (see Chs. 3 and 4).

While the legitimacy of the main points of critique of neoclassical theory is beyond question, there are two lines of response (or defence) that can be mounted with some confidence. The first is based on the normative role of neoclassical theory. Much of the critique concerns the failure of neoclassical theory to explain reality, yet this is neither its sole nor its principal purpose. To determine the optimum location for a plant, be it that of an independent entrepreneur or a branch of multinational corporation, is a task eminently suited to the analytical apparatus of neoclassical theory. As suggested in earlier parts of this chapter, both private and public location decisions can be assisted by models derived from neoclassical theory. Indeed, it is impossible to think of any alternative way to proceed in a world where optimality is so often conceived of as a resource-allocation problem – whether under capitalism or socialism. To build on non-material social considerations in some welfare extension is difficult, of course, but by no means impossible. If neoclassical theory is used retrospectively to reveal that real industrial plants are not at their economic or social optimum locations, this would hardly be surprising in a world of limited human ability and wisdom.

The second possible response to the critics of neoclassical location theory is that it may not be quite as impotent a positive framework as is often asserted. Indeed, there have been suggestions from leading adherents of the organizational approach that Weberian variable-cost analysis may be finding a new relevance (Smith 1981: 134–5). However much the modern multinational may be able to control costs and prices, to cross-subsidize plants and to take advantage of new technology, spatial differentiation still matters. Dicken (1977: 141) summarizes the position as follows:

> I believe that it is possible to argue that *least cost* location theory still has some relevance in helping us to understand the spatial organization of activities – not at the scale normally assumed, but rather at a greatly enlarged organizational and geographical scale. In other words, I would argue that such theory is relevant in precisely those circumstances of corporate organizational structure which are held to have rendered it impotent.

It is not difficult to envisage world-scale locational triangles, in which one corner might be a source of localized or cheaply-produced material, another the major market of northwest Europe or the USA, and the third a cheap-labour location in southeast Asia. Within Britain, location decisions in recent years may have been as simple as resolving the conflicting attractions of metropolitan market and cheap peripheral labour, at least in some industries like clothing manufacture and electronic components assembly. While transport costs are evidently far less significant than in Weber's time, cheap labour locations may well be more so.

Some support for the neoclassical perspectives comes incidentally from the structural school, which in some respects is the most profound source of criticism. Central to the Marxian analysis which informs the structural approach is the imperative of profit-seeking in the competitive world of capitalism. As Mandel (1978: 75) puts it, 'the problem of the extension of capital into new realms of production – whether technical or geographical – is ultimately determined by a difference in the level of profit'. What better framework for identifying such differences than neoclassical location theory? The single-minded determination with which capital extracts surplus value from labour in its unremitting quest for profits, at least as portrayed in some structural accounts, gives *homo marxicus* an uncanny resemblance to *homo economicus*. It seems only a matter of time before some elements of neoclassical location analysis are rehabilitated, within a broader theory which sees both capital and labour responding to basic structural imperatives which leave limited scope for freedom of action.

Affiliation with a particular perspective can be as much a matter of ideology as of scientific conviction. In location analysis, the supreme irony is that the neoclassical view (like neoclassical economics) is increasingly rejected in the capitalist world as devoid of social content, yet found useful in the socialist world by virtue of its normative properties and planning applications; Marxian theory is appealing to those who seek scientific penetration of capitalism, yet as a normative framework it is singularly unhelpful (as Soviet planners have found to their cost). The confusion between positive and normative theory, and between science and ideology, has prevailed over the field for years. To Lösch (1954: 4) the way out was clear: 'The real duty . . . is not to explain our sorry reality, but to improve it. The question of the best location is far more dignified than determination of the actual one.' While wishing to endorse such elevated sentiments, we now appreciate that how we judge what is best cannot be independent of how we understand what actually is.

References

Dicken, P. 1977 A note on location theory and the large business enterprise. *Area* **9**: 138–43.

Gregory, D. 1980 Alfred Weber and location theory. In D. R. Stoddart (ed.) *Geography, Ideology and Social Concern*. Blackwell, Oxford.

Haddad, P. R., Schwartzman, J. 1974 A space cost curve of industrial location. *Economic Geography* **50**: 141–3.

Hoover, E. M. 1937 *Location theory and the shoe and leather industries*. Harvard University Press, Harvard, Mass.

Hoover, E. M. 1948 *The location of economic activity*. McGraw-Hill, New York.

Hotelling, H. 1929 Stability in competition. *Economic Journal* **39**: 41–57.

Isard, W. 1956 *Location and the space economy*. MIT Press, Cambridge, Mass.

Kennelly, R. A. 1954–55 The Location of the Mexican Steel Industry. *Revista Geographica* **15**: 109–29; **16**: 199–213; **17**: 60–77.

Lewis, P. W. 1970 Measuring Spatial Interaction. *Geographica Annaler* **52**, Series B, 1: 22–39.

Lindberg, O. 1951 *Studier över pappersindustriens lokalisering*. Almqvist ond Wiksells, Uppsala.

Lösch, A. 1954 *The economics of Location*, trans. Woglom, W. H. from *Die räumliche Ordnung der Wirtschaft* (1940) Yale Univ. Press, New Haven, Conn.

Mandel, E. 1978 *Late Capitalism*. Verso, London.

McDermott, P. J. 1973 Spatial margins and industrial location in New Zealand. *New Zealand Geographer* **29**: 64–74.

Moses, L. N. 1958 Location and the theory of production. *Quarterly Journal of Economics* **73**: 259–72.

Palander, T. 1935 *Beitrage zur standorts theorie*. Almqvist ond Wiksells, Uppsala.

Rawstron, E. M. 1958 Three principles of industrial location. *Transactions*, Institute of British Geographers **25**: 132–42.

Robinson, P. A. 1979 *A study of the distribution and location of the electronics industry in Great Britain*. Unpublished PhD thesis, Birkbeck College, University of London.

Smith, D. M. 1970 The location of the British hosiery industry since the middle of the nineteenth century. In R. H. Osborne, F. A. Barnes, J. C. Doornkamp (eds) *Geographical Essays in Honour of K. C. Edwards*. Department of Geography, University of Nottingham, 719–29.

Smith, D. M. 1971 *Industrial location: an economic geographical analysis*. John Wiley, New York (page references in text are to 2nd edn, 1981).

Smith, D. M. 1977 *Human Geography: a welfare approach*. Edward Arnold, London.

Taylor, J. 1975 *Problems of minimum cost location: the Kuhn ond Kuenne algorithm*. Occasional Papers 4, Department of Geography, Queen Mary College, University of London.

Taylor, M. J. 1970 Location decisions of small firms. *Area* **2**: 51–4.

Tornqvist, G. 1962 *Transport costs on a location factor in manufacturing industry*. Lund Studies in Geography. Series B, 23.

Weber, A. 1909 *Uber den standort der industrien*, trans. C. J. Friedrich as *Alfred Weber's theory of industrial location*. University of Chicago Press, Chicago, 1929 (page references in text are to this edition).

CHAPTER 3

Behavioural approaches to industrial location studies

Peter A. Wood

3.1 The originals of the behavioural approach

The 'behavioural approach' to the study of industrial location first developed on both sides of the Atlantic after the mid-1960s. In America, such work was closely associated with the group of researchers based in the University of Washington, including Thomas, Krumme, Steed, Erickson, Le Heron and Rees (Carr 1983). Their interest was in the 'Geography of Enterprise' (Krumme 1969), exploring the impacts of key economic and technical changes on local areas, particularly through study of investment decisions by large industrial firms. They envisaged a process of continuous 'adaption' in the structure of organizations, of which location changes were a part, in response to a rapidly changing commercial environment (Alchian 1950; see also Townroe 1972). Study of firm–environment interaction, inherited from behavioural theories in economics and organizational studies, has thus been a primary focus for so-called 'behavioural' research into manufacturing location change over the past fifteen years.

In Britain, interest in behavioural studies was a rather separate development, in a planning environment dominated by strong, spatially differentiated regional and urban policies. Steed's work in Northern Ireland in the late 1960s, inspired by the Washington School, for example, although widely admired had limited impact, not only because of its geographical isolation from 'mainland' interests, but also because the comprehensiveness of its interpretation was perhaps too diffuse for pragmatic British taste (Steed 1968; 1971a; 1971b). As regional policy intensified after 1964, behavioural ideas became attractive because the idealized, static spatial theories of Weber, Hotelling, Lösch and Isard, which had so dominated theoretical thinking before, seemed to offer very limited practical guidance to urban and regional planners (Wood 1969). After earlier hints (Hill 1954; Luttrell 1962), behavioural research in Britain gathered momentum because it promised more realistic generalizations about the behaviour of firms in response to regional policy inducements. In the United States, the Washington School devised a theoretical framework of study intended to deal with both the growing domination of manufacturing by large firms and the increasingly rapid technical and market changes which they faced. In Britain, the aim was not so much a new theory of location as a practical guide to its significance in general investment decision-making by firms. Some attempts were made to adapt behavioural ideas from economics and organization theory to the locational problem, but nothing emerged resembling a body of 'behavioural location theory'.

Recently, as we shall see at the end of this chapter, more attention has been paid to the theoretical basis of behavioural studies in Britain. They have especially been criticized because of their apparent over-emphasis on the role of private firms in creating industrial change. Such an emphasis has been more evident in British work, however, than elsewhere, including the United States and Sweden, largely because of the preoccupation with the influence of regional policies on firms' decisions. Of course, as the limited effects of regional measures have come to be recognized, other aspects of change in the

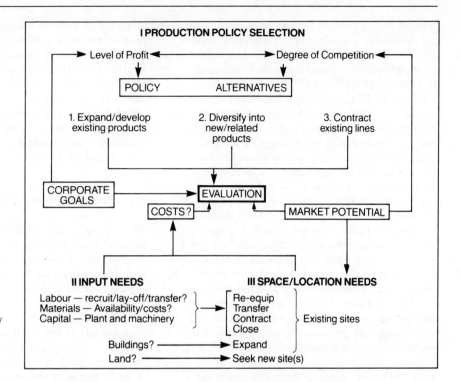

Fig. 3.1 Production policy selection, input needs and location

economic environment have been examined, including growing international competition, rapid technological change, concentration of ownership, and the spatial effects of other, 'non-spatial' government policies. The perspective has widened from a concern primarily with the characteristics of firms, as the complexities of the outside factors that influence their behaviour have emerged. The structuralist critique has also been influential since the late 1970s. Nevertheless, a balanced consideration of broad economic changes and the reactions to them by decision-makers has always been fundamental to behavioural theories more generally.

3.2 Elements of behavioural analysis

The term 'behavioural' was adopted when criticisms of classical theories focused on their simplifying assumptions about business behaviour (Pred 1967). Far from being all-knowing profit-maximizers, as these theories assumed, decision-makers usually act without perfect knowledge of all relevant circumstances, and in relation to 'suboptimal' goals, such as the achievement of satisfactory profits or even simply to survive. Even more significant, as industrial firms become larger, they also achieve a position from which they can control important aspects of their external environments, including the prices of materials, of labour, and of their goods in the market. The behaviour of at least some decision-makers thus exerts significant independent influence on the course of industrial change, in clear violation of classical assumptions (Dicken 1971; Hamilton 1974: Ch. 1). In economics, Cyert and March had developed a 'Behavioural Theory of the Firm' which attempted to model oligopolistic behaviour in relation to price and output decisions (Cyert and March 1963). Under the influence of the quantitative, theoretical enthusiasm of the late 1960s, a few geographers saw a similar analytical theory of large firm location decision-making as an appropriate goal. Again, although American workers have shown more interest in these developments, such an approach has not formed the mainstream of subsequent behavioural studies in Britain. This has remained predominantly empirical and descriptive in emphasis, with the implied aim of moving through greater understanding towards broader generalization about firm location behaviour.

Behavioural studies in Britain have examined the operations of all types of industrial organization, including the smallest as well as the largest. For private industry, as in North America, this has involved the study of 'enterprise'. More generally, interest has focused on the 'geography of organizations'. The ideas of Thompson, Lawrence and Lorsch, Emery and Trist, Burns and Stalker, and Katz and Kahn have been particularly influential (see Dicken 1976; Wood 1978; McDermott and Taylor 1982; Marshall 1982b). Structures of control adopted by firms have been examined, together with the ways these are manipulated in relation to goals such as making profits, achieving good market performance, innovating new products or processes, or surviving through rationalization. Spatial change, including the opening, closure expansion or contraction of plants, and functional changes within existing sites, have been analysed in the context of such corporate strategies. Figure 3.1 simply illustrates the interrelationships between I, the selection of alternative goals and market conditions; II, the inputs required to implement the chosen policies and the costs associated with them; and III, the adjustment of the location pattern of production associated with these input needs and, in some cases, with access to regional markets. The evaluation process requires each of these facets of change to be considered, but location needs are shown to be very much the by-product of broader production considerations.

Uncertainty about the future means that the process of entrepreneurial decision-making needs to strike a balance between the exploitation of commercial opportunities and the minimization of the risks of the failure. The urge to reduce risks is often evident in attempts to increase control over elements in the outside environment, including suppliers and customers, as well as to protect the essential functions of the firm's own activities from unexpected outside events. Nowadays corporate restructuring, including mergers and takeovers, is a more or less continuous process, driven by the need in ever-changing circumstances to achieve a pattern of control which will ensure profitable operations in the short and the long term. Internally, also, increasing resources are spent on specialist functions designed to monitor external change. Important trends in economic organization are therefore partly the consequences of the need to reduce or to anticipate unpredictable environmental changes, by shifting the firm – environment boundary.

Similar trends towards large size and centralized control have occurred in the public sector of the British economy, of course, including the long-term expansion of the nationalized industries, the growth of the Civil Service and the organization of health, education and other social services. Although the emphasis of work in industrial geography has been on the private sector, the organizational insights of behavioural studies have a wider relevance, and overlap into broader areas of social and political geography. Although their goals and motives may differ, the extent to which centrally-controlled organizations can effectively engage the talents of the workforce, or serve the diverse needs of customers, are equally important issues for both private and public sector agencies.

3.3 Firm-environment characteristics

We have seen that the basic framework of organizational studies rests on the complementary examination of two interacting elements: the firm and its environment. Modern *firms* are complex organizations, consisting of many individuals and groups who may influence the taking of decisions, as managers of strategic or routine functions, as shareholders or other capital suppliers, or as workers' representatives. Organizations are therefore coalitions of interest, whose 'behaviour' is the outcome of varied types of group activity, even though some groups or individuals may, of course, dominate. Much modern writing on organizations, whether economic or sociological in emphasis, concentrates on the internal characteristics of firms. Its explicit purpose is to help managers to be effective by defining clear goals, gaining agreement from others, exerting authority and coordinating functions efficiently (for example, see Child 1977; Silverman 1970; Dessler 1976; Galbraith 1977). The emphasis of the organizational literature is therefore unashamedly managerial and assumes that the best guide to success is the achievement of agreed corporate goals as effectively as possible. More generally, therefore, this view of the firm tends to be pervasive in behavioural studies, even though the appropriateness of the priorities of private firms may be questioned in the light of broader social goals.

One particular weakness of organization theory has been the somewhat confused analysis of what is meant by the '*environment*' of firms (McDermott and Taylor 1982: Ch. 2). Generally, it has been regarded as a set of external uncertainties, embodied in the behaviour of other organizations, which need to be anticipated and reacted to by business managers. Although we have seen that there is some validity in Carr's recent criticism that since the late 1960s, behavioural location studies

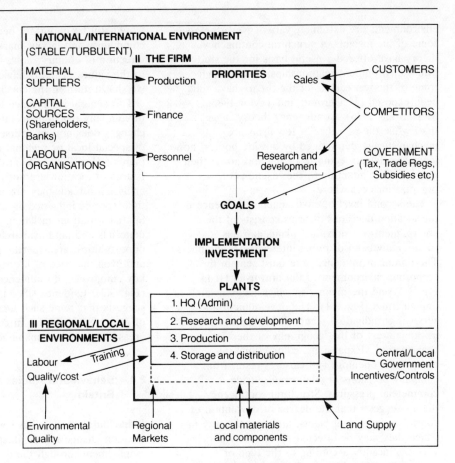

Fig. 3.2 Corporate environments of the firm

have neglected consideration of the economic environment, a number of commentators have discussed systematic ways of analysing its changes. Most notably through a long debate about the significance of industrial linkage and through attempts to adapt structural contingency theory to location analysis by McDermott and Taylor. Other more general contributions by Norcliffe, and by Hamilton and Linge have also systematized the influence of environmental developments on location patterns at regional and international scales (Norcliffe 1975; Hamilton and Linge 1979).

In summary, environmental influences need to be examined at two levels: that of the *firm*, reacting especially to national and international trends transmitted through the other organizations upon which it is dependent to varying degrees; and that of the *plants* owned by firms, the specific locations where the consequences of corporate decisions are felt. In Figure 3.2 the national and international environment of a firm (I) is seen as consisting of other organizations with which it has significant material or information contacts, including materials suppliers, customers, competitors, labour organizations, suppliers of capital such as banks, and national and local governments. Changes in inputs, prices, demand and competition affect home and overseas sales, levels and types of investment, wage rates and, crucially in capitalist firms, profits and returns on capital. One way of controlling for these variables in location studies has been to study firms in particular industries, within which those with similar products and commercial environments can be examined together. The application of 'structural contigency theory' of organizational behaviour has also attempted to explain the effect of environmental conditions on the behaviour of firms. It suggests that the organizational structures and strategies of large firms ideally adapt to the types of environment, whether stable or turbulent, in which they operate. More centralized control and decision-making is appropriate to firms serving a stable market. Devolved patterns of decision-making, on the other hand, work better for firms that are engaged in rapidly changing or highly competitive businesses. Empirical evidence suggests that the adaptive reactions of large firms to different operating environments are extremely varied. Nevertheless, some of the insights of structural contingency theory have offered a useful basis for the study of firm – environment relationships, even though some of the weaknesses of the theory have thus been exposed (McDermott and Taylor 1982; McDermott 1976). Contingency theory suggests that, while the activities of few firms are mechanistically determined by outside bodies, none have complete freedom to define and attain their goals without adapting to the reactions of organizations elsewhere.

The second level of environmental influence on the location decisions of firms consists of the environments of individual plants and has traditionally been of prime interest to geographers. The dominant influence here must remain the corporate environment of the firm itself (II in Fig. 3.2) and the manner in which its goals are implemented (Fig. 3.1). The development and survival of individual plants should be understood in the context of the 'geography of the organization'. Expansions, contractions, re-equipment, openings and closures result from company decisions in relation to outside commercial pressures. Structural contingency theory suggests that the degree of devolution of control to individual plants, and thus the way in which they may be affected by company policies, vary significantly according to the type of environment faced by the firm (Wood 1978). In fact, the theory, like most organization theory, is not conceptualized in a sense which can be directly related to the control of physical plants. It is concerned with purely functional groupings within organizations, and the influence of the spatial arrangement of activities is only occasionally considered as an element in organizational design (see Child 1977: Ch. 3). Location is seen as a relatively minor complication for effective management, compared with the complexities of coordinating diverse technical functions, achieving appropriate levels of supervision, or planning and coordinating future investment. Nevertheless, there is plenty of evidence from geographical and other research that the functional specialization of manufacturing locations, for example into control, administrative, research and development, production and distribution activities, has grown over the past two decades. Investment and production trends have encouraged a growing pattern of spatial differentiation in industrial functions. Of course, even if these trends were not discernible, industrial changes at the plant level would still be important because of their community and labour-market impacts. In this direction of enquiry, however, there may be limits to the values of the behaviour approach, to which we shall return in the conclusion.

The general significance for particular locations of company reactions to international and national changes does not, of course, rule out the effects of special local or regional influences on individual plants (III in Fig. 3.2). Traditional 'geographical' factors of location are not neglected by behavioural analysis; instead, they are judged in relation to their specific influence on corporate decisions. Spatial variations in labour supply and costs, materials and land availability, regional market opportunities, area-specific government incentives and even the style of life sought by managers and key employees, do influence location decisions (Fig. 3.2). Evidence since the mid-1960s in fact suggests that some such local attributes have been of growing importance in determining whether areas retain or lose manufacturing investment.

3.4 Behavioural studies of location change in Britain

Behavioural analysis has sought explanation of location change in manufacturing investment and employment through the decisions made by the organizations that own plants in different areas. These decisions are guided by general corporate goals in relation to market and other opportunities, but may also be influenced by the working environments found in particular areas. Empirical studies based upon these assumptions have adopted one of two approaches:

1. Study of aggregate patterns of location change, usually from secondary data, as a basis for inferring explanations for those changes.
2. Study of detailed processes of, and influences on location decision-making and behaviour.

This distinction is not a matter of principle but more one of convenience, dictated by the availability of information. Both approaches are empirically based, starting from observed patterns and changes at the local level. It is, however, axiomatic in behavioural research that the reasons for location change are not to be found locally, nor are they merely locational in character; much wider corporate and environmental influences are at work.

Analysis of aggregate changes has been based

almost entirely on secondary data on employment from official or other sources. The pursuit of other measures of local economic performance, such as output, investment or productivity has been discouraged by inadequate monitoring of these attributes. Detailed studies of location change, on the other hand, have usually been based on primary questionnaire surveys of sample firms in particular areas or industries. These have often been supported by relevant but often rather indirect secondary information, for example on the location patterns of the headquarters of major firms, government research and development activities, and various measures of technical innovation or environmental attractiveness. Our understanding of the processes of industrial location change in Britain has thus developed from the interpretation of disparate evidence, varying in scale of analysis (aggregate or micro-level), quality of source (secondary/primary) and directness in relation to specific decisions made by identifiable organizations. Much of this evidence is reviewed elsewhere in this volume. Nevertheless, examination of individual aspects of behavioural work does not necessarily reveal the degree of cumulative understanding that has emerged from the diverse approaches. Here the emphasis will therefore be on the relationships between the results of various type of behavioural work since the late 1960s.

The interpretation of aggregate patterns of location change

British industrial location research between 1965 and 1984 has taken place against a backdrop of profound change, the most important feature of which is the decline of manufacturing employment at an accelerating rate. Related to this, the geographical pattern of industrial change has taken on a new form. Aggregate analysis of the new industrial geography of Britain has therefore been especially significant in directing attention towards particular trends, providing a framework for more detailed study. In interpreting aggregate patterns of employment change themselves, of course, inferences must be made about industrial behaviour based on *area* characteristics. The measurable attributes of different locations are assumed to have been interpreted and acted upon consistently by decision-makers. What such studies offer in the breadth of their description of employment change therefore needs to be balanced against the indirectness of their behavioural interpretations.

The most comprehensive analysis has been undertaken by Keeble (1976: Chs. 2 to 5). As a result of the mapping and multivariate statistical analysis of manufacturing employment and floorspace changes in 62 subregions between 1959, 1966 and 1971, he was able to demonstrate that during the 1960s the long-established division between the 'core' of prosperity and growth in and around London and the West Midlands conurbations, and the 'periphery' of the North and West, became less dominant. Instead, the trend changed to what Keeble described as a 'periphery – core' pattern, in which industry declined rapidly in all the large conurbations, including London and the West Midlands, and grew only in small – medium size towns and relatively rural sub-regions. Keeble explained this shift on the basis of other characteristics of the sub-regions, chosen to test various hypotheses about industrial location decision-making during the period. He showed that throughout the period and later, in another study which included the recession conditions up to 1976, the poor performance of the large industrial areas became increasingly dominant (Keeble 1980). This seemed to represent a progressive rejection of the advantages of geographical proximity to other firms and industries which had supposedly been so influential in sustaining the industrial conurbations since the nineteenth century. In the 1960s, Keeble had also identified the significant positive influences on manufacturing growth of residential attraction and the growing impact of assisted area status as the decade proceeded. After 1971, however, with decline in manufacturing employment, neither of these influences seemed to persist. Regional policy, in particular, exerted an increasingly unclear effect on an urban – rural division of growth which cut across the old distinction between assisted and growth regions.

The patterns analysed by Keeble were confirmed by Fothergill and Gudgin, employing a longer and more detailed data set which they compiled, from 1952–79 (Fothergill and Gudgin 1982). In 1966, manufacturing employment had reached a peak of nine millions. After this a growing range of industries, as well as those traditionally regarded as declining since the 1930s, began to contract. Thus neither the inherited structure of local industry nor the possession of factories moved earlier from elsewhere any longer insured areas against job losses. Instead, other area attributes, especially those which contrasted the qualities of the cities against small towns, began to tell. Fothergill and Gudgin's interpretation of these patterns emphasizes the increasingly extensive use of industrial floorspace and the shortage of suitable sites for expansion in large urban areas. In behavioural terms, their conclusions suggest that increasing pressures towards higher labour productivity and more extensive use of space rule out consideration by firms of new locations in large urban areas, and call into question the retention of their existing plants in such areas.

As well as these general studies of aggregate geographical patterns of employment change in Britain, others have used secondary data to examine particular aspects of change, especially those related to the effects of regional policy. For example, in the early 1970s several studies traced patterns of industrial movement to the Development Areas and, in some cases, out of the cities to new expanded towns (Keeble 1971; 1972; Sant 1975; Townroe 1979; Fothergill, Kitson and Monk 1983). This work was further developed through attempts to evaluate the employment impacts of different aspects of regional policy, including development grants, the Regional Employment Premium and Industrial Development Certificate restrictions in growth areas (for a review, see Schofield 1979; House of Commons 1981). Most of this work neglected the organizational context of decisions to move, generally begging the question of whether other forms of policy might have been more effective in persuading firms to invest in particular areas. Such aggregate analysis also could not establish how much movement might have taken place even without official encouragement, or comment on the status and long-term permanence of much assisted area investment in relation to the subsequent patterns of firms' operations.

More recently, in response to changing economic conditions, analysis of the local impacts of redundancies has attracted more attention. The decisions of major firms to close plants or reduce employment in particular areas have attracted a good deal of public attention, and this work has generally shown a greater awareness of the behavioural context of local employment change. It has highlighted the need to understand more fully the options open to industrial decision-makers and the reasons why they adopt particular rationalization strategies (Townsend 1981; 1982; 1983b).

The inevitable limitations of any explanation of location change based on the interpretation of aggregate employment trends alone imply that such studies must be complemented by detailed examination of particular firms or groups of firms. Another approach, however, has been to improve the quality of aggregate data on local industrial change by compiling establishment-level data banks and undertaking 'components of change' studies for particular areas (for a review, see Healey 1983). In these studies, employment change in individual plants over a period is categorized according to whether it has resulted from plant closures or openings, movement, or from on-site changes in established plants. A number of local plant-level data banks were established during the 1970s and have provided valuable insights into the diversity of processes that underlie the aggregate spatial patterns identified by Keeble and by Fothergill and Gudgin. The significance of plant closures in large urban areas, whether by large firms or through the collapse of the small firm sector, has been particularly noteworthy, placing in context the more modest impact of the out-migration of firms encouraged by various planning policies. The on-site contraction even of surviving plants has also been a significant phenomenon, seemingly confirming the operation of the processes suggested by Fothergill and Gudgin. Another feature neglected by both aggregate employment studies and small-scale sample surveys is the appreciable gross level of new plant establishment and plant expansions in the cities even during periods of net decline. Urban economies are thus revealed as being much more dynamic than the impression gained from aggregate trends (for example, see Whitelegg 1976; Dennis 1978; Dicken and Lloyd 1978; Firn and Swales 1978; Lloyd and Mason 1978; Lloyd 1979; Lloyd and Mason 1979; Mason 1980a; 1980b; 1981). Even where such diversity is analysed, however, the full behavioural context of plant closures, openings, contractions and expansions can still only be inferred. Nevertheless the potential of plant-level data banks in providing a framework for behavioural studies has been demonstrated by the addition of information on plant ownership, and a few examples of comprehensive research on the company policies which have led to particularly significant local changes (e.g. Lloyd and Reeve 1982).

Detailed studies of location decision-making and behaviour

Within the context of the monitoring of aggregate trends and the questions which it raises about the reasons for change at the local level, the main volume of enquiry in British industrial location studies in recent years has been into groups of firms, investigating particular influences on patterns of location change. This work has been reviewed regularly elsewhere (Keeble 1977–79; Wood 1980–82; Malecki 1982; Hayter and Watts 1983; Carr 1983), and other chapters in this volume offer further details. The cumulative insight derived from this research can, however, be summarized in terms of three inter-related questions:

1. The most fundamental is, 'How are industrial location decisions taken in relation to the general processes of investment appraisal and change by firms?' The answers to this vary widely according to the types of firm engaged in particular changes and the external circumstances which they face.
2. The second question is, therefore, 'How does the ownership of industrial plants at particular locations, whether by single- or multiplant firms, based locally, elsewhere in the UK or abroad, influence their activities?' Change in ownership, as well as change in the economic environment more generally, may thus influence the prospects of particular industrial sites.
3. The third element of behavioural research concerns the environment of firms: 'What are the most critical external relationships which govern the activities of the firm, through both material and information exchanges, and what role do particular plants play in sustaining these relationships?

More specific, policy-oriented research has generally attempted to answer one or more of these broader questions, pursuing the interaction between them in different ways. For example, the effectiveness of area-based regional or urban policies must obviously be assessed in the light of the types of firm affected, their general investment strategies, and the status of particular plants in their operations (Cooper 1976; Harrison 1982; McGreevy and Thomson 1983). Similarly, recent studies of the regional impact of rapid technological change have adopted a firm-based approach, in which such change is seen as both an external contingency and a crucial element influencing internal patterns of investment allocation (Oakey 1979; Thwaites 1982; Oakey, Thwaites and Nash 1982; Rothwell 1982; Oakey 1983).

Location decision-making

A preoccupation with regional policy, and its efforts to direct 'mobile' investment to areas of high unemployment, led to an early emphasis in the study of location decision-making on relocations and branch plant establishment. The work of Townroe, in particular, pioneered analysis of the organizational context of such decisions from samples of firms moving with various types of government assistance (Townroe 1969; 1971; 1972; 1979). More conventional questionnaire studies of the reasons for movement were also undertaken, the most comprehensive being by the government itself (Department of Trade and Industry 1973). The assumption of this type of enquiry seemed to be that locational change was a relatively unusual, strategic decision for manufacturing firms, although aggregate studies of movement such as those by Howard (1968), Keeble (1971) and Sant (1975) showed that even these changes had become increasingly frequent during the post-war period. North's study of all types of locational change in the plastics industry, however, demonstrated that a much wider range of locational adjustments was taking place as a result of the routine processes of investment, and that in many cases major moves to new sites were regarded almost as 'last resorts' after cheaper and organizationally less disruptive options had been rejected (North 1974). These included on-site changes within established plants, expansions and contractions, and takeover activity. North also related such locational adjustments to various types of non-locational strategies, through a linked sequence of decisions (cf. Fig. 3.1). It is true that the orientation of much work at this time, as in North America, was towards the allocation of *growth* to various locations. This reflected the perspective of both commercial and public planning at the time. Nevertheless, the framework of study was as applicable to the conditions of recession and contraction which followed in the late 1970s.

The variety of local experience, including the closure and contraction as well as the opening and expansion of plants, became clearer as the results of components of change studies were made available. One comprehensive survey of the types of locational change undertaken by a wide variety of firms was later published for a similar period to that of North's, 1960–72; by Hamilton (1978). Almost 1,500 medium-sized firms were asked to describe how they had changed their geographical patterns of production. No less than 944 of the firms had set up at least one new plant and over 400 had established more than one. To balance this high mobility, 617 firms had closed 724 plants, rationalizing production, often at the expense of old sites in congested urban areas. Many firms made more than one type of change during the period, opening plants in some areas, closing others, and relocating production or administrative facilities as part of a process of continuous adjustment. Forty-three per cent of the sample had undergone some sort of change through mergers or acquisitions. Locational flexibility at a time of increasing government control over new factory building must have been one important benefit of such activity. Hamilton also reported that *all* of the multilocational firms in the sample had altered the relative importance or functions of established units and these changes, like the moves and closures, had favoured outer suburban and small town locations, rather than inner cities and larger towns. Almost 1,000 firms expanded the floorspace of established plants, a strategy generally favoured over new branch plant construction. An elusive but no less important element of change was the substantial alteration of production lines within existing factories. Almost 500 firms claimed to have made such changes and most other firms pointed out that designs, components and material inputs had altered, even where production lines appeared to be similar, especially through the introduction of synthetic fibres, plastics and alloys.

Thus by the mid-1970s the almost routine nature of location change as the result of a wide variety of corporate strategies was established. North also outlined some general circumstances in which particular types of locational adjustments in the plastics industry took place during the expansion conditions of the 1960s. For example, site extensions and takeovers or mergers were most commonly undertaken by well-established medium or large, capital-intensive multiplant companies. In most cases, these changes were part of a planned development of production. On the other hand, small firms or firms in recently-created groups frequently underwent more rapid, unplanned change at established locations, and were often more or less forced to relocate production because of operational difficulties there. Few firms undertook objective analyses of their locational options – inertia towards existing locations was strong, keeping the distance and disturbance of moves to a minimum. Later in the 1970s, Healey applied a similar methodology successfully to a declining industry (Healey 1981a; 1981b; 1982), reaffirming the significance for locational studies of exploration, on an industry basis, of the relationship between a wide variety of firm types, their goals and their locational responses over a particular period of general economic change.

Ownership and location

The second area of basic behavioural research has devoted most attention to the influence of multiplant ownership on locational change and, in particular, the growing economic control of many areas by large outside firms. A complementary interest in recent years has been the remaining role of the locally-based, small firm sector. The volume of writing on these topics forbids a comprehensive review here (Watts 1981), but some general conclusions arising from the work can be outlined.

In the early 1980s, it is clear that growing external control of regional industrial economies is an inevitable consequence of the domination of manufacturing by large firms and the decline of locally-based capital in specialized sectors (Firn 1975; Townroe 1975). The balance of short- and long-term benefits and losses to particular regions arising from this situation remains unclear. Access to outside capital, management expertise and innovative capacity must be regarded as potential advantages, at least in the short term, while the loss of local control and high-level managerial and technical employment may bring longer-term disadvantages (McDermott 1976). The spectre of the 'branch plant economy', dominated by routine, low-wage production activities and vulnerable to market and technological changes, certainly became a reality in some areas during the recession after 1978. Nevertheless, the definition and mode of operation of regional external control are much more varied than this caricature would suggest. Large multiplant firms obviously have greater flexibility than smaller firms to adjust the size and role of particular sites in relation to non-local, corporate needs. The nature of their control may, however, vary from the loosest 'conglomerate' relationship, concerned only with major financial decisions, to direct day-to-day production management in close coordination with other sites. Even when local autonomy is high, this offers no particular guarantee of stability. In summary, therefore, the pattern of ownership alone is a poor guide to the prediction of behaviour (Dicken 1976).

The recognition of this fact, together with the difficulty even of satisfactorily defining external control, has directed attention towards three special aspects of multiplant ownership and firm behaviour.

1. The first is ownership by overseas firms. Regional external ownership by UK firms is almost impossible to define and measure. If all branch

plants, subsidiaries of outside companies and companies with significant outside representation on their boards of directors are regarded as externally controlled, only firms owned by locally-resident entrepreneurs remain. Even these firms are usually heavily dependent upon large firms based elsewhere. With the whole notion of regional external control founded on such definitional quicksand, the identification of foreign ownership seems more secure. Of course, only direct ownership is usually acknowledged, excluding more subtle financial, contractual and licensing relationships between UK and overseas firms. This almost certainly neglects the increasing links between medium-sized companies in different countries, especially within the European Economic Community. It also cannot account for the effects on 'home' investment of the high and increasing involvement of UK companies abroad (Owens 1980).

The regional effects of foreign ownership patterns, studied in isolation from particular economic conditions, are highly diverse and unpredictable. It does appear, however, that overseas firms, whose plants are generally larger, more productive and have grown more rapidly than the average UK company (Dunning 1974; Dicken and Lloyd 1976), have favoured peripheral regional development in recent decades. Nevertheless there remains a legacy of activity in East Anglia and the South East arising from pre-war investment by US companies and more recent merger activities with EEC-based firms (Dicken 1980; Dicken and Lloyd 1980; Law 1980; Watts 1979; 1980; Dicken 1983).

2. The second aspect of research into the effects of multiplant ownership has offered rather more substantial evidence of its regional impact. This has investigated the emerging pattern of functional specialization, especially associated with the concentration of non-routine high-level activities in London and the South East region. The simplest and most accessible evidence for this has come from studies of the distributions of company headquarters, which carries two circumstantial implications. The first is that, since headquarters are centres of *control*, their uneven distribution implies a disadvantage to some regions in making claims for a fair share of *investment* opportunities. Secondly, if headquarters are concentrated in one region, their high-level *employment* opportunities in management, research, and in specialized commercial and technical services will be denied to other regions.

While, for reasons already summarized, evidence of the first of these assumptions is not clearly established, the effects of the second of these influences seems to have been confirmed. It has also been suggested, of course, that these two aspects of regional functional specialization are not independent and that the long-term influence of high-level employment trends will adversely affect the quality and quantity of investment in peripheral regions.

In 1972, 51 per cent of the headquarters of the top thousand manufacturing companies in Britain, responsible for over three-quarters of their turnover, were located in the Greater London area. The larger the company, the more likely it was to be based in the capital (Evans 1973; Westaway 1974a). Between 1972 and 1977, Greater London and the rest of the South East increased their shares of headquarters, mainly at the expense of the North West, the Midlands and Yorkshire and Humberside (Goddard and Smith 1978). This growing domination by the South East occurred because a higher proportion of large firms in this region expanded and were engaged in more company acquisitions. Although London remained pre-eminent, by 1977 the rest of the South East possessed more 'top thousand' headquarters than any other region. The most powerful agents of these trends during the period, according to Goddard and Smith, were UK manufacturing companies. The influence of firms in non-manufacturing and of foreign-owned firms was neutral in the redistribution process.

Westaway (1974b) suggested that a number of undesirable economic and social effects would follow what he saw as a shift towards the 'information rich' environment of the South East. The availability of specialist commercial and technical information is the key to modern economic growth. The trends are cumulative, with those regions gaining the headquarters of major companies becoming thereby more desirable for other companies wishing to be 'in the know'. Other work, by Buswell and Lewis (1970), had already demonstrated the concentration of 58 per cent of all commercial and government-sponsored research and development in the South East and West Midlands (see also Parsons 1972; Howells 1984). The implications are clear; although it may not be possible to demonstrate a bias in *productive* investment towards the South East, further growth in the expanding *information-processing* sectors of companies, especially in the growth industries, seems inevitable. Decision-makers located in the rest of the country, unless in areas of particular industrial specialization (which are in decline in any case), are likely to be starved of the knowledge they require to compete successfully. Commercial and industrial advisory services are also likely to be less well developed and less effective in such areas than in the South East (Marshall 1979). In employment terms, a growing dichotomy between the middle-class South, with its high quality of life for skilled managerial, administrative and technical workers, and the 'working-class' North and West seems to have been emerging even in the early 1960s (Westaway 1974a). Much of this trend can be explained as a result of developments in the geographical patterns of service industry employment, but there is also little doubt that the evolution of multiregion manufacturing organizations has made a significant contribution to disparities in the provision of managerial and technical employment.

3. Some studies of the behaviour of multiplant firms have particularly focused on the impacts of takeover and merger activity. Such activity is, of course, only one aspect of the process of corporate restructuring which dominates modern industry. It has nevertheless offered useful opportunities to examine the locational consequences of major events in this process. The reasons for takeover and merger activity are very varied, and their outcomes complex and unpredictable. Even industry-wide studies can give only a partial view, as one effect of the merger boom since the mid-1960s has been the grouping of diverse activities into multiproduct conglomerates (Smith and Taylor 1983). The most useful general starting point for explanation, as Massey and Meegan (1979) have demonstrated, is the *purpose* of capital restructuring, for example to rationalize inefficient production, acquire new technology, consolidate managerial or sales expertise, reduce competition or generally to gain a larger market share. Many acquisitions are not carefully planned in advance, however, and may result from unexpected events such as the impending bankruptcy of a key supplier or the financial weakness of a competitor. Large firms nevertheless plan fairly consistently to seek out small, innovative firms in need of capital and marketing expertise as vehicles for moving into new technical or product areas. In all such cases, spatial changes form an important element of the post-merger adjustment, including the relocation of production, distribution and administrative functions, setting up new patterns of material and information exchange, the closure of some sites and the consolidation of others. Leigh and North undertook what is so far probably the most comprehensive study of such regional impacts arising from restructuring in a group of industries during 1973–74 (Leigh and North 1978a; 1978b). Like Goddard and Smith, they found that the pattern of acquisition favoured increasing control from the South East. On the other hand, in about one-third of the cases, appreciable local independence was retained, at least for the two-year period after the mergers that they examined. In only a quarter of the cases were plants fully assimilated into the new parent company. After the takeover, expansion of output, although not of employment, was more common than plant closures. On balance, Leigh and North saw acquisition activity as being at least partly beneficial, allowing regional assets to be concentrated into more efficient and competitive firms, but they acknowledged that the longer-term negative effects might include the loss of local control and managerial staff, and the declining use of local service industries.

The potentially damaging effects of takeover activity have been emphasized by Smith in the North (Smith 1979). Local ownership of plants with more than 100 employees in this region fell to 21 per cent in 1973 from 46 per cent a decade earlier. This trend was caused more by the takeover of sucessful local firms than by the setting up of branch plants by large outside organizations. The result was a loss of regional employment and growth potential, through a high closure rate of plants or lower rates of employment growth after acquisition. Although acknowledging Leigh and North's evidence that some benefits accrue to regions as a result of acquisition activity (in terms of productivity and good management practices), Smith's conclusions about the effects of acquisition on employment in the short term were largely negative.

The external relationships of plants
Consideration of merger and acquisition activity illustrates how various aspects of behavioural research (concerned with locational adjustment, multiplant ownership, regional impacts) offer complementary evidence about spatial trends. A final component to be added to this variety of approaches is conventionally described as the study of 'industrial linkage'. This explores geographical patterns of materials supply, product distribution, and the exchange of information for routine and strategic functions that support the operations of

individual plants. Such work relates these external patterns of exchange to plant functions and organizational status. In the past, debate about the measurement and significance of linkages has often neglected their importance as the channels through which both macro-economic and regional trends are transmitted to local plants, and thus their role in the study of the effects of plant ownership and plant – environment relationships more generally (Taylor 1975; McDermott and Taylor 1982).

This significance is well illustrated by Marshall's evidence, which complemented Smith's study of growing external ownership in the North (Marshall 1979). Marshall examined the relationship between local or non-local factory ownership and the use of local materials and information sources. He demonstrated that the levels of local *materials* exchange were not particularly affected by the ownership of plants; the types of technology employed and scale of operations were much more important. The *communications* links and business services of multiplant firms, however, were heavily

Table 3.1 The segmented economy

---

1. **Large business organizations** (corporate sector, including emerging global corporations, based on the international movement of finance within a wide network of affiliates):

Consist of a number of companies (see (i) to (iv) below); multiplant; often multinational; involved in diverse sectors.

- (i) Leaders: Innovative products, services or investment opportunities.
High growth potential and risk.
Depend on technical know-how and skilled labour.

- (ii) Intermediate: Established/mainstream activities of the corporation, serving developed markets.
High, steady profits. Moderate growth potential, relatively secure.
Depend on management quality and capital to succeed.

- (iii) Laggards: Fading or saturated markets, increasing competition.
Low profits.
Dependent on sales performance, automated production methods, unskilled labour to survive.

- (iv) Support: Provides services to the rest of the corporation (e.g. management, finance, computing).
Competition from 'in-house' provision of services, and from 'independent' outside firms.

2. **Smaller firms**

Single companies; few sites (often single-site); nationally/regionally based; dependent on large firms for survival; independent ownership.

- (i) Leaders: Based on personal innovation.
Young, opportunistic.
High birth rate and death rate.

- (ii) Intermediate: Older firms, occupying established product or service 'niche'. Depend on large organizations in various ways, e.g.:
  (a) Satellites –
  Subcontractors, providing manufacturing or service facilities; franchises.
  High birth and death rate.
  (b) 'Loyal opposition' –
  Complementary to large organizations in products, or offering specialist goods/services not offered by large organizations.
  Relatively stable.

- (iii) Laggards: No growth planned; survivors. Include:
  (a) Craftsmen –
  Skilled proprietors.
  Limited capital/technology.
  (b) Satisfied –
  'Family firms'; limited lifespan.
  Size deliberately constrained.

Source: After Taylor and Thrift 1982

orientated towards their headquarters outside the region, rather than to local sources. Marshall argued that this behaviour reduces the demand for local industrial services and thus affects the level and quality of provision for locally-based industrial firms at a time when specialist financial, legal, market consultancy, computer and other information is of growing significance to the success of small business (Marshall 1982a). An indirect impact of external ownership is therefore to degrade the 'information environment' of peripheral regions.

The vulnerability of the small firm sector to the operations of large firms is also made clear from this work. This is particularly significant since the growth of small firms is often regarded as offering better prospects for future employment as large firms reduce their workforce needs. In fact, evidence from various parts of the country suggests that small firms are highly volatile and, at best, can be expected to create a significant net addition of employment only over a long period of time. While there may indeed be some scope for a small firm revival in various areas of Britain, detailed enquiry into their roles and operating environments in recent years has offered little evidence to suggest that small firms provide a truly independent source of employment growth (Storey 1981; 1982; 1983; Cross 1981; Walker and Green 1982; Mason 1983).

The most comprehensive and convincing recent formulation of the positions of small firms in relation to different types of large firms has been proposed by Taylor and Thrift (1982; 1983). Building on a critical review of earlier organizational and linkage studies, they develop a framework for 'particularizing the environment' of firms through the study of sets of competing, controlling and complementary organizations. Industrial linkages are re-interpreted as the spatial manifestations of the power networks within which organizations operate. Fundamental to an understanding of the environment within which firms operate is the uneven distribution of resources and influence. Thus in Table 3.1, a strong dualism is illustrated between two sets of enterprise; the large business enterprise, in the corporate sector, and the smaller firm. The former are often multinational, commanding various sectors of operations. The latter are not only small in size, often operating from single plants, but also work in an environment dominated by the actions of large firms, which are generally able to exercise much greater flexibility of operations. In both 'segments', individual organizations may be further sub-divided in relation to their general role in the power structure, and especially whether they lead or follow other firms. Continuous change, through takeovers, licensing, sub-contracting, sale and labour transfers, occurs between the segments as large organizations seek to generate new products and to enter new markets. Small firms are therefore unlikely to grow into large organizations because of the financial domination of established large firms and the specialization and instability of small firms markets. The attraction of Taylor and Thrift's framework is that, while integrating many aspects of past behavioural work (for example on linkages, large and small firms, technical innovation, takeover and merger activity), it also points to new areas of research. A comprehensive approach to all types of firm is suggested, in groups ranging from highly localized small firms to the global activities of multinational corporations. Taylor and Thrift suggest that different aspects of classical, behavioural and structuralist location theory may be appropriately applied to different segments. The framework particularly emphasizes the inter-relations of business organizations through the operations of networks of power or control, and the processes of change which they undergo.

In Fig. 3.3, firms in both segments are shown to change their role through time. In the corporate sector, as laggard firms fall behind, capital is transferred to new leading firms and to more profitable intermediate firms. The dynamism of the small firm sector is sustained by high-risk leading firms which are often generated by motivated individuals breaking away from large firms. They are also often absorbed into large firms once they are successful, through takeover activity. Other types of small firms are described in Table 3.1; some act primarily as sub-contracting satellites to large organizations, while others specialize in products or services which large organizations do not find worth offering (the 'Loyal Opposition'). Many small firms remain small because they are tied to the technical and managerial skills of individuals, as craftsmen, or in family firms, who have no wish to expand into larger firms after setting up independently. Thus small, local firms operate in an environment that is largely determined by the operations of large organizations while, within large corporations, leader, laggard and intermediate firms make different contributions to both the current and the future operations of the organization to which they belong. This framework therefore offers a basis for the study of

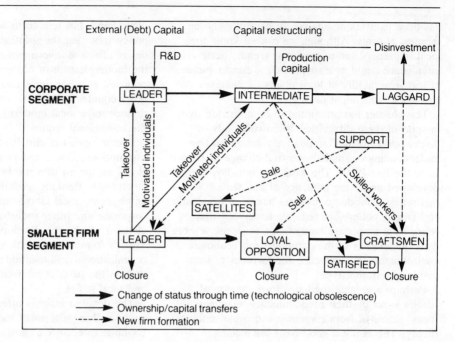

Fig. 3.3 Major directions and forms of transition within and between segments of the economy

critical elements in the environments of all types of firms, linking the national and multinational to the local perspective.

3.5 Conclusion

In the period reviewed in this chapter, behavioural research has anticipated and charted the growing functional specialization of manufacturing in various regions, offering many insights into the spatial effects of the development strategies of large firms in circumstances of both growth and decline. Fundamental changes in the manufacturing geography of Britain have been identified and measured, stimulating a debate on the contemporary nature and future of manufacturing in different types of area. Behavioural studies have challenged many older, simple notions about the causes of local economic change, including the still-prevalent assumptions that sectoral composition and unemployment indices are adequate guides to policy. Through components of change studies, the complexity and variety of local economic adjustments and their relationship to varied types of corporate change have been thrown into stark relief. The wider decision-making context of locational change has been explored, revealing the importance of 'non-locational' decisions, including acquisitions, mergers and other forms of financial restructuring in creating a continuous stream of locational adjustments. The implications of growing external control, including that by both foreign and UK multinational firms, for regional development have been explored and discussed. One valuable outcome of this work has been recognition of the declining role of local or even regionally-based manufacturing and producer service organizations and the cumulative impact on the remaining locally-based firms of increased external control. Behavioural research into industrial location has offered by far the most thorough critique of conventional assumptions about the role of small firms in stimulating employment growth. Other important areas of research not reviewed in this chapter in detail, but which have challenged popular views about local economic change, concern the nature and impact of technical innovation at the local level, and the effects of regional and industrial policy incentives.

If this work has appeared to be somewhat disjointed in its coverage, weak in its predictive power and diffuse in its policy impact, one crucial reason has been the inadequacy of the working data base (Healey 1983). For example, evidence on the productivity and efficiency of local activities would offer a much better basis for assessing their future prospects than the available information on employment alone. In most cases, commercial confidentiality denies researchers access to the best evidence for the operating efficiency and corporate status of plants in various areas. The alternative has been painstaking analysis of carefully chosen samples of firms. Although we cannot know the local impacts of future economic trends, better intelligence could at least provide a clearer picture of the vulnerability of local plants to whatever strategies firms might pursue.

This chapter has attempted to demonstrate that, in spite of these difficulties, a growing body of diverse evidence has cumulatively enhanced our understanding of spatial industrial change in Britain since the late 1960s. The work has not, however, developed under the influence of any strong theoretical orthodoxy; progress has been pragmatic and largely policy-oriented. We have seen that renewed attempts are being made to devise a more dynamic analysis of the corporate and economic environments that influence the fortunes of local plants of different types.

Perhaps a greater challenge for behavioural studies is set by their predominantly 'managerial' focus, inherited from economic and organizational theory. This is not a good basis for placing industrial location trends in an alternative context of evaluation, for example viewed from the perspective of the workforce or local community welfare. The most vociferous expressions of this perspective in Britain were incorporated in the various Community Development Project studies of the mid-1970s, and have continued to be heard from workers and community groups in areas which have suffered major job losses. The structuralist interpretation of these patterns of employment change, such as that offered by Massey (1979a; 1979b), emphasizes developments at the broadest level in the organization of capitalism itself. The reasons for the decline of manufacturing employment, the movement of investment out of the conurbations, and the growth of female and part-time employment in non-conurbation areas are summarized in terms of a 'new spatial division of labour', reflecting the strategies of large firms in the face of increasingly intense international competition and rapid technological change.

The insights of behavioural research are thus made to appear, at best, like intricate descriptions of how a machine is working, offering no view about the value of what it produces. Formally, of course, structural analysis offers no more guidance on such normative issues, although its exponents argue that the priorities of private industry will continue to create geographical inequalities which conventional policies will have little power to reduce. Behavioural research in Britain, on the other hand, has tended to accept the predominant policy view that the solution to these inequalities lies in effective investment in new technology and the modernization of old industries, as have occurred at other times and, more recently, in other countries. Regional and social policies exist to encourage local modernization and to alleviate the transitional strains.

Events since the mid-1970s have shaken these assumptions. One consequence will certainly be a closer alignment of some behavioural research with structuralist thinking. Another should be a greater emphasis on local labour market reactions to corporate and other industrial changes. A further element will be the development of Taylor and Thrift's framework to link various types of organizational and locational change more formally than in the past, at international, national, regional and local scales.

The detailed insights offered by behavioural analyses into patterns of location change will continue to occupy a central position in regional research. Aggregate employment trends will still be analysed and interpreted, ideally augmented by other measures, for example of investment or output. The locational decisions of large and small firms, and the influence on them of patterns of ownership and changes in the external environment will continue to be investigated. As in the past, the *context* of these enquiries may shift in relation to the changing policy environment. In the 1970s, the growing influence of regional external ownership, the emergence of urban decline and patterns of redundancy attracted new attention. In the 1980s, the spatial impact of technological change and the quality as well as the quantity of employment demand are among the issues of central interest. The employment role, nationally and locally, of service industries is at last also being given more serious attention. Whatever debate take place about the epistemological basis of locational research, the economic conditions and policy priorities of the late 1980s will guide inquiry as actively then as in the past, even though perhaps in different directions.

References

Alchian, A. A. 1950 Uncertainty, evolution and economic theory. *J. Pol. Econ.* **58**: 211–21.

Buswell, R. J., Lewis, E. W. 1970 The geographical distribution of industrial research activity in the UK. *Reg. Stud.* **4**: 297–306.

Carr, M. 1983 A contribution to the review and critique of behavioural industrial location theory. *Prog. Hum. Geog.* **7**: 386–401.

Child, J. 1977 *Organization: A guide to problems and practice*. Harper and Row.

Cooper, M. J. M. 1976 Government influence on industrial location. *Town Planning Rev.* **47**: 384–97.

Cross, M. 1981 *New firm formation and regional development*. Gower.

Cyert, R. M., March, J. G. 1963 *A behavioural theory of the firm*. Prentice-Hall, Englewood Cliffs.

Dennis, R. 1978 The decline of manufacturing employment in Greater London, 1966–74. *Urban Studies* **15**(1): 63–74.

Department of Trade and Industry 1973 Memorandum on the inquiry into location attitudes and experience. *Minutes of Evidence*, Trade and Industry Sub-Committee of the House of Commons Expenditure Committee, 4 July, Session 1972–73, HMSO: 525–668.

Dessler, G. 1976 *Organisation and management*. Prentice-Hall, Englewood Cliffs.

Dicken, P. 1971 Some aspects of the decision making behaviour of business organizations. *Econ. Geog.* **47**: 426–37.

Dicken, P. 1976 The multi-plant business enterprise and geographical space: some issues in the study of external control and regional development. *Reg. Stud.* **10**: 401–12.

Dicken, P. 1980 Foreign direct investment in European manufacturing industry: the changing position of the UK as a host country. *Geoforum* **11**: 289–313.

Dicken, P. 1983 Japanese manufacturing investment in the United Kingdom: a flood or a mere trickle? *Area* **15**: 273–84.

Dicken, P., Lloyd, P. E. 1976 Geographical perspectives on US investment in the UK. *Env. and Planning (A)* **8**: 685–705.

Dicken, P., Lloyd, P. 1978 Inner metropolitan industrial change, enterprise structures and policy issues: case studies of Manchester and Merseyside. *Reg. Stud.* **12**(2): 181–98

Dicken, P., Lloyd, P. E. 1980 Patterns and processes of change in the spatial distribution of foreign-controlled manufacturing employment in the UK, 1963–75. *Env. and Planning (A)* **12**: 1405–26.

Dunning, J. H. 1974 *United States industry in Britain*. Financial Times, London.

Evans, A. W. 1973 The location of the headquarters of industrial companies. *Urb. Stud.* **10**: 387–95.

Firn, J. R. 1975 External control and regional development: the case of Scotland. *Env. and Planning (A)* **7**: 393–414.

Firn, J. R., Swales, J.K. 1978 The formation of new manufacturing establishments in the central Clydeside and West Midlands conurbations 1963–1972: a comparative analysis. *Reg. Stud.* **12**: 199–214.

Fothergill, G., Gudgin, S. 1982 *Unequal growth: urban and regional employment change in the UK*. Heinemann.

Fothergill, S., Kitson, M. Monk, S. 1983 The impact of the New and Expanded Town programmes on industrial location in Britain, 1960–78. *Reg. Stud.* **17**: 251–60

Galbraith, J. R. 1977 *Organization design*. Addison-Wesley, Reading, Massachusetts.

Goddard, J. B., Smith, I. J. 1978 Changes in corporate control in the British urban system. *Env. and Planning (A)* **9**: 1073–84.

Hamilton, F. E. I. 1974 *Spatial perspectives on industrial organisation and decision making*. Wiley.

Hamilton, F. E. I. 1978 Aspects of industrial mobility in the British economy. *Reg. Stud.* **12**: 153–66.

Hamilton, F. E. I., Linge, G. J. R. 1979 *Spatial analysis, industry and the industrial environment, Vol. I: Industrial Systems*. Wiley: Ch. 1, 1–24.

Harrison, R. T. 1982 Assisted industry, employment stability and industrial decline. *Reg. Stud.* **16**: 267–86.

Hayter, R., Watts, H. D. 1983 The geography of enterprise: a reappraisal. *Prog. Hum. Geog.* **7**: 157–81.

Healey, M. J. 1981a Product changes in multi-plant enterprises. *Geoforum* **12**: 359–70.

Healey, M. J. 1981b Locational adjustment and the characteristics of manufacturing plants. *Trans. Inst. Brit. Geog.* (New Series) **6**: 394–412.

Healey, M. J. 1982 Plant closures in multi-plant enterprises – the case of a declining industrial sector. *Reg. Stud.* **16**: 37–51.

Healey, M. J. 1983 *Urban and regional industrial research: the changing UK data base*. Geobooks.

Hill, C. 1954 Some aspects of industrial location. *J. Ind. Econ.* **3**: 184–92.

House of Commons, Committee of Public Accounts 1981 *Measuring the effectiveness of regional industrial policy*. House of Commons Paper 206, HMSO.

Howard, R. S. 1968 *The movement of manufacturing industry in the United Kingdom, 1945–65*. HMSO.

Howells, J. R. L. 1984 The location of R and D: some observations and evidence from Britain. *Reg. Stud.* **18**: 13–30.

Keeble, D. E. 1971 Employment mobility in Britain. In Chisholm, M. and Manners, G. *Spatial policy problems of the British economy*. Cambridge University Press, 24–68.

Keeble, D. E. 1972 Industrial movement and regional development. *Town Planning Rev.* **43**: 3–25.

Keeble, D. E. 1976 *Industrial location and planning in the United Kingdom*. Methuen.

Keeble, D. E. 1977–79 Industrial geography. *Prog. Hum. Geog.* **1**: 304–12; **2**: 318–23; **3**: 425–33.

Keeble, D. E. 1980 Industrial decline, regional policy and the urban–rural manufacturing shift in the UK. *Env. and Planning (A)* **12**(8): 945–62.

Krumme, G. 1969 Towards a geography of enterprise. *Econ. Geog.* **45**: 30–40.

Law, C. M. 1980 The foreign company's location investment decision and its role in British regional development. *Tijd. Econ. Soc. Geog.* **71**: 15–20.

Leigh, R., North, D. 1978a Acquisition in British industries: implications for regional development. In Hamilton, F. E. I. *Contemporary industrialisation.* Longman, 158–81.

Leigh, R., North, D. 1978b Regional aspects of acquisition activity in British manufacturing activity. *Reg. Stud.* **12**: 227–46.

Lloyd, P. E. 1979 The components of industrial change for the Merseyside inner area: 1966–1975. *Urb. Stud.* **16**: 45–60.

Lloyd, P. E., Mason, C. M. 1978 Manufacturing industry in the inner city: a case study of Greater Manchester. *Trans. Inst. Brit. Geog.* New Series **3**(1): 66–90.

Lloyd, P. E., Mason, C. M. 1979 Industrial movement in north-west England, 1966–75. *Env. and Planning (A)* **11**: 1367–85.

Lloyd, P. E., Reeve, D. E. 1982 North West England 1971–1977: a case study in industrial decline and economic restructuring. *Reg. Stud.* **16**(5): 345–60.

Luttrell, W. F. 1962 *Factory location and industrial movement.* 2 vols. N.I.E.S.R.

McDermott, P. J. 1976 Ownership, organisation and regional dependence in the Scottish electronics industry. *Reg. Stud.* **10**: 319–25.

McDermott, P. J., Taylor, M. J. 1982 *Industrial organisation and location.* Cambridge University Press.

McGreevy, T. E., Thomson, A. W. J. 1983 Regional policy and company investment behaviour. *Reg. Stud.* **17**: 347–58.

Malecki, E. J. 1982 Industrial geography: introduction to the special issue. *Env. and Planning (A)* **14**: 1571–6.

Marshall, J. N. 1979 Ownership, organization and industrial linkage: a case study in the Northern region of England. *Reg. Stud.* **13**: 531–58.

Marshall, J. N. 1982a Linkages between manufacturing industry and business services. *Env. and Planning (A)* **14**: 1523–40.

Marshall, J. N. 1982b Organisational theory and industrial location. *Env. and Planning (A)* **14**: 1667–84.

Mason, C. M. 1980a Industrial decline in Greater Manchester, 1966–75: a components of change approach. *Urb. Stud.* **17**(2): 173–84.

Mason, C. M. 1980b Intra-urban plant relocation: a case study of Greater Manchester. *Reg. Stud.* **14**(4): 267–84.

Mason, C. M. 1981 Manufacturing decentralization: some evidence from Greater Manchester. *Env. and Planning (A)* **13**(7): 861–84.

Mason, C. M. 1983 Some definitional difficulties in new firms research. *Area* **15**: 53–60.

Massey, D. 1979a In what sense a regional problem? *Reg. Stud.* **13**: 233–44.

Massey, D. 1979b A critical evaluation of industrial location theory. In Hamilton, F. E. I., Linge, G. J. R. *Spatial analysis, industry and the industrial environment, Vol. I: Industrial Systems.* Wiley, Ch. 4, 57–72.

Massey, D., Meegan, R. A. 1979 The geography of industrial reorganisation. *Prog. in Planning* **10**(3): 155–237.

Norcliffe, G. 1975 A theory of manufacturing places. In Collins, L., Walker, D. F. *Locational dynamics of manufacturing activity.* Wiley, 19–58.

North, D. J. 1974 The process of locational change in different manufacturing organisations, in Hamilton, F. E. I. (ed.) *Spatial perspectives on industrial organisation and decision making,* 213–44.

Oakey, R. P. 1979 The effect of technical contacts with local research establishments on the location of the British instruments industry. *Area* **11**: 145–50.

Oakey, R. P. 1983 New technology, government policy and regional manufacturing employment. *Area* **15**: 61–5.

Oakey, R. P., Thwaites, A. T., Nash, P. A. 1982 Technological change and regional development: some evidence on regional variations in product and process innovation. *Env. and Planning (A)* **14**: 995–1138.

Owens, P. R. 1980 Direct foreign investment – some spatial implications for the source economy. *Tijd. Econ. Soc. Geog.* **71**: 50–62.

Parsons, G. 1972 The giant manufacturing corporations and regional development in Britain. *Area* **4**: 99–103.

Pred, A. R. 1967 Behaviour and location, Part I. *Lund Studies in Geography,* Series B, No. 27.

Rothwell, R. 1982 The role of technology in industrial change: implications for regional policy. *Reg. Stud.* **16**: 361–70.

Sant, M. E. C. 1975 *Industrial movement and regional development: the British case.* Pergamon.

Schofield, J. A. 1979 Marco-evaluations of the impact of regional policy in Britain: a review of recent research. *Urb. Stud.* **16**: 251–71.

Silverman, D. 1970 *The theory of organisations.* Heinemann.

Smith, I. J. 1979 The effect of external takeovers on manufacturing employment change in the Northern region between 1963 and 1973. *Reg. Stud.* **13**: 421–38.

Smith, I. J., Taylor, M. J. 1983 Takeover, closures and the restructuring of the UK iron foundry industry. *Env. and Planning (A)* **15**: 639–62.

Steed, G. P. F. 1968 The changing milieu of a firm: a study in manufacturing geography. *Ann. Assn. Amer. Geog.* **58**: 506–25.

Steed, G. P. F. 1971a Changing processes of corporate-environment relations. *Area* **3**: 207–11.

Steed, G. P. F. 1971b Plant adaptation, firm environments and locational analysis. *Prof. Geog.* **23**: 324–7.

Storey, D. J. 1981 New firm formation, employment change and the small firm: the case of Cleveland County. *Urb. Stud.* **18**: 335–46.

Storey, D. J. 1982 *Entrepreneurship and the new firm.* Croom Helm.

Storey, D. J. 1983 Job accounts and firm size. *Area* **15**: 231–7.

Taylor, M. J. 1975 Organisational growth, spatial interaction and location decision-making. *Reg. Stud.* **9**: 313–24.

Taylor, M. J., Thrift, N. J. 1982 Industrial linkage and the segmented economy: I some theoretical proposals. *Env. and Planning (A)* **14**: 1601–14.

Taylor, M. J., Thrift, N. J. 1983 Business organisation. segmentation and location. *Reg. Stud.* **17**: 445–65.

Thwaites, A. T. 1982 Some evidence of regional variations in the introduction and diffusion of industrial products and processes within British manufacturing industry. *Reg. Stud.* **16**: 371–82.

Townroe, P. M. 1969 Locational choice and the individual firm. *Reg. Stud.* **3**: 15–24.

Townroe, P. M. 1971 *Industrial location decisions: a study in management behaviour*. University of Birmingham, Centre for Urban and Regional Studies, Occasional Paper No. 15.

Townroe, P. M. 1972 Some behavioural considerations in the industrial location decision. *Reg. Stud.* **6**: 261–72.

Townroe, P. M. 1975 Branch plants and regional development. *Town Pl. Rev.* **46**: 47–62.

Townroe, P. M. 1979 *Industrial movement: experience in the US and UK*. Saxon House.

Townsend, A. R. 1981 Geographical perspectives on major job losses in the United Kingdom 1977–80. *Area* **13**: 31–8.

Townsend, A. R. 1982a *The impact of recession*. Croom Helm.

Townsend, A. R. 1982b Recession and the regions in Great Britain, 1976–80: analyses of redundancy data. *Env. and Planning (A)* **14**: 1389–1404.

Walker, S., Green, H. 1982 The role of the small firm in the process of economic regeneration: the evidence from Leeds. In Collins, L. (ed.) *Industrial decline and regeneration*. Department of Geography, University of Edinburgh.

Watts, H. D. 1979 Large firms, multinationals and regional development: some new evidence from the UK. *Env. and Planning (A)* **11**: 71–81.

Watts, H. D. 1980 The location of European direct investment in the United Kingdom. *Tijd. Econ. Soc. Geog.* **71**: 3–13.

Watts, H. D. 1981 *The branch-plant economy*. Longman.

Westaway, J. 1974a Contact potential and the occupational structure of the British urban system, 1961–66: an empirical study. *Reg. Stud.* **8**: 57–73.

Westaway, J. 1974b The spatial hierarchy of business organisations and its implications for the British urban system. *Reg. Stud.* **8**: 145–55.

Whitelegg, J. 1976 Births and deaths of firms in the inner city. *Urb. Stud.* **13**(3): 333–8.

Wood, P. A. 1969 Industrial location and linkage. *Area* **1**: 32–9.

Wood, P. A. 1978 Industrial organisation, location and planning. *Reg. Stud.* **12**: 143–52.

Wood, P. A. 1980–82 Industrial geography. *Prog. Hum. Geog.* **4**: 406–16; **5**: 414–19; **6**: 576–83.

CHAPTER 4

Structural approaches to industrial location

Martin J. Boddy

So-called 'structural' approaches to location theory attempt to relate the changing geography of jobs and industry to the underlying structure of capitalist society, economic and class relations and to the social and spatial organization of commodity production rooted in the contradictory relationship between, primarily, capital and labour. They place particular emphasis on the analysis of production itself rather than simply on 'locational factors', they see production as an essentially social process structured by capital – labour relations and the wider political and ideological context, they situate empirical analyses in their wider national, international and historical contexts, and finally, they place particular emphasis on the role of labour and the nature of the labour process as a key element in the production process. Recognizing the importance of spatially and geographically unique outcomes of more general underlying processes, the aim of theoretical and conceptual development is to provide a framework through which to undertake empirical analysis of specific situations, rather than to generate universally applicable, formal models.

Examples would include, in particular, Massey and Meegan's (1982) analysis of the processes of economic restructuring and the geography of job loss and Massey's subsequent (1984) extended analysis of social processes and the spatial organization of production. While not wishing to deny the variety of work which might go under this banner, the latter is the fullest account of the general approach and it brings together and develops many of the themes explored in earlier analyses and critiques of traditional approaches (Massey 1977; 1979; Massey and Meegan 1979).

Similar issues have been explored at a theoretical and empirical level in the US, in particular by Storper and Walker (1981; 1983) and Storper (1981). Regional uneven development has been explored by, for example, Carney et al. (1980), Perrons (1981) and Morgan and Sayer (1983). Analyses of particular sectors include Morgan's (1983) analysis of the restructuring of the steel industry, Morgan and Sayer's (1984) analyses of the electronics industry in South Wales and the M4 Corridor, and Cooke et al. (1984) on the semiconductor industry. Combined with this, there has been an increasing concern both with locality-based studies and with sectoral analyses which have a regional or locality-level focus. Sayer and Morgan's analysis of electronics falls into this category (see Morgan and Sayer 1984); the many other examples include Murgatroyd and Urry (1983) on the Lancaster economy, Boddy and Lovering (1985a; 1985b) on the Bristol region and, most explicitly, the collection of studies initiated in 1985 by the Economic and Social Research Council on 'The Changing Urban and Regional System'. Several studies have focused on individual companies or have built methodologically on this scale of analysis (Peck and Townsend 1984; Boddy and Lovering 1985a). Locality-focused sectoral analyses have also, in a more policy-oriented sense, underlain the sort of economic and employment strategies developed by local authorities such as Sheffield, the Greater London Council and West Midlands County Council in the early 1980s (Boddy 1984).

The development of the 'structural' perspective in part reflects changes since the 1960s in the focus of attention – the increasingly complex organization of production with the growth of multiplant and multinational firms and the 'branch plant' phenomenon; regional differences based increasingly on the functions performed (e.g. headquarters, R and D, basic assembly) and related employment characteristics (e.g. basic assembly and supervisory, professional and technical, managerial) rather than sectoral specialization; decentralization of employment away from the major urban areas and the central regions reflected in inner-city decline; and major transformations in industrial and employment structure with the slump in manufacturing, and service expansion, in the context of deepening recession. These concerns are, of course, shared with other approaches. It is worth noting, however, that structural analyses were developed in the context of recession, internationalization, increased functional differentiation and the generalization of economic decline and the rundown of regional policy. In contrast, much of traditional industrial geography was concerned, in its formative years with, in particular, the location decisions of mobile industry and the impacts of regional policy.

The structural perspective is also, of course, rooted in the transformation of social science since the late 1960s and the development within geography of 'radical' approaches grounded in Marxist analysis. In industrial geography, radical approaches were built up from a specific critique of the dominant neoclassical traditions of location theory and behavioural alternatives and their inherent explanatory shortcomings, increasing in the face of the evident transformations in industrial and employment structure (Massey 1977; Storper 1981). In brief, formal neoclassical models saw individual profit-maximizing firms locating in relation to market forces in the context of perfect competition and static equilibrium. This is oversimplified, but the spatial distribution of 'factors of production' and of markets was the key to the location of industry and employment. Operationally, even allowing for such assumptions, they were incapable of handling any but the simplest of idealized situations abstracted from reality. Considerable refinements have been achieved, the approach has provided many insights and, as Smith argues in Chapter 2, it can be important in a normative, planning context. Many of the fundamental problems remain, however, and its explanatory value is strictly limited.

Behavioural approaches attempted to overcome some of the disabling assumptions of neoclassical approaches by, for example, recognition of sub-optimal goals. Much more important, however, was the growing concern to relate locational behaviour and employment impacts to organizational structure and corporate strategy – what Storper (1981) has called the 'systems' approach, drawing on organization theory and systems theory as well as behavioural analysis. This contributed, together with a large volume of essentially empirical analysis, to a great mass of largely empirical analysis of urban and regional change in industry and employment (e.g. Keeble 1976; Goddard and Champion 1983), contributing significantly to our knowledge of the patterns of change and suggestive at least of the underlying processes. Some of this work, particularly on the structures of business organizations, starts, moreover, as Wood indicates in Chapter 3, to overlap with the empirical concerns of structural analysis grounded in the Marxist tradition (viz. Taylor and Thrift 1983). It would be wrong, therefore, to overemphasize the differences. There are, however, important contrasts even at the empirical level and much more so in terms of explanation and implications for action, as will become apparent. In particular, organizational approaches remain focused largely on locational factors and the relation between spatial patterns; they fail to recognize the social character of production based around the (conflictual) relationship between labour and capital, focusing instead on managerial decision-making and corporate strategy within a context in which labour is a 'factor of production'; and they have frequently mis-specified the organizational structure of production and processes of change, generating chaotic conceptions rather than coherent explanation (Sayer 1984).

4.1 Social structure and the organization of production

The starting point of structural analysis is, according to one definition, that: 'It aims to set the changing geography of industry and employment within the wider context of the development of capitalist society and to examine the particular developments in Britain through the prism of the evolution of class and economic relations both within the country and internationally. Specifically, it aims to explore the geography of industry and of jobs through an interpretation of the spatial organization of the social relations of capitalist production' (Massey 1984: 6). Expanding on this, we can set out some of the implications, looking in turn at the central importance of production, its essentially social character, the relation between structure and real world outcomes, the issue of scale and, finally, the key role of labour and the labour process.

4.2 The production process

First, in order to understand the changing geography of jobs and industry, we have to look at the production process and at the spatial organization of production rather than simply at 'location factors' such as labour supply, availability of premises or regional policy incentives. Traditional, empirical analyses have generally sought to relate the geographically differentiated pattern of industrial or employment change to the characteristics of areas, to the spatial distribution of supposed 'location factors'. They have largely ignored the fact that spatial patterns do not simply or necessarily reflect spatial causes, but reflect, also, the locational requirements of industry, and these can change as a result of changes in the production process. For example, decentralization of employment towards the peripheral regions and convergence of indicators of interregional inequality from the mid-1960s coincided with an increase in the strength of regional policy, leading to claims that the policy succeeded. However, as Massey (1979) has pointed out, the shift may also have reflected the changing locational requirements of industry due to changes in the production process. In fact, it is argued, convergence, in part at least, reflected a newly-emergent 'spatial division' based on a geographically differentiated 'spatial hierachy of functions; (Hymer 1972; Westaway 1974) in which basic production activity was increasingly decentralized while higher management and technical operations were increasingly concentrated in the centre.

Geographical characteristics, factors of production, are not unimportant. Their relevance depends, however, on the locational requirements of production. Locational requirements, moreover, vary between different elements of the 'functional hierachy', they vary within as well as between sectors. Analysis of the spatial organization of production and the processes of economic restructuring is therefore crucial:

> 'the kinds of locational characteristics which will influence a company's choice of where to build a new factory, expand, run down or close will depend on the nature and demands of the process of production. In turn, changes in production, and consequently in location factors, were argued to be the result, not of some autonomous choice by management, but of wider economic and political forces, shop-floor relations within the firm, and of the company's reaction to both. (Massey 1984: 14)

4.3 Production as a social process

Central to the analysis of production is the process of capitalist commodity production and the accumulation of capital in pursuit of private profit. Integral to this is the basic class division between capital and labour, between those who own and control the means of production and those who must sell their labour as a commodity. We need, of course, to elaborate on this rather bald statement. Production is essentially a social process involving, in particular, relations between capital and labour. These are not, however, undifferentiated categories; social structure, moreover, develops historically with specific national and subnational characteristics (Cooke 1984). It reflects, in part, the spatial organization of production itself: 'what are called "interregional relations" and geographical differences in type of employment are in large part the spatial expression of the relations of production and the divisions of labour within society' (Massey 1984: 39). Capital itself, for example, varies in terms of social form, place in the economic structure and scale, while between the 'bourgeoisie' amd 'proletariat' one can recognize a range of intermediate class locations ranging in occupational terms from, say, top managers to foremen and line supervisors (Wright 1976). Thus: 'the study of industry and of production is not just a matter of "the economic", and economic relations and phenomena are themselves constructed within a wider field of social, political and ideological relations. A real exploration of industrial geography takes one into historical shifts in national politics, into the vast varieties of social forms of capital, into the whole arena of gender relations . . .' (Massey 1984: 7). Central to the actual processes of restructuring and the outcome of external pressures on businesses are relations between capital and labour, capital's attempt to accumulate and to maintain control and labour's to resist.

4.4 Structure and outcome

While some versions of Marxist theory have verged on the mechanical – arguing, for example, for the inevitability of spatially uneven development (discussed in Browett 1984) – recognition of the social and historically-specific character of the production process and the relations of production emphasizes the non-deterministic nature of this approach and the necessity for empirical research. Massey and Meegan (1982: 12), for example, in their analysis of job loss, argued that 'Under the present economic system in the UK, it is the process of production for profit to which employment decline ultimately needs to be related. But this relationship is not simple and undifferentiated. The "imperatives of capital accumulation" can produce, in different economic and political circumstances and in different industries, very different responses.' Job loss, as their analysis demonstrates, has taken place in the context of different processes of economic restructuring but, as they demonstrate, 'these varying kinds of change within production could each be related to the exigencies of production for profit – to the system's overall dynamic' (Massey and Meegan 1982: 12). More generally, 'the way in which a firm copes with the pressures upon it depends on the kind of capital involved, on the kind of labour which it faces, and on the battle between them. It also depends on how those pressures are defined and on how they are translated through the wider political and social context' (Massey 1984: 5). Moreover, the overlaying of successive rounds of investment leads, in specific localities, to the superimposition of different spatial divisions of labour. The 'factors of production', such as labour and infrastructure, are themselves largely the product of previous spatial divisions of labour and rounds of investment, the product, again, of a social process. In this context, empirical analysis is concerned, then, with geographically and historically specific unique outcomes.

4.5 The national and international context

Structural analyses have frequently been focused at the regional or locality level. This obviously reflects the central concern with spatially uneven development, with regional issues and the inner city, and with locationally and historically unique outcomes. It has reflected also recognition of the need to handle simultaneously the social, political, ideological and cultural levels as well as the economic. This is partly because these levels and their combination do have an element of locational specificity and partly for practical reasons, given the complexities of the subject matter. This has therefore reinforced a concern with the subnational level. There has, however, been a specific concern to situate regional or locality-oriented analyses within the national and international context. Again, this is implied in the emphasis given to the spatial organization of production which obviously takes the analysis well beyond regional confines. It has, however, been strongly reinforced by the increasing scale, internal differentiation, mobility and internationalization of the modern corporation and their markets, and the importance of the new international division of labour (Froebel et al. 1980).

4.6 The importance of labour

The internationalization of capital coupled with the importance placed on the social character of production, centred around relations between capital and labour, has led to an increasing concern with labour, labour markets and the nature of the labour process. As Storper and Walker (1983) have argued, the globalization and vastly increased 'locational capability' of much of capital has decreased the importance of non-labour locational factors. The relative importance of labour has increased as a consequence and 'As capital develops its capability of locating more freely with respect to most commodity sources and markets, it can afford to be more attuned to labour force differences' (Storper and Walker 1983: 3–4). Capital is therefore able to take advantage of spatial differentiation in labour, and functionally-differentiated capital, with different labour requirements, will tend to become spatially dispersed. Urry (1981) has similarly drawn attention to the increased importance of specifically local labour market characteristics: 'large national or international capital will seek those labour markets which fit its requirements and, following the new international division of labour thesis, it will look for a variety of markets, each being appropriate to its different fractional operations' (Urry 1981: 468).

4.7 The labour process and the relations of production

Added to this, the emphasis on production as a social process in which capital and labour combine and conflict has increasingly generated direct concern with the labour process and the relations of production. Labour is no longer seen simply as a passive 'factor of production'. At a general, historical level, Dunford et al. (1981) have related the changing structures of spatial inequality to successive regimes of capital accumulation rooted in broadly defined phases in the development of the labour process (see also Perrons 1981). It is argued that 'Studies of regional and local economies taken together provide necessary knowledge about the variety of outcomes produced by specific processes of accumulation operating in particular contexts. But these outcomes and the underlying causes of regional and urban problems cannot be understood without being related both to the broader macroeconomic and political framework and to the general laws of capitalist development' (Dunford et al. 1981: 404). Drawing on Aglietta (1979), they distinguish between nineteenth-century 'machinofacture'; 1930s scientific management and 'Fordism' (assembly-line production as in car manufacture); and finally the collapse of 'Fordist' growth, leading to more diffused and geographically peripheral growth in the attempt to restore profitable capital accumulation through 'neo-Fordist' labour processes based on automation and new management practices. It is the latter, according to Dunford et al., which underlies the observed convergence of indicators of regional inequality and is the basis for the new geographical division of labour – the gap between centres of control and localities characterized by externally-controlled economic development and by relatively low-paid, unskilled employment. While developing an analysis of the relations between spatial inequalities and the processes of capital accumulation, these approaches have been criticized for the somewhat mechanistic links which have been drawn from the general laws, the 'needs' of capital accumulation to the dominant form of labour process and related patterns of regional imbalance, perhaps somewhat unfairly given their emphasis on the variety of outcomes quoted earlier in the paragraph. The problem is that they pay little attention to the processes which generate that variety, the different forms of job loss and spatial reorganization of production. In particular, it has been argued that they have underplayed the role of worker resistance and the active role of labour in shaping the labour process and the spatial organization of production. Furthermore, the attempted periodization into phases marked by the predominance of broad processes of capital accumulation and labour process underestimates the variety of relations established in the contest between capital and labour in the workplace, including the degree of variation in management strategy demonstrated in other studies (Edwards 1979; Zimbalist 1979; Morgan and Sayer 1982).

Developing this theme, it has been argued that production changes, including the spatial organization of production, are actively shaped by the interaction of capital and labour in the workplace, by management strategy and worker resistance (Cooke 1983; Taylor and Thrift 1983). Such interaction is shaped by the context of labour market and product market conditions, which influence the relative strength of management and labour. Friedman (1977; 1983), for example, draws the relatively simple distinction between strategies of 'direct control' and 'responsible autonomy' in analysing the characteristics of inner-city labour markets and urban decay. In the case of direct control, managers reduce the responsibility and discretion of the individual worker by means of close, hierarchical supervision and by setting out specific tasks to be performed in great detail. With responsible autonomy, workers are given responsibilty and status, supervision is lighter and employees are encouraged to identify their own interests with those of management and 'the company' as a whole. Friedman argues that firms have tended to differentiate their workforces into 'central' workers essential to long-run profitability towards whom management develops strategies of responsible autonomy, and 'peripheral', increasingly de-skilled, workers who can more easily be replaced and whose work conditions are characterized by direct control. It is the laying off of peripheral workers, subject to direct control, which underlies inner-city unemployment and related problems. In fact, Friedman's analysis slides from arguments about labour market segmentation and the segmentation of workforces within the workplace (Loveridge and Mok 1979; Gordon et al. 1982) to the segmentation of business organizations themselves (Averitt 1968; Taylor and Thrift 1983). The two are obviously related, particularly given the increasing importance of spatial hierarchies of function within sectors. Failure to distinguish between the two can, however, obscure the importance of segmentation within large business organizations themselves which to a large extent underlies the new spatial division of labour, and of internal labour market segmentation within the individual workplace itself, which is crucial in understanding the impacts of employment change on different groups of workers and potential workers.

Management strategies and labour–management relations are in practice liable to be much more complex than the simple 'responsible autonomy–direct control' continuum would suggest. Morgan and Sayer (1984), for example, look at the re-making of management–labour relations in the electrical engineering industry in South Wales. This has included the introduction of flexible work practices, single union plants, no strike agreements, simplified pay structures, less rigid job descriptions, decentralized management and the marginalization of formal union structures. An essential element in this process has been the context of recession, the influx of foreign capital, the introduction of new technology and a new management philosophy. The latter has emphasized the securing of consensus and consent rather than direct control but is considerably more refined than the simple notion of 'responsible autonomy'. The study emphasizes the fact that firms do not simply locate in relation to the characteristics of 'local labour', but are actively involved in the making and re-making of capital – labour relations. This in turn may be reflected in the recomposition of the labour force in terms of, for example, age, gender, skill or ethnic group – changes in the geography of employment and unemployment and their impacts on different groups in the labour market. Employers' estimations of the possibilities for establishing new work practices, and how these work out in practice, then become an important element in determining the changing spatial division of labour and the geography of industry and employment.

4.8 The geography of production

Analysis of forms of economic restructuring and the geography of production change are central to structural approaches. As already indicated, Massey and Meegan's (1982) analysis of economic restructuring and job loss, and Massey's (1984) exposition which concentrates more on spatial structures of production and the links between social relations and spatial organization are key examples, worth looking at in a bit more detail.

Massey and Meegan set out to identify the different processes of production change which underly job loss at a sectoral level within the UK. They then translated this into the spatial patterning of job loss. Emphasizing initially that job loss cannot be divorced from changes in output and productivity, they distinguish three main forms of production change which may underly job loss. Intensification involves 'the reorganization of an existing production process, without abandoning capacity and without major investment in new forms of production, in order to increase labour productivity' (Massey and Meegan 1982: 31). It might involve, for example, reductions in 'manning levels' or changes in work practices, resulting in increased output per worker. Investment and technical change defines job loss in the context of significant investment, particularly in the process of production but possibly also in product innovation where the latter reduces labour needs. Employment decline is not, then, associated simply with declining industries but may occur in the context of major new investment and innovation in terms of production techniques and products. Rationalization, finally, generates job loss simply through the partial or total scrapping of productive capacity.

These three processes represent 'pure' forms of restructuring which in practice are to varying degrees combined. Different forms of restructuring are pursued by different sectors and firms depending on the circumstances which confront them in their search for profitable production: the need, for example, to increase the rate of profit or to counter the lack of profitability, to preserve or expand market shares, or to counter product obsolescence and market decline in a context of market change and shifting competition. The analysis goes on to demonstrate the need to understand the different mechanisms of job loss in order to explain the spatial structure of employment decline. For 'the form of production change through which jobs are lost may be at least as important for an understanding of the geographical pattern as the locational characteristics, and even locational characteristics may vary in the way they operate, depending on the kind of production change' (Massey and Meegan 1982: 123). With intensification, for example, unequal effects between regions simply reflect the differential distribution of job loss. Rationalization, on the other hand, may involve redistribution of capacity and employment between sites, when output is concentrated on to remaining plants. The effects of new investment and technical change are more variable, involving the possibility of job gain from investment at new sites as well as employment displacement. Different forms of production change do not determine spatial outcomes, particularly when they combine in practice. They do, however, have different implications for the possible spatial outcomes of economic restructuring.

Massey and Meegan's analysis convincingly demonstrates the real importance of production change in understanding the geography of job loss, and has been increasingly drawn on in empirical work (Peck and Townsend 1984; Boddy and Lovering 1986b). In fact, it can be argued that this approach needs extending. Empirically, like far too much of 'industrial geography', they confine themselves to manufacturing industry and ignore the whole range of service activities. Second, they focus only on job loss rather than employment change as a whole, including job gain. This artificially limits the scope of their analysis – it becomes, for example, a bit ragged when considering 'investment and technical change' which can, as they acknowledge, result in job gain. Third, the range of possible forms of restructuring can usefully be extended (which they again acknowledge). Other types of restructuring include, for example, simple expansion without significant change in technique, product or work organization (the inverse of rationalization); relocation of functions, product lines or entire installations, in order maybe to restructure input costs or open up new markets, a strategy often combined with other forms of production change. Finally, the simple notion of 'job loss' needs to be carried forward by considering labour market processes and impacts on different groups in the labour market. Numerical 'job loss' impacts very differently on different groups, depending on the dominant form of restructuring and the distributional mechanisms operating in the labour market. Moreover, 'labour market strategy' in terms of both recruitment and internal labour market management is itself a central element in production change. This would include for example, attempts to break down traditional craft organization and barriers to flexible working, deskilling of certain groups, the differential development of 'responsible autonomy' and 'direct control' among different groups of workers, the search for labour with little experience of union organization and traditional work practices, or the 'feminization' of particular tasks (Friedman 1983; Morgan and Sayer 1984).

More explicit attention is given to spatial structures of production in Massey's subsequent exposition of the relations between social structures and the geography of production. Extracting this particular issue from her much broader discussion, it is argued that the spatial distribution of employment 'can be interpreted as the outcome of the way in which production is organized over space' (Massey 1984: 67). Three particular examples of spatial structures are discussed, each with different implications in terms of interregional dependence. In locationally-concentrated spatial structures, all the elements of production are located within each region, including ownership, control, administration and production. Interregional relations consist of market relations between firms or between firms and markets, while interregional differentiation is commonly based on sectoral specialization. Spatial structures based on cloning, on the other hand, replicate production activity in each region but headquarter administration and control are centralized. Interregional relations involve external control and location within a hierarchy of management and control. Finally, part-process structures distribute elements of the production process itself between different regions. Here, interregional relations involve not only relations of control but also dependency on other regions within the total production process. Within cloning and part-process structures, regional differentiation relates more directly to the production process and production change within the enterprise. Possibilities for change are, however, conditioned by the spatial structure of production. Capacity cuts, for example, can be pursued within a cloning structure by the closure of branch plants within particular regions, concentrating job loss; interregional dependencies within part-process structures, however, mean that, despite branch–plant structures, outright closure is less likely and the impacts are likely to be distributed across several regions.

As with the analysis of production change, the forms of spatial structure discussed are pure forms, unlikely to be found as such in practice. They illustrate the fact that production can be organized spatially in a variety of ways and offer a means through which to approach the complexities of the real world without collapsing into unstructured empiricism. The key point is that different spatial structures of production imply different forms of geographical differentiation and spatial inequality.

'Geographical uneven development does not vary only in degree, as some of the arguments about increasing uneven development imply, it varies also in nature. And this variation reflects the plurality of ways in which the relations of production can be organized spatially' (Massey 1984: 82). As has already been suggested, this emphasis on business organization starts to overlap with the descriptive analyses of systems and organizational approaches seen, for example, in the typology developed by Taylor and Thrift (1983), discussed in Chapter 3. One major contrast between structural analyses and much of this work is that structural analyses seek to relate business organization to the interrelated social and spatial organization of production grounded in capital – labour relations and situated in its wider political and ideological context. It is from this that structural approaches derive their explanatory power and their utility as a basis for action. Analyses of business organization based solely at the empirical level with little concern for conceptual structure are unlikely to develop coherent accounts of the processes at work.

4.9 Methodological issues

Structural approaches have, therefore, a number of methodological implications. Leaving aside the more mechanistic applications of Marxist analysis, Massey (1984: 49) concludes that 'what we are concerned with in empirical analysis is geographical and historical specificity. Indeed, what lies behind the whole notion of uneven development is the fact of highly differentiated and unique outcomes.' From this perspective, the point of theory in this context is not then to divide the world into that which fits with some *a priori* model and the rest which can be described but not explained – a situation which led to the breakdown of formal neoclassical approaches and the dominance of purely descriptive methodologies. The point of theoretical concepts and of frameworks such as those described above 'is to provide tools for the analysis of specific situations, analyses on which action can be based' (Massey 1984: 50). The need for theoretically-structured empirical analysis is firmly established. The particular relationship between theory, observation and reality suggested here relates to a broader philosophical critique of traditional approaches to explanation in the social sciences. While not necessarily espousing the pure 'realist' perspective as set out by Sayer (1984), it does nevertheless reject positivist approaches which attempt to derive generally applicable regularities or 'laws' on the basis of empirical observation.

More specifically, theoretically-grounded structural analyses have seriously undermined traditional empirical abstractions in the field of industrial and employment change and the mechanisms at work. Sayer (1982; 1984) makes the formal distinction between 'rational abstraction' and 'chaotic conceptions'. Rational abstraction 'isolates a significant element of the world which has some unity and autonomous force' whereas chaotic conception 'combines the unrelated or divides the indivisible' (Sayer 1982: 71). The problem of branch plants illustrates the point. As indicated earlier, the impact of 'branch plants' as a form of spatial structure on the geography of job loss may be different depending on whether cloning or part-process structures are involved. The category 'branch plant' is in this context a chaotic rather than rational abstraction since it does not relate to the processes of change operating.

The same problem arises with the widely-used technique of 'components of change' analysis, for example, which disaggregates employment change in an area into that due to closures and openings, moves in or out, and *in situ* change. As Massey and Meegan's analysis of job loss makes clear, however, the same component can reflect several different processes: partial relocation of capacity, for example, would be recorded as *in situ* change although analogous to, and possibly responding to the same factors, as the out-movement of a complete plant; *in situ* job loss, on the other hand, might reflect (and conflate) several different processes of job loss – intensification and rationalization, as well as partial relocation of capacity away from the area. So 'not only may the same stimulus produce similar responses in a number of different components . . . but also each of the components may relate to a whole number of different processes' (Massey and Meegan 1982: 192). Similarly, shift-share analysis, which continues to be widely used, purports to divide up employment change within an area into that due to local economic structure and reflecting sectoral change at the national level, and that due to the particular 'locational characteristics' of the area. The observed pattern of locational change may, however, reflect the changing spatial organization of production and the production characteristics of local industry rather than geographical locational factors particular to the area. Employment loss through rationalization may be concentrated in an area because plant and machinery there is older and less productive, or simply because a company decides to pull out of a particular product line for reasons of corporate strategy, regardless of the locational characteristics of the area itself. As observed earlier, spatial outcomes are not necessarily related to spatial causes.

The same point was made earlier in relation to regional policy – the observed convergence of indicators of regional inequality in part at least reflected changes in the spatial organization of production rather than a simple response to spatial policies. Similar problems exist with Fothergill and Gudgin's (1983) attempt to explain the observable urban to rural shift in manufacturing employment by its correlation with a single factor, the supposed lack of room for physical expansion in the major urban areas, in total disregard for the range of processes underlying the spatial restructuring of employment. There is, of course, no reason why an apparently general spatial pattern of employment change should reflect a single causal mechanism, nor that such a causal mechanism be 'spatial' in character. It is likely, for example, that the character of productive capacity itself in the older industrial areas rather than the characteristics of these areas themselves (lack of space for expansion) has contributed to their differential decline. Productive capacity in these areas is likely to be older and less productive and therefore more likely to be written off. New investment in turn is likely to reflect the emerging new spatial division of labour based on the interaction of the production process and location factors, the changing spatial organization of production. In Sayer's terms, techniques like components of change or shift-share analysis, and attempts to tie spatial patterns to simple spatial causes are likely therefore to represent 'chaotic conceptions'.

4.10 Policy consideratons

Much of mainstream urban and regional policy has followed on from analyses which relate problems to location factors and to the characteristics of the 'problem areas'. It follows, from this perspective, that it is largely a question of making these areas more attractive by means of subsidies and other forms of financial intervention, provision of land and premises, environmental and infrastructure improvements, and information and promotional activity. The alternative is to direct locational decisions despite the perceived characteristics of different areas, as with the 'stick' element of earlier regional policy.

From a structural perspective, the characteristics of different localities are obviously relevant. But the analysis points to the need in policy terms to focus as well on issues related to production and the spatial organization of production. For it is at this level and in the interaction between the organization of production and the characteristics of local areas that the changing geography of industry and employment is to be understood. Moreover, the social organization of production and the ownership and control over productive capacity and investment flows exercised by capital is fundamental to the changing spatial structure of productive activity and patterns of inequality. Policy issues in this sense rapidly become 'political' rather than 'technical' in character, since they tie back urban and regional problems to social organization rather than spatial distributions. As Massey has observed, the problem of 'external control' for the majority of the population is not 'geographical'. It is not the fact that capital is owned and controlled from outside of their particular region that lies at the root of inequality and job loss. The problem for labour lies in the social structure of production, exclusion at the social level from control over investment and production. The bases of urban and regional problems and the possibilities for change are in this sense rooted in the conflict between capital and labour, its influence on forms of production change and on the changing spatial organization of production.

The role of the state and of 'policy' as conventionally understood (as opposed to the conflicting strategies of capital and labour) depend then on the role of the state in relation to this conflict. The state itself is not viewed as in some sense neutral, intervening in a technical sense in order to counter identifiable 'urban' or 'regional problems'; it is itself seen as a locus of political power and a site of class conflict. This, therefore, conditions the scope of 'policy response' to such problems and the capacity of the state to intervene effectively to secure real change, and defines the way in which 'policy' has to be understood. There are, of course, issues of policy implementation and effectiveness. But it is largely economic and political power, expressed through the state apparatus, which from a structural perspective determines the broad scope of policy and structures the agenda (Benson 1978: Whitting 1986).

In this sense it is the nature of the state which

conditions the traditionally dominant forms of urban and regional policy, the primary concern with attempts to remodel spatial factors of production, rather than with the organization of production itself. And it is in this sense that local authorities such as Sheffield, West Midlands County and the GLC took tentative steps to develop policies which take into account, and aimed to intervene in, the organization of production, rather than simply restructuring the locational environment within which firms operate (Boddy 1984). The structural perspective thus has specific implications as much for policy and action as for explanation.

References

Averitt, R. T. 1968 *The dual economy: the dynamics of American industry structure*. Norton, New York.

Aglietta, M. 1979 *A theory of capitalist regulation – the US experience*. New Left Books.

Bassett, K. 1983 Corporate structure and corporate change in a local economy: the case of Bristol. *Environment and Planning* **A 16**: 879–900.

Benson, J. K. 1982 A framework for comparative analysis. In Rogers, D., Whettem, D. *Interregional coordination*.

Boddy, M. 1984 Local economic and employment strategies. *Local Socialism?* Macmillan.

Boddy, M., Lovering, J. 1986a High technology industry in the Bristol sub-region: the aerospace/defence nexus. *Regional Studies* **20**(3): 217–31.

Boddy, M., Lovering, J., Bassett, K. 1986b *Sunbelt city? A case study of economic restructuring in Britain's M4 Growth Corridor*. Oxford University Press.

Browett, J. 1984 On the necessity and inevitability of uneven spatial development under capitalism. *International Journal of Urban and Regional Research* **8**(2): 155–76.

Carney, J., Hudson, R., Lewis, J. 1980 *Regions in Crisis*. Croom Helm.

Cooke, P. 1983 Labour market discontinuity and spatial development. *Progress in Human Geography* **7**(4): 543–65.

Cooke, P. 1984 Recent theories of political regionalism: a critique and an alternative. *International Journal of Urban and Regional Research* **8**(4): 549–72.

Cooke, P., Morgan, K., Jackson, D. 1984 New Technology and regional development in austerity Britain: the case of the semiconductor industry. *Regional Studies* **18**(4): 277–89.

Dunford, M., Geddes, M., Perrons, D. 1981 Regional policy and the crises in the UK: a long-run perspective. *International Journal of Urban and Regional Research* **5**(3): 377–410.

Edwards, R. C. 1979 *Contested Terrain: the transformation of the workplace in the twentieth century*. Basic Books, New York.

Fothergill, S., Gudgin, G. 1983 Trends in regional manufacturing employment: the main influences. In Goddard, J. B., Champion, A. G. (eds.) *The urban and regional transformation of Britain*. Methuen. 27–50.

Friedman, A. 1977 *Industry and Labour: class struggle at work and monopoly capitalism*. Macmillan.

Friedman, A. 1983 Social relations at work and the generation of inner city decay. Regional inequality and the state in Britain. In Anderson, J., Duncan, S., Hudson, R. *Redundant spaces in cities and regions?* Academic Press, 51–66.

Fröbel, F., Heinrichs, J., Kreye, B. 1980 *The new international division of labour*. Cambridge University Press.

Goddard, J. B., Champion, A. G. 1983 (eds.) *The urban and regional transformation of Britain*. Methuen.

Gordon, D. M., Edwards, R. C., Reich, M. 1982 *Segmented work, divided workers; the historical transformation of labour in the United States*. Cambridge University Press.

Hymer, S. 1972 The multinational corporation and the law of uneven development. In Bhagwati, J. N. (ed.) *Economics and the world order*. Macmillan, New York.

Keeble, D. 1976 *Industrial location and planning in the United Kingdom*. Methuen.

Loveridge, R., Mok, A. C. 1979 *Theories of labour market segmentation: a critique*. Martinus Nijhoff, The Hague.

Massey, D. 1977 Towards a critique of industrial location theory. In Peet, R. (ed.) *Radical geography*. Methuen, 181–9.

Massey, D. 1979 In what sense a regional problem? *Regional Studies* **13**: 233–43.

Massey, D. 1984 *Spatial divisions of labour: social structures and the geography of production*. Macmillan.

Massey, D., Meegan, R. 1979 The geography of industrial reorganisation: the spatial effects of the restructuring of the electrical engineering sector under the Industrial Reorganisation Corporation. *Progress in planning* Vol. x: Part 3.

Massey, D., Meegan, R. 1982 *The anatomy of job loss*. Methuen.

Morgan, K. 1983 Restructuring steel: the crises of labour and locality in Britain. *International Journal of Urban and Regional Research* **7**(2): 175–201.

Morgan, K., Sayer, A. 1983 Regional inequality and the state in Britain. In Anderson, J., Duncan, S., Hudson, R. (eds) *Redundant spaces in cities and regions?* Academic Press, 17–50.

Morgan, K., Sayer, A. 1984 A 'modern' industry in a 'mature' region: the re-making of management – labour relations. *Urban and Regional Studies Working Paper* 39, University of Sussex.

Murgatroyd, L., Urry, J. 1983 The restructuring of a local economy: the case of Lancaster Regional inequality and the state in Britain. In Anderson, J., Duncan, S., Hudson, R. *Redundant spaces in cities and regions?* Academic Press, 67–98.

Peck, F., Townsend, A. R. 1984 A contrasting experience of recession and spatial restructuring: British Shipbuilders, Plessey and Metal Box. *Regional Studies* **18**(4): 319–38.

Perrons, D. 1981 The role of Ireland in the new international division of labour: a proposed framework for analysis. *Regional Studies* **15**(2): 81–100.

Sayer, A. 1982 Explanation in economic geography: abstraction versus generalisation. *Progress in Human Geography* **6**: 68–88.

Sayer, A. 1984 *Method in social science: a realist approach*. Hutchinson.

Storper, M. 1981 Toward a structural theory of industrial location. In Rees, J., Hewings, G., Stafford, H. (eds). *Industrial location and regional systems*. Croom Helm, 17–40.

Storper, M., Walker, R. 1981 Capital and industrial location. *Progress in Human Geography* **5**: 473–509.

Storper, M., Walker, R. 1983 The theory of labour and the theory of location. *International Journal of Urban and Regional Research* **7**(1): 1–43.

Taylor, M., Thrift, N. 1983 Business organisation, segmentation and location. *Regional Studies* **17**(6): 445–65.

Urry, J. 1981 Localities, regions and social class. *International Journal of Urban and Regional Research* **5**(4): 455–74.

Westaway, J. 1974 The spatial hierarchy of business organisation and its implications for the British urban system. *Regional Studies* **8**: 145–55.

Whitting, G. 1986 Central-local state relations in UK land supply. In M. Goldsmith, S. Villadsen (eds.) *Urban political theory and the management of fiscal stress*. Gower, 161–78.

Wright, E. O. 1976 Class boundaries in advanced capitalist societies. *New Left Review* **98**: 3–41.

Zimbalist, A. 1979 *Case studies on the labour process*. Monthly Review Press.

Introduction to the Factors of Production

The development of theories of industrial location and decision-making as outlined in Chapters 2–4 forms a sequence in which an increasing number of factors are taken into consideration. In the four chapters which follow, five of these factors are dealt with in turn: labour, capital, space, technology and industrial linkages. In Chapter 5, Bill Lever examines the interaction of capital and labour within the manufacturing sector's production function. Neither appears to play a major role in determining the location of industry, but as the price of each varies, both within the UK and relative to other countries, companies substitute one for another in an attempt to sustain their competitiveness. The long-run effect of this has generally been to reduce the labour input into any given volume of output and to increase the capital input in the form of plant and machinery (and the associated land and premises). The consequence of this trend, and of demand deficiency, has been to reduce the aggregate demand for labour, particularly in manufacturing, and hence to increase recorded levels of unemployment. Unemployment, in its turn, does not impact uniformly, with certain groups (young, ethnic minority, unskilled) and certain areas (inner city, Development Areas) suffering much higher rates than others. Lever thus picks up the theme developed by Martin Boddy in Chapter 4 that manufacturing production can be seen as a conflictive outcome of two forces – capital, seeking to maximize profits by treating labour as a commodity to be dealt with solely in terms of quantities and prices, and labour, seeking to maximize the return on the one asset which it has for sale. The return to capital in UK industry has fallen dramaticaly since about 1970, and changes in legislation have permitted major investors and companies to place more of their investment outside the UK in countries where rates of return are better. Small enterprises, of course, do not have this option and still remain very dependent upon personal capital and locally-secured loans against the security of personal assets.

In Chapter 6, Stephen Fothergill, Michael Kitson and Sarah Monk provide a full exposition of what David Keeble in Chapter 1 described as the 'constrained location' explanation of the urban to rural shift in manufacturing employment. Their argument is that when enterprises wish to expand output, or merely to remain competitive by substituting more capital equipment for labour, they require more space and there is a tendency for this to be less readily available in inner London and the centres of the other conurbations than elsewhere. Thus in the inner city, growth is either constrained or deferred, or firms seek new locations elsewhere, thus achieving the urban to rural shift in the location of industry. This pattern of change is exacerbated when the study is extended to take into account not only the availability of space but also its age and quality. Further analysis, however, leads them to the conclusion that it is more the sort of site and premises which an enterprise occupies than whether it is located in a large city, a town or a rural area *per se* which influences the nature of its economic change.

In Chapter 7, John Goddard and Alf Thwaites examine the role of technological innovation in industrial change in Britain, distinguishing between product innovation and process innovation. The

South East region particularly appears to benefit from innovations, although there is some evidence that foreign multinationals have been instrumental in introducing a significant share of innovations into the Development Areas. Studies of individual establishments show how important research and development have become in keeping firms competitive, and once again the South East region seems particularly well-served both by in-house research and development and the availability of bought-in services. Finally, employment change in a small number of sectors is shown to have been powerfully influenced by the introduction of new products and process technologies, leading the authors to speculate that in the period 1978–81 the rate of job decline in the South East would have been as serious as that in the Northern region had the South East not had the benefit of the enhanced rate of innovation.

Lastly in this part, Neill Marshall in Chapter 8 looks at the changing role of industrial linkages – both material and service – in the explanation of industrial change and regional development. From what he traces as the high point of linkage analysis in the 1960s, Neill Marshall follows its decline in importance which he correlates with the recession in the British manufacturing economy, but then argues that there exists a much wider role for linkage analysis. It represents the mechanisms by which national economic changes are transmitted to the firm or enterprise and also by which the impact of changes at firm level on the regional/local economy can be measured.

There are two respects in which the four chapters anticipate the major themes of the second half of the volume. Firstly, while all four chapters look back towards the location theory chapters (it is not difficult, for example, to see the links between Neill Marshall's analysis of linkages in Chapter 8 and David Smith's neoclassical location theory in Chapter 2), they also have to make increasing use of explanations which are specific only to certain types of industrial establishment. Bill Lever shows how different sizes of enterprise draw investment capital from different sources while Stephen Fothergill *et al.* describe how different sizes of enterprise face up to the problem of the need for space for expansion in different ways. John Goddard and Alf Thwaites show how ownership type plays an important role in explaining a company's prospensity to innovate: small independent companies, especially those located outside the South East, place fewer resources into research and development than do large multiplant or multinational companies, and are thus less likely to enjoy the benefits of product or process innovation. Neill Marshall, too, shows how multiplant companies have a different pattern of buying in services and material inputs to those of single-plant independents. Using variables such as size and ownership, rather than product market or industrial sector, conforms to what Peter Wood in Chapter 3 described as the geography of enterprise – an approach which assumes that large companies in an industry, chemicals for example, will have more in common with large companies in other industries – say, motor vehicle manufacture or heavy engineering – than they will with small chemical companies.

Secondly, all four chapters begin to look ahead to the role of public sector policy, and this is dealt with in the final three chapters of this volume. For example, Bill Lever points out that the continued use of capital subsidies in regional policy, while sustaining output, may be removing jobs; Stephen Fothergill *et al.* argue cogently that industrial growth and employment in the large cities may be being lost because of the inadequacy in the supply of suitable premises; John Goddard and Alf Thwaites query whether public policy can enhance the technological capacity of the lagging regions of Britain or whether the South East has a comparative advantage for innovation which would be impossible or at least extremely expensive to overcome; and Neill Marshall argues that indigenous industry with its local linkages and consequent high negative multiplier effects would benefit from enhanced educational and management training programmes.

CHAPTER 5

Labour and capital

William F. Lever

5.1 Introduction

Classical economic theory identifies three factors of production – land, labour and capital: the value of the output of any enterprise is equal to the value of the amounts of each of these factors which are utilized. Geographers, however, have tended to underestimate the importance of the factors of production in explaining the locational dynamics of manufacturing industry. Locational theory initially was dominated by distance/transport costs, concentrating on the optimum location as defined by the point at which aggregate transport costs were minimized (see Ch. 2). Of the three factors of production, only land has played a full role in locational theory, and this more often at the urban and metropolitan scales than at the regional or national scales. This may represent a tacit belief that the costs of labour and capital do not vary so significantly over space as to represent an important element in the locational decision. It can be argued that, unlike land, the price of which may vary very sharply over comparatively short distances, the prices of labour in the form of wages and of capital in the form of interest tend to be nationally determined and in consequence do not enter into the locational decision. Studies of capital in manufacturing have tended to stress not the price of capital but its availability to enterprises of different types and sizes. Studies of labour have tended to stress not wage rates so much as the skill composition, productivity and the quality of industrial relations, and the relative availability of labour in the form of registered unemployed workers and activity rates. When attempts have been made to assess the relative importance of a wide range of locational factors, aspects of capital have rarely achieved much prominence. For example, the ILAG study (Department of Trade and Industry 1973) of the interregional movement of manufacturing enterprises makes virtually no mention of capital availability or price. Labour availability, however, was the most frequently mentioned influence on locational choice in the ILAG study, and discriminated significantly between firms moving to the Development Areas where labour was plentiful and those moving to regions such as the South East and West Midlands which, over the period covered by the survey, were relatively 'tight' labour markets. Labour availability in the survey, however, did not correlate with the price of labour: few ILAG firms indicated that lower wages were a significant factor in locational choice and a substantial minority of firms moving to Scotland and Merseyside indicated that they paid higher wages after the move.

The importance of labour and capital in the explanation of industrial change in Britain is largely twofold. Firstly, as the technology of production in manufacturing has changed throughout the post-war period, firms have increasingly substituted capital, in the form of machinery, for labour. Secondly, where the government has intervened in the location of manufacturing in Britain at both the urban and the regional scales, subsidies to capital and assistance with labour supply have been major instruments of such policy.

This chapter begins by assessing the roles of labour and capital in location theory and then deals separately with the operation of labour and capital markets in British industry. It concludes by examining the role of capital and labour subsidies in policies for British industry and describes the potential for conflict between labour and capital within the current debate on restructuring in British industry.

5.2 Labour, capital and location theory

Location theory suggests that firms will select locations in which, on the basis of spatially-variant factor costs and revenue levels, profits will be maximized. In consequence, different industrial sectors will tend to locate in different regions according to the theory of comparative advantage (Moroney and Walker 1966; Estle 1967). Industrial sectors which are capital-intensive will tend to select capital-rich regions; labour-intensive industries will tend to locate in regions where labour is available and cheap. Most of the empirical work was undertaken in the United States where, for example, New England was a region which was relatively abundant in labour and was therefore tending to attract industries which were relatively labour-intensive. Evidence on the American South was ambiguous in that it has traditionally been regarded as a labour-abundant region but showed evidence of a concentration of capital-intensive industries. The answer lay in the difficulty of estimating capital vintage, for much of the industry in the South is more recent in origin than that in the North East and in consequence has much more recent capital invested: hence the disparity.

In the UK, a similar analysis also yields ambiguous results. The Census of Production data for 1981 indicate that expenditure on labour (both wages and salaries but excluding non-wage labour costs such as National Insurance) equalled 20.6 per cent of the value of gross output in that year for all manufacturing. At major industrial order level the range of values of labour costs expressed as a percentage of gross output extended from 28.5 in man-made fibres to 8.3 per cent in the food, drink and tobacco industries. The Census of Production lists net capital expenditure for the year on plant, buildings and vehicles, but these data, unlike those on labour costs, are not tied to production levels and the amount of net capital expenditure in any one year may vary substantially as firms choose to advance or delay capital investment in response to changing market conditions. In the depressed state of the economy in 1981, figures for net capital investment are likely therefore to have been well below average. The Census of Production does not provide data on the value of *all* capital investment discounted for age, or total capital vintage, which would be a more reliable measure of capital intensity than expenditure in a single year. For the whole of manufacturing industry, net capital expenditure in 1981 amounted to 3.3 per cent of gross output.

On the basis of the Heckscher – Ohlin theorem, labour-intensive industries should be drawn to those regions where labour is available, cheap or both, while capital-intensive industries should be drawn to those regions where capital is available. In the British context, we can divide the standard regions into those with a labour surplus (North, Wales, Scotland), those with relatively tight labour markets (South East, East Anglia, West Midlands) and intermediate (East Midlands, South West, Yorkshire and Humberside, North West) according to their long-run unemployment rates and status for regional assistance. Seven sectors have high proportions of wage expenditure: man-made fibres, miscellaneous metal goods, mechanical engineering, electrical machinery, textiles, footwear and clothing, and instrument engineering. Within these seven sectors there are eleven location quotients (LQ) in excess of 1.5 when measured at the regional level. Contrary to our expectation, these are not clustered in the labour-surplus regions, but rather in the intermediate regions. Only one LQ, textiles in Scotland, exceeds 1.5 in an Assisted Region, whereas there are three (metal goods and electrical machinery in the West Midlands, and instrument engineering in the South East) in the labour-shortage regions and seven in the intermediate regions. It would therefore seem that at the regional level there is no association between labour intensity and labour availability as measured by unemployment rates. Alternative explanations might include the fact that some of the sectors involved in the study have long-run histories of location in the regions concerned (e.g. textiles in the North West, clothing and footwear in the East Midlands) so that twentieth-century measures of labour market conditions are inappropriate, and the fact is that some of these sectors place heavy stress upon labour quality rather than availability or price as a locational determinant.

Although it has been suggested that capital is less likely to be a locational factor than are attributes of the labour force, the comparable analysis does suggest that capital-intensive industries are located in the capital-rich regions of

Britain, namely the South East and the West Midlands. Allowing for the reservation that a single year's net capital investment may not be an accurate measure of overall capital intensity, the sectors in which net capital investment in 1981 was significantly above the average for all manufacturing industry in that year were energy production, chemicals, office machinery, electrical machinery, motor vehicles, paper and printing, and rubber and plastics. Between them, these seven sectors produced eight location quotients greater than 1.5. Of these eight, five were in the capital-rich regions (office machinery and paper and printing in the South East, and electrical machinery, motor vehicles and rubber and plastics in the West Midlands). There were two in the intermediate regions, and only one in the peripheral regions. This last LQ linked Scotland with energy production – a reflection of the exploitation of North Sea oil.

The role of capital and labour in industrial location theory has been more extensively studied at the urban metropolitan scale as prices may vary much more significantly over these shorter distances than interregionally, and the local labour market is felt to be a more appropriate spatial unit of analysis than the region (Goodman 1970; Cheshire 1979). Scott (1982), in an important review article, examines the role of capital, labour and land in determining intrametropolitan location decisions. Scott deals with capital within the metropolitan region as commodity capital (Sraffa 1960) in the form of both sunk capital (i.e. spatially fixed and unconvertible) and circulating commodity capital (flows of goods and services). By adopting this definition and concentrating upon circulating commodity capital, Scott focuses his analysis upon linkages. In a classical Weberian sense, capital in this form tends to locate productive enterprises close to sources of material inputs or to markets. If services are included in the linkage pattern, then agglomeration economies tend to cause enterprises to locate in the largest urban centres and close to their central points (Walker 1977; Taylor 1978; Townroe 1970). A number of researchers (e.g. Wood 1969) have stressed that it is not merely the volume of linkages which influences locational choice but quality, complexity and variability. Any flow of goods, services or information involves both the fixed costs of the negotiation of the linkage and the direct cost of the actual linkage. The fixed costs tend to decrease in unit terms with increases in the volume of the linkage, thus enabling larger enterprises to enjoy scale economies and in consequence to select less central locations. This trend is reinforced by the tendency for larger enterprises to undertake standardized, mass-production stages of product manufacture which require little access to rapidly changing information and which can thus be conducted well away from the information-rich centres of the largest urban settlements. By contrast, many small enterprises whether in manufacturing or services undertake small batch or individual client-specified products and thus require close access to large urban centres and their associated markets (Gilmour 1974).

Theoretical treatments of labour markets and wage rate differentials have tended to stress interurban rather than intraurban comparisons. Vipond (1974) argued that unemployment, and hence wage rates, is related to urban size although in both positive and negative ways. Large cities may have lower unemployment rates (and hence higher wages) because their wider range of employment opportunities makes it easier to match up redundant workers with appropriate job opportunities, thereby reducing job search and recruitment times and in consequence reducing frictional unemployment. Conversely, the longer distances and difficulties in communication in large cities may impede the matching up of workers and jobs, thereby raising the unemployment rate. Using data from the 1966 Census of Population, Vipond was able to show a positive correlation between *male* unemployment and city size, although she did indicate that industrial structure might also influence the result. Conversely, there was a negative relationship between *female* unemployment and city size, with industrial structure again as a confusing element. The facts that large cities tend to have a higher proportion of their total employment in services, and that services employ a higher proportion of females than does manufacturing, tend to confuse the identification of a simple bivariable relationship between urban size and unemployment. Studies from the United States (Fuch 1967) and Britain (Mackay 1970) have, once regional and structural effects are allowed for, shown a positive relationship between wage rates and urban size. There are two broad reasons for this relationship. Firstly, living costs, particularly rent and commuting costs, are known to be higher in large cities than in smaller urban centres (although the relationship breaks down at the other end of the urban – rural continuum where very small urban centres in remote rural locations may have high costs of living), and workers resident in these largest centres will seek higher wages to compensate for these higher costs. Secondly, for a number of reasons, the marginal product of labour is thought to be higher in large cities and thus employers are able to afford higher wage rates because labour is more productive. At the intraurban scale, it is hypothesized that commuting costs have to be offset by additions to basic wages and thus that in a classical monocentric urban labour market, wage rates will be high in the city centre which is some distance from the suburban dwelling labour force, and will decline as employment opportunities are found closer to the suburbs (Muth 1969). However, as Ravallion (1979) points out, where the classical assumption of all employment being located in the Central Business District is relaxed, employers who locate in the suburbs where residential densities are much lower are likely to have to attract their labour force to commute over longer distances and to do so will have to pay higher wages. In this latter case, the types of enterprises attracted to peripheral locations will be more capital-intensive and require less labour than those located closer to the city centre.

Analyses of the locational response to spatial differences in the cost of capital and the cost of labour are made more complicated by the fact that enterprises do not require a fixed capital : labour ratio but substitute one for the other as their relative prices change. It was Predöhl (1925) who first drew attention to the role of substitution in locational analysis. He argued that in a city where labour was ample and cheap close to the centre but expensive at the city edge, whereas land, which is used in conjunction with capital, is cheap and plentiful at the city edge, then substitutions between capital and labour will have the effect of moving the optimum location. As the price of labour in Britain, as in most developed countries, has risen faster than the price of industrial raw materials and components and plant, then there has been a general tendency for industry to reduce the amount of labour used to generate a given volume of output and to increase the amount of capital (and land). This tendency has been increased by the fact that labour is regarded as a direct cost, more or less proportional to output, whereas capital and land are regarded as a fixed cost in that plant, premises and land have to be acquired but then do not relate closely to the volume of production. In a theoretical sense, therefore, the suburbanization of manufacturing industry can be regarded as the locational adjustment necessitated by capital – labour substitution.

In an analysis of 1,231 manufacturing establishments in Glasgow and Birmingham, Lever (1974) attempted to link patterns of decentralization to the relative importance of capital and labour and to agglomeration economies on a sectoral basis. Using Census of Production data, the study ranked the 117 manufacturing sectors according to the proportion of net output made up by wages and salaries and that made up by capital expenditure and land costs. It was theorized that those sectors in which the labour : capital ratio was most heavily weighted on the labour side would be those that (a) were located initially close to the city centre and that (b) over the study period were decentralizing most rapidly. The study, over the period 1959–69, did show three things. Industries which were capital- and land-intensive were already quite suburbanized in their locations before 1959 and therefore had low rates of mobility during the study period. Industries which were labour-intensive were tending to capital – labour substitute during the study period and showed the highest rates of plant mobility. Some industries in which both capital *and* labour were small cost elements relative to the cost of external economies (bought-in services, subcontracted manufacturing processes etc.) tended to have the most centralized locations at the start of the period and not to decentralize significantly during the study period.

5.3 Labour

Chapter 1 has already described how the process of de-industrialization has altered the demand for labour in Britain. The major movements in the labour market are contained in Table 5.1. The working population of the UK defined as the employed labour force plus the registered unemployed has grown from 25.36 m. in 1968 to 27.00 m. in 1984: the economy would therefore have required to produce a net growth of about 100,000 jobs per annum in order to keep pace with the growth of the workforce and to keep unemployment at its 1968 level of 2.4 per cent. This growth in the working population is attributable to the growth in absolute numbers in the population of working age and to the growth in the number of females in work or registered as unemployed. The number of employees in

Table 5.1 Working population, UK* ('000)

	Employees in employment			Self-employed	Employed labour force†	Unemployed	Working population
	Males	Females	All				
1966	14,779	8,375	23,154	1,684	25,256	336	25,592
1968	14,242	8,327	22,569	1,780	24,756	608	25,364
1970	14,052	8,434	22,486	1,890	24,750	637	25,387
1972	13,532	8,499	22,031	2,003	24,405	938	25,343
1974	13,621	8,996	22,617	2,005	24,971	602	25,573
1976	13,345	9,071	22,416	1,960	24,713	1,242	25,955
1978	13,312	9,259	22,571	1,904	24,796	1,379	26,175
1980	13,325	9,629	22,953	1,984	25,258	1,376	26,634
1982	12,222	9,197	21,419	2,172	23,919	2,821	26,740
1984	11,921	9,296	21,217	2,313	23,856	3,143	26,999

Source: Employment Gazette
* Includes members of HM Forces
† The figures are for March in each case and are thus seasonally unadjusted

employment fell from 23.15 m. in 1966 to 21.22 m. in 1984, a decline of 8.5 per cent, but almost all of this fall occurred between 1980 and 1984. Within the total number of employees in employment, there has been a substantial change in the sexual composition of the labour force: in 1966, 63.8 per cent of all employees were male; by 1984 the proportion had fallen to 56.2 per cent. The number of males in work fell almost without interruption so that there were almost 3 m. jobs fewer for males in 1984 than there had been in 1966. For females, the number of jobs rose by about 1.25 m. between 1966 and 1984, although there was a sharp fall between 1980 and 1982. This change in the sexual composition of the employed labour force is due to two processes, as Table 5.2 indicates. Firstly, the growth in service employment generally has increased the number of jobs occupied by females, the proportion rising from 47.8 per cent in 1974 to 54.1 per cent in 1984. There has been no rise in the proportion of females employed in manufacturing and allied sectors despite the widely-held impression that manufacturing industry has become 'lighter' in character with the introduction of more machinery. Indeed, the proportion of jobs in manufacturing held by females actually declined between 1974 and 1984 from 27.8 to 24.8 per cent. What the figures in Table 5.2 do not show is the growth of part-time female employment. Of all females in employment in 1974, 38.3 per cent were in part-time employment; by 1984 the proportion was 50.4 per cent. This growth is largely due to the high proportion of females employed part-time in retail distribution (59.2 per cent) and in hotels and catering services (74.5 per cent).

Table 5.2 Sex composition of the workforce, 1974–84 ('000)

	Manufacturing + primary + const.	Services	Total
1974			
Total	8,617	13,681	22,298
Male	6,220 (72.2%)	7,144 (52.2%)	13,364
Female	2,397 (27.8%)	6,537 (47.8%)	8,934
1984			
Total	7,416	13,334	20,750
Male	5,577 (75.2%)	6,125 (45.9%)	11,702
Female	1,839 (24.8%)	7,209 (54.1%)	9,048

Source: Employment Gazette

The growth in the number of self-employed persons recorded in Table 5.1 can be explained by the very large numbers of people leaving employment in manufacturing industry, often with some lump sum of redundancy payment, and setting up in their own business. The large number of forms of government assistance, such as the Enterprise Allowance Scheme, and of agencies offering advice has increased this volume.

Finally, Table 5.1 records the growth in the number of registered unemployed which rose from 376,000 (1.5 per cent) in 1965 to 3,143,000 in March 1984 (11.6 per cent) with the largest increase occurring between March 1980 when the rate was 5.2 per cent and December 1981 when it was 10.3 per cent. The reasons for the growth in unemployment are many. In the mid-1960s the normal pattern of rising and falling unemployment in a four- to five-year cycle was broken.

Labour and capital

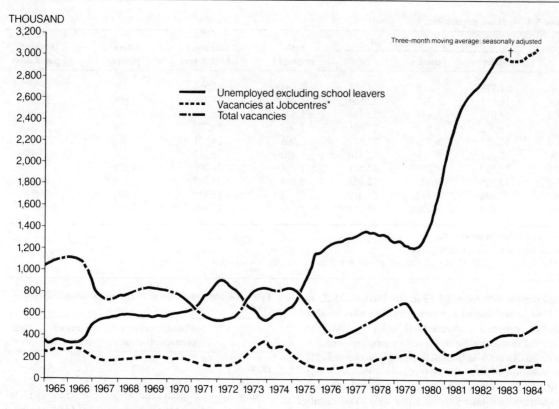

*Vacancies at Jobcentres are only about a third of total vacancies.
†Figures affected by the Budget provisions for men aged 60 and over.

Fig. 5.1 Unemployment and vacancies: UK 1965–84

Restructuring in the economy created new jobs in expanding industries but they were unlikely to be filled by workers released by declining industries who went on to the unemployed register. In the 1970s the world's developed nations moved into deep recession in part because of the dramatic rises in the price of crude oil in 1973 and 1979. Since 1979, the British government has pursued policies with the objective of reducing the rate of inflation, sustaining a balance of trade equilibrium and reducing levels of public sector expenditure rather than seeking to move back towards full employment. Where, with the lessening of the recession since 1982, there have been increases in demands for goods and services, there has been a tendency for employers to use more capital or to make more intensive use of their existing labour force (through overtime, for example, which rose from 8 m. hours per week in 1981 to 12 m. hours per week in 1984) rather than to take on additional workers, in order to avoid paying the fixed non-wage costs of the additional labour. Thus, by October 1984, there were 3.1 m. registered unemployed in Great Britain, or 13.2 per cent of the workforce; when disaggregated by sex, the rate for males was 15.6 per cent, that for females was 9.9 per cent. In addition to this registered unemployment, a number of other groups, school-leavers, young adults on various training schemes and females looking for employment but not formally registered, should be included amongst the unemployed, and the true total may be closer to 4.5 m. (ASTMS 1982).

Figure 5.1 shows how registered unemployment has risen since 1965 with a particularly dramatic rise between early 1980 and late 1981. Labour economists divide unemployment into three types – frictional, structural and demand deficient. Frictional unemployment occurs when there are workers without jobs but jobs do exist for their skills although they do not know of them, or the jobs are in locations different to those of the workers. Structural unemployment exists where there are workers without jobs and there are no jobs anywhere for their skills, but there are jobs available which require different skills. In both these cases, policies are required to match up jobs and workers, either by improving information flows in the labour market, or by assisting the relocation of the redundant workers, or by retraining them to provide them with skills for which there is still a demand. The third type of unemployment, demand deficient, occurs when there are no jobs of any kind for the unemployed. Figure 5.1 shows the number of vacant jobs registered with the Department of Employment, a figure which generally oscillated between 300,000 and 100,000 between 1965 and 1984, and stood at 164,000 in the third quarter of 1984. It is generally recognized (Manpower Services Commission 1984) that only about one-third of all vacancies are so notified currently and that this proportion has risen from about one-quarter over the past twenty years. If the notified vacancy figures are grossed up to take account of this fact to give total vacancies, there is clearly a different picture before 1975 and after 1975 as Fig. 5.1 shows. Before 1975, there was generally an excess of vacancies over registered unemployed, with the exception of a short period in 1971–72: it is therefore possible to regard most of the unemployment prior to 1975 as frictional and/or structural. After 1975 the numbers of unemployed climb way above the estimated numbers of vacancies, which reached an absolute low in early 1981. The gap between the two lines conceptually measures the amount of demand deficiency in the total volume of unemployment. By mid-1984, on this basis it is possible to say that only about 540,000 of the registered unemployed are either frictionally or structurally unemployed, and the remainder, or just over 2.5 m., are unemployed because of the depth of the current economic recession.

It is clear, not only that there has been a massive rise in the numbers of registered unemployed, but that unemployment hits certain groups within the national labour force much harder than others. We have already indicated that unemployment rates for males are higher than for females. This may, however, reflect lower registration rates for females rather than the true state of the labour market. Age correlates very strongly with unemployment. When unemployment rates reached their peak in 1982, the rate for those under twenty years of age was about 26 per cent compared with the national average of 13.6 per cent. The rate for those in their early twenties was 19.5 per cent, and the rate fell successively for each ten-year age band, reaching a minimum of 8.4 per cent for those in the 45–54-year-old band.

Thereafter, the unemployment rate rose again, reaching 14.4 per cent for those over sixty years old. The growth in the number of young unemployed has been one of the most marked features of the current recession. Evidence suggests that although changes in youth unemployment are closely linked with fluctuations in the overall rate, they are greatly exaggerated. It has been demonstrated that when the overall rate rises by 1 per cent, the rate for those under twenty rises by 1.7 per cent (Makenham 1980).

There are clear differences in unemployment rates by ethnic origin. The 1983 Labour Force Survey showed that unemployment amongst non-whites was considerably higher than amongst the white population. The highest unemployment rates were found amongst West Indian men (28 per cent) and both men (22 per cent) and women (18 per cent) of Asian origin, compared with rates for whites for males of 12 per cent and females of 10 per cent. However, the overall figures disguise wide variations between age groups. Young men and women aged 16–24 of all ethnic groups have high unemployment rates, but the rates are particularly high for young West Indian men and women, and young Asian women. The age structure of the non-white population is much younger than that of the white population and this accounts for some of the differences in unemployment rates.

Thirdly, different occupational groups within the labour force have different unemployment rates. The Labour Force Survey shows a distinct concentration of the unemployed in the manual, and particularly semi- and unskilled manual, groups. Of those unemployed who had had a job less than three years prior to the 1983 survey, 68.1 per cent were in manual occupations and 31.9 per cent in managerial, professional, clerical and other non-manual occupations. Not only were manual workers more likely to be unemployed, but they were more likely to be unemployed for longer periods: 76.7 per cent of those unemployed for more than twelve months were manual workers.

Lastly, location within Britain influences a worker's probability of being unemployed. At the regional level by October 1984, unemployment rates ranged between 21.1 per cent in Northern Ireland and 9.9 per cent in the South East. At the local level, the spatial disparities are even greater. In terms of the Department of Employment's travel-to-work areas, by October 1984 the unemployment rate range extended from 39.9 per cent in Strabane and 34.5 per cent in Cookstown,

Table 5.3 Unemployment ratios, standard regions

	1978	1979	1980	1981	1982	1983	1984
S.E.	0.69	0.64	0.62	0.67	0.70	0.72	0.75
E. Ang.	0.83	0.79	0.78	0.80	0.80	0.79	0.76
S.W.	1.05	1.02	0.94	0.88	0.88	0.87	0.86
W. Mid.	0.89	0.98	1.07	1.20	1.22	1.21	1.18
E. Mid.	0.86	0.83	0.90	0.92	0.90	0.91	0.94
Yorks. & Humb.	0.98	1.02	1.07	1.10	1.09	1.09	1.09
N.W.	1.27	1.22	1.25	1.22	1.21	1.22	1.23
North	1.51	1.57	1.53	1.41	1.36	1.37	1.39
Wales	1.34	1.38	1.38	1.30	1.27	1.23	1.23
Scotland	1.41	1.40	1.34	1.19	1.16	1.16	1.16
N. Irel.	2.04	2.02	1.88	1.61	1.55	1.57	1.63
s.d.	0.41	0.41	0.37	0.26	0.27	0.27	0.28

Source: *Employment Gazette*

both in Northern Ireland, to 4.8 per cent in Winchester, 5.5 per cent in Crawley, both in the South East, and 5.3 per cent in Clitheroe in the North West.

There has been considerable discussion in the literature on British regional labour markets as to whether the regions' unemployment rates are converging towards or diverging from the national rate over time. The reason for the concern is based upon the need to assess the efficacy of, or the need for, a regional policy which seeks to stimulate growth or at least to slow decline in the least prosperous regions for reasons of equity. Frost and Spence (1983), and earlier Manners (1976), argued that between the mid-1960s and the mid-1970s the differences between the areas of traditionally high unemployment in Scotland, Wales, the North and Northern Ireland and those areas characterized by lower rates in the rest of the country narrowed. On this basis there has been a call for a reconsideration of regional policy (Townsend 1980), drawing attention to the needs of the inner areas of all large cities. In the early 1970s there was a slight widening of unemployment rate differentials, with the situation in Scotland worsening markedly. By the mid-1970s there was some convergence as the more prosperous regions such as the South East Anglia showed unemployment rates worsening faster than the national average. By the late 1970s (Gillespie and Owen, 1981) convergence continued but a number of changes in the performance of certain regional economies began. Table 5.3 lists the ratios between regional unemployment rates and the national rates from 1978 to 1984 and therefore covers the deepest recession. The most noticeable feature of the table is the West Midlands' performance whereby a traditionally prosperous region with unemployment rates lower than the national average developed, in 1980–81, rates more than 20 per cent above the national average. Moving in the opposite direction, the South West goes from above-average to below-average unemployment rates. By calculating the standard deviations for these ratios it is possible to demonstrate that considerable convergence did occur between 1978 (s.d. = 0.41) and 1981 (s.d. = 0.26). Much of this convergence can be attributed to three regions: the rapid worsening in the West Midlands position, and the sharp decline in the levels of unemployment disadvantage in the two most depressed regions, Northern Ireland and Scotland. It may be possible from Table 5.3 to discern that the situation since 1982 shows some signs of renewed divergence: there has been a marginal increase in the standard deviation. Perhaps more significantly, in a number of prosperous regions (East Anglia, South West, West Midlands) the ratio values are beginning to decline whereas in the four depressed regions they are either worsening (N. Ireland and the North) or at least not continuing their earlier improvement (Wales and Scotland).

The more frequently used scale of analysis is the local labour market, which is defined by journey-to-work data (Smart 1974). Definitions usually require that journey-to-work flows across the boundary both inward and outward are below a certain predetermined percentage limit and that within the boundary there is a high level of integration. Within such a local labour market there are three groups: the stock of employed workers, the stock of registered unemployed, and the stock of non-working population who are of working age who may be either not in the labour force at all or are unregistered unemployed (Armstrong and Taylor 1983). Persons move between these three groups in several ways. Most simply, people move from the stock of employed workers by redundancies and voluntary quits, or by retirement or illness, and to the stock of employed workers when they are hired or rehired. People who are not working may register or deregister as unemployed, or age may take them into or out of the workforce. Labour economists have argued that it is at least as important to examine these flows on to and off the register as it is to look at the stock on the register at any one time. Fowler (1968), for example, posed the question: where two localities have the same unemployment rate but in one case it consists of a small group of persons each of whom is unemployed for a long time and in the other it consists of a relatively large number of persons each of whom is unemployed for a short spell, which has the most serious problem? Layard (1981), by way of answer, indicated that it would be fairer to 'spread' the unemployment amongst a large number of people for relatively short periods. However, data on duration of unemployment for Great Britain as a whole show that while the rate of unemployment has risen rapidly, the rate of long-term unemployment (greater than one year) has grown at a more rapid rate. In a study of twenty-eight local labour markets in the North West region, Armstrong and Taylor (1983) were able to demonstrate that the composition of unemployment at the local scale differs considerably. In some cases, high rates of both male and female unemployment were caused primarily by high inflow rates; these typically were coastal resort towns such as Blackpool, Southport, Lancaster/Morecambe where there is a strong seasonality to unemployment. In other cases, high rates of unemployment were caused primarily by a relatively large core of permanently or long-term unemployed, and these typically were south Lancashire industrial towns such as Liverpool, St Helens, Birkenhead, Widnes and Wigan.

In addition to the composition of unemployment at the local labour market level, there has been interest in both the amplitude of the swings between high and low levels of unemployment and the phenomenon of leading or lagging labour markets relative to the national trend. Frost and Spence (1981) compared the sensitivity of local unemployment rates to the changes in the national rate at the level of economic subregions. The level of local sensitivity is strongly linked to the overall regional level of unemployment: local labour markets in regions with high rates of unemployment had the widest fluctuations in unemployment rates. The highest sensitivity coefficients were found in areas such as the industrial north east, industrial south Wales, Glasgow, northwest Wales, all of which are local labour markets in the assisted areas. Some labour markets, however, did not conform to this pattern, showing signs of high sensitivity despite locations in the relatively prosperous regions: these include the West Midlands conurbation, much of Yorkshire and Lancashire, and Kent. In these cases, high sensitivity is explained by the presence of volatile industries such as textiles, vehicles, metals and heavy engineering. An alternative approach to measuring the violence of short-run changes in local unemployment rates was used by Lever (1981) who for 301 local labour markets in Great Britain fitted long-run unemployment trend curves and measured the extent to which individual quarterly rates diverged from this trend. Using this measure, the widest fluctuations were found in the intermediate regions of the East Midlands, Yorkshire and Humberside, the North West and the West Midlands, and the narrowest in the depressed of Wales, Scotland and the North. In this respect, Lever's findings correlate closely with the earlier studies based on 1950s data by Brechling (1967) and Thirlwall (1966), although analysing the industrial structure of the more volatile labour markets does confirm Frost and Spence's finding that textiles, metal goods and heavy engineering are associated with wider than average swings in local unemployment.

Bassett and Haggett (1971) began the study of those local labour markets which recorded changes in their unemployment rate significantly earlier or later than the nation as a whole but their study was focused solely on the South West region. Brechling (1967) found that at the regional level the Midlands, in the period 1952–63, tended to lead national trends by one-quarter whereas Scotland and the North lagged by one-quarter. A more recent study (Lever 1981) used turning-points in the unemployment rates for 295 local labour markets in July 1966 and January 1974 when unemployment began to rise nationally, and July 1972 when it began to fall. For the July 1966 turning-point there was a range of five quarters lead (Stroud and Yeovil) to one quarter lag (Fraserburgh and Stratford). For the July 1972 downturn there was a range of five quarters lead (Milford Haven) to one quarter lag (99 local labour markets). For the January 1974 upturn in rates, the range extended from two quarters lead (Barrow-in-Furness and Warminster) to three quarters lag (seven local labour markets). No simple regional model distinguishing between prosperous and depressed regions provides an explanation of the incidence of local leads and lags. The South East, the South West and the East Midlands appeared persistently to lead into and out of recession, whereas the West Midlands and East Anglia are amongst the last regions to experience rising unemployment rates. By size of local labour market, the medium-sized group (with populations in the 50,000–200,000 range) were the first to record unemployment upturns, whilst the large centres (>200,000 inhabitants) were the last. When unemployment began to reduce in July 1972, it was the smallest urban centres which led the trend.

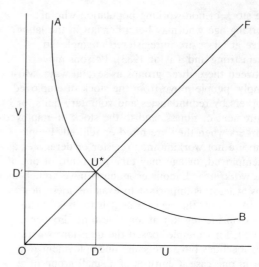

Fig. 5.2 The UV curve in a local labour market

The last characteristic of local labour markets which has been studied extensively is that of efficiency. Figure 5.1 shows how at the national level there is a tendency for vacancies and unemployment to be negatively correlated: when unemployment is high, there are few vacancies, and vice versa. The same phenomenon obviously occurs at the level of local labour markets. If the numbers of vacancies and unemployed in a local labour market for a large number of points in time are graphed against one another, the result will be a diagram such as that in Fig. 5.2, known as the UV curve. At a point U^* on the UV curve such that $U = V$, and where the UV curve intersects the line FO, we have the point at which, according to our earlier definition, all unemployment may be regarded as frictional or structural and the labour market is in equilibrium. At points to the left and above U^* on the arc AU^* there are more vacancies than unemployed and the labour market is regarded as tight. At points to the right of and below U^* unemployment exceeds vacancies and therefore by definition demand-deficient unemployment is present. The size of U^*D in Fig. 5.2 measures the number of vacancies and unemployed which can co-exist in a local labour market and therefore represents a measure of the inefficiency in the system. A study of British local labour markets (McPhail 1982) shows that there is a slight correlation between the size of the local labour market and inefficiency. For both small (fewer than 50,000 inhabitants) and medium-sized local labour markets (50,000–150,000 inhabitants) the value of U^*D'' when expressed as a rate was 2.54 per cent: in other words, where there is an unemployment rate of 2.54 per cent or less, only frictional and structural unemployment is present; where it exceeds that figure, the additional unemployment can be regarded as locally demand deficient (although there may of course be compensating vacancies elsewhere outside the local labour market available for the unemployed were they to move). For local labour markets with populations over 150,000, the value of U^*D was 3.55 per cent. The assumption must be that in larger cities the size of the labour market makes it less easy to bring vacancies and unemployed workers together.

One factor which does increase the efficiency of a local labour market is the presence within it of a single large employer or a concentration upon a single industry. The existence of a single dominant employer reduces frictional unemployment simply because a large proportion of vacancies are located in one establishment and it is thus relatively easy for an unemployed worker to gain information. For British employer-dominated labour markets (Lever 1978; 1980), the average value of U^*D is 2.18 per cent. Those local labour markets which are dominated by a single industry based on several enterprises are likely to have lower levels of structural unemployment because the skill needs of that industry are known to the local labour force and labour shortages are more likely than surpluses (Manpower Services Commission 1983). For industry-dominated labour markets, the value U^*D was 1.63 per cent and for other local labour markets it was 2.58 per cent.

5.4 Capital

We have already indicated that capital availability and price of capital are not powerful locational determinants for British industry, although there are some recorded examples of spatial differentials in capital markets. Keeble (1976), for example, reports an example of higher interest rates being charged for risky investment in the North and Wales (9 per cent) than in the South East (7.8 per cent), and Crum and Gudgin (1977) described the advantages of location close to London's capital markets. Nevertheless capital is important in two respects. Firstly, the level of capital investment in the economy as a whole is clearly crucial to those industries which supply capital goods such as plant and machinery, industrial vehicles and industrial construction. Unless these sectors are able to diversify their output into consumption goods, and this is not often the case, a decline in industrial investment inevitably means a reduction in demand for their products and a consequent reduction in employment. Secondly, if the substitution of labour by capital is undertaken by firms to make themselves more competitive through a reduction in production costs, a slowing in the rate of capital investment must result in a reduction in relative competitiveness.

The recession of the late 1970s and early 1980s had led to weak levels of private sector investment in the UK. Manufacturing investment declined sharply from the 1979 peak so that by late 1981 it was down to the same level as in 1965. This reflects the sharp decline in output after 1980 and the consequent increase in unutilized capacity accompanied by a severe squeeze in company profits. By 1981, pre-tax real rates of return had fallen to just over 2 per cent compared to 11 per cent in 1965, 8 per cent in 1973 and 7 per cent in 1978. After taking into account the sharp rise in interest rates in 1980 and 1981, and higher local authority taxes, the rate of profit probably fell considerably more than is suggested by these rate of return figures. Industrial and commercial companies outside manufacturing and North Sea oil operations also recorded falls in profit rates, but they remained above those for manufacturing. This explains the switch of investment towards services and distribution by 1982, and the strong rise in investment in 1982 reflects a marked rise (12 per cent) in private sector house investment and in office development. Investors, by 1982, were

Table 5.4 UK capital investment

	Net capital expenditure (£m.)			NCE per employee (£)		
	1980	1981	1982	1980	1981	1982
Metal manufacture	347	292	224	1,185	1,235	1,095
Non-metallic mineral prod.	404	267	286	1,559	1,179	1,356
Chemicals	1,002	837	815	2,865	2,628	2,686
Metal goods	313	219	210	671	551	561
Mech. engineering	632	558	557	692	687	763
Office machinery	56	62	60	1,218	1,621	1,678 *
Electric. engineering	477	479	501	712	806	901 *
Motor vehicles	626	469	461	1,400	1,283	1,436
Other vehicles	282	311	235	733	825	661
Instrument eng.	59	47	58	586	516	653 *
Food, drink, tob.	958	877	1,012	1,365	1,340	1,620 *
Textiles	141	114	129	450	424	518 *
Footwear, clothing	83	76	86	216	225	281 *
Timber	162	108	115	687	499	554
Paper, printing	556	464	424	1,063	959	924
Rubber and plastic	280	200	205	1,047	910	995
Other	42	30	49	521	528	771
Manuf. indust.*	6,537	5,512	5,531	1,002	951	1,029

Source: Annual Census of Production, BSO
*Includes some smaller sectors such as man-made fibres, leather etc.

clearly looking to sectors other than manufacturing to generate profits at least at the level of, and preferably higher than, the prevailing rate of inflation (OECD 1983). Table 5.4 shows the decline in investment in manufacturing industry in the period 1980–82. Overall net capital investment fell from £6.5 bn in 1980 to £5.5 bn in 1982, although there was some evidence of an upturn in investment in the latter part of 1982. Very few manufacturing sectors increased their level of investment between 1980 and 1982 as Table 5.4 shows: only office machinery, electrical engineering, foodstuffs, and footwear and clothing (excluding the miscellaneous category) increased investment in absolute terms and in no case, once inflation was taken into account, was there a *real* rise in investment. We have argued that faced with increasing competition in world markets, there is a tendency for industries to replace labour by capital. The data on net capital investment per employee do show that in some sectors such as office machinery, electrical engineering, instrument engineering, foodstuffs, textiles and clothing this is true, but these sectors remain in the minority.

The declining rate of profit is attributable to two factors. Statistical analysis has shown that the rate of capital accumulation has been much greater than that of income distribution. In the consumer goods sector, these two factors were equally important in the decline of profitability. Only in the investment goods sector, excluding construction, was the decline in the rate of profit mainly attributable to increasing wages (Reati 1984). The rate of profit by industry varies substantially. Tyler (1984), using Census of Production data for 1980, showed that while all manufacturing industry in the UK had a *gross* profit margin of 10.5 per cent, the rate for individual sectors ranged between −0.1 per cent in motor vehicles, 1.8 per cent in fish processing, 2.1 per cent in textile machinery, and 28.3 per cent in pharmaceuticals, 25.6 per cent in soft drinks, and 22.2 per cent in tobacco processing machinery. Geographers and others have made a number of analyses of the manner in which profit rates vary over space. Moore *et al.* (1980) used Census of Production data for 1958–68 to show that in terms of profits per employee, these were generally lower in most of the conurbations than in their surrounding hinterlands. The exception was London where not only were profits higher than in the rest of the South East but they also exceeded the national average. Once industrial structure was standardized for, it was still clear that the rate of profit growth was lower in the conurbations than elsewhere. A study by Fothergill and Gudgin (1982) based solely on two industrial sectors – mechanical engineering and clothing – confirmed these findings for a more recent time period. As Table 5.5 shows, the gross profit rate in the clothing industry rises as the urban size decreases. With the exception of London, which is seen to be a profitable location for the mechanical engineering industry, the same pattern is true. This is attributed to lower sales and higher labour costs per unit of capital in the conurbations rather than the level of other costs (Fothergill *et al.* 1982). The most recent study (Tyler 1984) relates profit not to urban size but to distance from London. He shows that for one sector, telecommunications equipment manufacture, profits are lowest closest to London, rise sharply up to about 65 miles from London and thereafter continue to rise but at a steadily decreasing rate.

Urban location theory has stressed the distinction between labour-intensive inner-city industries and capital-intensive peripheral industries. In a sample survey of matched pairs of firms, Lever *et al.* (1981) showed that at least in one year, 1978, investment per employee was higher in the inner city (£363 per head) than in the outer conurbation (£189 per head), but both these figures were substantially below the UK national average (£831

Table 5.5 Average profitability and type of area

	Mechanical engineering 1970–75		Clothing 1968–74	
	No. of firms	Profitability (%)	No. of firms	Profitability (%)
London	35	14.0	43	11.2
Conurbations	67	11.1	64	11.4
Free-standing cities	55	12.7	30	14.1
Industrial towns	32	13.7	15	14.7
County towns + rural	45	16.4	28	18.5

Sources: NEDO; Fothergill and Gudgin (1982)

Table 5.6 Net capital investment per head: Clydeside

Sector*	Inner Clydeside (£)	Outer Clydeside (£)
Food, drink, tobacco	1,109	2,338
Chemicals	1,381	970
Mechanical engineering	521	713
Instrument engineering	768	435
Electrical engineering	614	1,037
Other metal goods	223	514
Textiles	526	219
Clothing, footwear	154	465
Non-metal mineral prod.	2,192	2,397
Timber	451	633
Paper, printing	659	801

Source: Business Statistics Office
* Sectors with small numbers of establishments excluded

Table 5.7 Sources of capital for investment

	Small (<50 empl.) (%)	Medium (50–200 empl.) (%)	Large (>200 empl.) (%)
Group finance	25.8	32.4	76.5
Retained profits	56.2	68.6	55.9
Personal capital	28.1	8.8	0
Share issue	3.2	5.7	15.2
Bank loan	56.2	42.9	15.2
Finance house	19.4	20.6	0
Insurance	3.2	2.9	0
Govt. grants	62.5	65.7	94.1
ICFC	0	8.6	6.1
Other	6.7	11.4	5.1

Source: Lever *et al.* (1981)

per head). More aggregated data for 1981 (Table 5.6) show that in the majority of sectors, outer-city firms invested more per employee than their inner-city equivalents. In only three of eleven sectors surveyed was the reverse true and in one of these three, instrument engineering, the figure for the inner city was biased by a very large investment programme by a single company. Size of plant also plays a key role in determining the source of capital for investment. As Table 5.7 shows, based on a survey of 240 plants in Clydeside, the most commonly used capital sources are government grants, retained profits, corporate group financing and bank loans. However, whereas 94.1 per cent of investing large plants used available government grants, only 62.5 per cent and 65.7 per cent of small and medium-sized establishments did so. Large plants were also more likely to use group financing from within the company. By contrast, small plants were significantly more dependent both upon personal capital and upon loans from banks and finance houses, often using domestic property as security. Over 70 per cent of investment projects in small plants were financed by these commercial loans (although not necessarily in their entirety) compared with 60 per cent of projects in medium-sized plants and only 15 per cent of large plant projects.

Exogenously-owned plants are inevitably more likely to make use of corporate group financing in the form of retained profits or share issues. This degree of external control has led to considerable discussion on the potential disadvantage to a region's economy of placing a high reliance upon inward investment. For example, in a study of the Manchester and Merseyside conurbations, contrasting locally-controlled and externally-controlled establishments, Lloyd and Dicken (1983) showed that while entry rates are significantly higher for locally-controlled firms in both cities, thus creating a substantial proportion of new jobs, the same sector also had a much higher closure rate and was therefore responsible for vast job losses.

Table 5.8 Inward and outward capital flows, UK, 1972–83 (£m)

	1972	1973	1974	1975	1976	1977	1978	1979	1980	1981	1982
Total overseas invest. in UK	772	1,497	2,204	1,514	2,091	4,399	1,877	4,336	5,240	3,362	3,459
Total UK invest. abroad	1,402	1,760	1,148	1,367	2,269	2,334	4,604	6,544	8,146	10,671	10,768
Balance	−630	−263	+1,056	+147	−178	−2,065	−2,727	−2,208	−2,906	−7,309	−7,309

Source: Annual Abstract of Statistics, 1984

While the discussion about the merits and demerits of attracting inward, foreign investment into regions and cities continues, although the volume of such flows has reduced very sharply form the peak of the 1960s, increasing concern is now being expressed about the flow of British capital abroad. As Table 5.8 shows, inward investment into Britain peaked in absolute terms in 1980, but in real terms in 1977. Outward investment accelerated rapidly after 1979 when foreign investment restrictions were lifted. In consequence, from a position in 1974–75 where Britain was a net importer of capital, the country now experiences very substantial net outflows. While it is possible to criticize this flow on the grounds that such capital outflows do not create employment within Britain, given the declining profit rates in British industry they are scarcely surprising and it is worthy of note that by 1983–84 more than half the recorded profits of British industry were being earned abroad.

5.5 Labour versus capital

In a simple model of the productive process where revenue and material input costs are fixed by external forces, a firm has a fixed amount to spend upon land, labour and capital. Within these three production factors, a simple equation inversely links labour costs (wages) and capital costs (interest and/or profits). In this model, therefore, a rise in labour costs may well result in lower profits: conversely, if labour costs can be reduced either through reductions in wage rates or by redundancies, then more capital is available either for investment or for profits. However, increasing profits is not the sole reason for companies seeking to reduce their labour costs: the increasing pressure of external competition may force companies to reduce their labour force (wage *rates* rarely decline) in order to sustain existing profit levels or to reduce the rate at which profit levels fall (Broadbent 1977).

Faced with these kinds of pressures, companies restructure their production systems in three ways: (i) by intensification, defined as instituting changes which increase labour productivity without resort to massive new capital investment; (ii) investment and technical change where increases in labour productivity are achieved through significant investment and possibly changes in production technique; and (iii) rationalization, defined as a simple reduction in total capacity (Massey and Meegan 1982). Using national data on employment change, changes in output and changes in productivity, Massey and Meegan (1979) were able to categorize fifty-eight industrial sectors according to their type of restructuring. Thus intensification was experienced in the textile finishing, men's tailored clothing and footwear sectors, investment and technical change was the most common form of restructuring in general chemicals, iron and steel, textile machinery and scientific instruments, and rationalization was the process adopted in grain milling, ferrous castings, electrical machinery, woollens and paper and board.

Intensification, as defined above, can be achieved by a multiplicity of small changes – increasing fragmentation of work tasks, speeding up of conveyor belts, reduction of restrictive practices by labour, reorganizing the production equipment, for example. Such changes can, of course, be adopted to meet a shortrun increase in demand but are more likely to be introduced during recession and therefore be accompanied by job loss, and in some cases by small firm closure. The employment consequences of this process, as exemplified by the men's outerwear industry, are a fairly uniform rate of job loss over the country as a whole. By establishment type, it did seem that the large and medium-sized plants were performing rather better and in some cases taking on additional workers whereas small firms were contracting or closing. Regional variations in employment change were therefore reflections of plant size. Where restructuring takes the form of substantial capital investment and technical change, a fairly major reduction in the amount of labour required was usually involved for a given level of output. In virtually every sector examined by Massey and Meegan, major investment was accompanied by productivity increases (over the period 1968–73) generally of the order of 25 per cent over five years. However, given that they choose to use the fletton brick industry as their typical example, it is difficult to discern a regional pattern of employment shifts due to this form of restructuring. Lastly, rationalization or the reduction of capacity involves disinvestment through the scrapping of plant and cutbacks in the labour force. The spatial pattern of the job loss depends upon the criteria used to identify which capacity is to be cut. Profitability, or lack of it, is usually the key element in the multiplant company's prioritization of plants for closure. This lack of profitability may be attributed to the age of the capital stock and consequent low productivity of labour, or size of plant, by which smaller plants may fail because of lack of scale economies. Massey and Meegan in an earlier study (1978) of the electrical equipment industry used these arguments to explain how this form of restructuring generally operated to the disadvantage of the inner cities. As an example of rationalization, the ferrous castings industry shows patterns of employment change at the regional level which reflect different product demands. Thus regions such as the West Midlands and Wales did relatively well with their buoyant demand for car castings and ingot moulds whereas the North East with its base in shipbuilding did badly. These studies are important not for their ability to produce a coherent national model of employment change by sector, but for their illustration of the ways in which defensive restructuring by capital in the face of increasing external competitive pressure leads to an increasingly weak position on the part of labour.

In a more recent study, Lever (1984) has argued that one of the explanations of the urban – rural shift in Britain's manufacturing industry is that the balance between labour and capital may run more strongly in capital's favour in smaller urban centres. Faced with external pressures and risks resulting from international competition, interest and exchange rates changes, one company policy is to seek to gain as much control over the local labour market as possible. A single large plant in a relatively small local labour market is able to exercise control over labour in several ways. Firstly, it may be able to drive down prevailing wage rates to levels lower than would have been the case in a competitive labour market, through the use of monopsony power. For three types of labour, skilled male manual, other male manual and female manual, wage rates in 1981 were between 4 and 12 per cent lower in these controlled labour markets than in a control group of non-dominant firms. Secondly, these more dominant establishments experience lower labour turnover and voluntary quit rates than do non-dominant plants, thereby achieving cost savings on labour training. Lastly, industrial relations appear to be better in smaller local labour markets, where there is one or a small number of comparatively large firms. This implies that these types of establishments have a capacity to use monopsony power solely to exploit a relatively captive labour force in a context where there are few alternative employment opportunities. This is not universally true and there are a number of paternalist capitalist enterprises which recognize their obligations as employers in this type of situation and seek to manage their labour forces accordingly.

5.6 Conclusion

This chapter has examined a number of aspects of capital and labour in determining the operation and location of British industry. Although labour in particular is important to manufacturing, neither capital nor labour has played a major role in formal location theory. In terms of government intervention in the location of industry in the form of both regional and, more recently, urban policy, a heavy emphasis has been placed on the use of capital subsidy programmes and, less consistently, labour subsidy programmes and labour training. The switching between these two types of incentives has also led analysts to query whether using subsidies of this type to influence the location of industry has not had the effect of changing the production function of industry, substituting subsidized capital for unsubsidized labour when employment creation and maintenance was the key objective of such policies. These substitutions, their associated rises in productivity per worker and the global recession have all had the effect of radically transforming the national labour market. While there were substantial numbers of unemployed prior to the mid-1970s, their problem was often thought to be soluble in terms of better job search information, relocation, retraining or a combination of these factors. As unemployment has climbed since the mid-1970s and particularly in 1980 and 1981, a different situation has developed. There are now large numbers of workers who have either never had a job or who have been unemployed for a long period of time and for whom there is little prospect of employment. What has developed is a 'dual labour market' (Bosanquet and Doeringer 1973; Wachter 1974) in which the economy is seen as divided into two sectors, a primary sector characterized by high wages, specific training and structured internal labour markets, and a secondary sector characterized by low wages, uncertainty of employment and no internal career or promotion structure. Disadvantaged groups such as the young, the elderly, females and ethnic groups tend to become trapped in the secondary sector. As the national aggregate demand for labour falls and almost every skill category experiences excess labour supply, more and more workers take jobs below their ability or qualifications, a process known as deskilling, thereby making it even more difficult for those without a job to re-enter employment of any kind. This is then reflected in both higher unemployment and the growth of the informal sector (Dicken and Lloyd 1981: 205) where output and income tend to be missing from official accounts. In addition to the fact that unemployment tends to adhere to certain groups within the labour force, it tends to concentrate in certain locations. Indeed, at an analytic level it becomes difficult to disentangle whether a person is unemployed because of his or her personal characteristics or because of where they live, when where they live is influenced or even dictated in the case of public sector housing tenants by their personal characteristics including their unemployment. An unambiguous answer to this question would undoubtedly make it easier for policy-makers to intervene effectively in the local or the national labour market.

References

Armstrong, H., Taylor, J. 1983 Unemployment stocks and flows in the travel-to-work area of the North-West region. *Urban Studies* **20**: 311–25.

ASTMS 1982 *Quarterly Economic Review* August. Association of Scientific, Technical and Managerial Staff.

Bassett, K. A., Haggett, P. 1971 Towards short term forecasting for cyclical behaviour in regional system of cities. In Chisholm, M. D. I. *et al.* (eds) *Regional forecasting*. Butterworth.

Bosanquet, N., Doeringer, P. B. 1973 Is there a dual labour market in Great Britain? *Economic Journal* **83**: 421–35.

Brechling, F. 1967 Trends and cycles in British regional unemployment. *Oxford Economic Papers* **19**: 1–21.

Broadbent, T. A. 1977 *Planning and profit in the urban economy*. Methuen.

Cheshire, P. C. 1979 Inner areas as spatial labour markets: a critique of the inner area studies. *Urban Studies* **16**: 29–43.

Crum, R. E., Gudgin, G. 1977 *Non-productive activities in UK manufacturing industry*. Commission of the European Communities Regional Policy, Brussels (Regional policy series No. 3).

Department of Trade and Industry 1973 Memorandum on the enquiry into location attitudes and experience. *Minutes of evidence*, Trade and industry Sub-Committee of the House of Commons Expenditure Committee, 4th July session, 1972–73. HMSO.

Dicken, P., Lloyd, P. E. 1981 *Modern western society*. Harper and Row.

Estle, E. F. 1967 A more conclusive regional test of the Heckscher – Ohlin hypothesis. *Journ. Polit. Econ.* **74**: 573–86

Fothergill, S., Gudgin, G. 1982 *Unequal growth: urban and regional employment change in the UK*. Heinemann.

Fothergill, S., Kitson, M., Monk, S. 1982 *The profitability of manufacturing industry in the UK conurbations*. Industrial Location Research Project, Working Paper 2, Department of Land Economy, University of Cambridge.

Fowler, R. F. 1968 *Duration of unemployment on the register of wholly unemployed*. Studies in Official Statistics, Research Series 1. HMSO.

Frost, M. E., Spence, N. A. 1981 Unemployment, structural economic change and public policy in British regions. In Diamond, D. (ed.) *Progress in planning* **16**: 1–103.

Frost, M. E., Spence, N. A. 1983 Unemployment change. In Goddard, J. B., Champion, A. G. (eds) *The urban and regional transformation of Britain*. Methuen, 239–59.

Fuch, V. R. 1967 *Differentials in hourly earnings by region and city size*. Columbia Press, New York.

Gillespie, A. E., Owen, D. W. 1981 Unemployment trends in current recession. *Area* **13**: 189–96.

Gilmour, J. 1974 External economies of scale, interindustrial linkages and decision-making. In Hamilton, F. E. I. (ed.) *Spatial perspectives on industrial organisation and decision-making*. Wiley, 335–62.

Goodman, J. F. B. 1970 The definition of local labour markets: some empirical problems., *British Journ. Indus. Relat.* **8**: 179–96

Keeble, D. E. 1976 *Industrial location and planning in the United Kingdom*. Methuen.

Layard, R. 1981 Measuring the duration of unemployment: a note. *Scottish Journal of Political Economy* **28**: 273–7.

Lever, W. F. 1974 Manufacturing decentralization and shifts in factor costs and external economies. In Collins, L., Walker, D. F. (eds) *Locational dynamics of manufacturing activity*. Wiley, 295–324.

Lever, W. F. 1978 Company dominated labour markets: the British case. *Tijd. voor Econ. en Soc. Geog.* **69**: 306–12.

Lever, W. F. 1980 The operation of local labour markets in Great Britain. *Pap. Proc. Regional Sci. Assoc.* **44**: 37–55.

Lever, W. F. 1981 Employment change in urban and regional systems: the United Kingdom case. In Buhr, W., Friedrich, P. (eds) *Regional development under stagnation*. Nomos Verlag, Baden-Baden.

Lever, W. F. 1984 Industrial change and urban size: a

risk theory approach. In Barr, B. M., Walters, N. M. (eds) *Regional diversification and structural change*. Tantalus, Vancouver.

Lever, W. F., Danson, M. W., Malcolm, J. F. 1981 *Manufacturing and service industries in inner cities*. Report to the Department of the Environment. University of Glasgow.

Lloyd, P. E., Dicken, P. 1983 The components of change in metropolitan areas: events in their corporate context. In Goddard, J. B., Champion, A. G. (eds) *The urban and regional transformation of Britain*. Methuen, 51–70

Mackay, D. I. 1970 Wages and labour turnover. In Robinson, D. (ed.) *Local labour markets and wage structures*. Gower, 68–99.

McPhail, C. I. 1982 The impact of plant dominance on employer personnel policy and local labour market behaviour. Unpublished doctoral dissertation, University of Glasgow.

Makenham, P. 1980 the anatomy of youth unemployment. *Department of Employment Gazette* **88**: 234–6.

Manners, G. 1976 Reinterpreting the regional problem. *Three Banks Rev.* **111**: 35–5.

Manpower Services Commission 1983 *Engineering labour shortages in the North West*. Institute of Manpower Studies, University of Sussex.

Manpower Services Commission 1984 *Labour market quarterly report: Great Britain*. November. MSC.

Massey, D., Meegan, R. 1978 Industrial restructuring versus the cities. *Urban Studies* **15**: 273–88.

Massey, D., Meegan, R. 1979 Labour productivity and regional employment change. *Area* **11**: 137–45.

Massey, D., Meegan, R. 1982 *The anatomy of job loss*. Methuen.

Moore, B., Rhodes, J., Tyler, P. 1980 *New developments in the evaluation of regional policy*. SSRC urban and regional economics study group conference, May. University of Birmingham.

Moroney, J. R., Walker, J. M. 1966 A regional test of the Heckscher – Ohlin hypothesis. *Journ. Polit. Econ.* **74**: 573–86.

Muth, R. F. 1969 *Cities and Housing*. University of Chicago Press, Chicago.

OECD 1983 *Economic Survey 1982–83: the United Kingdom*. OECD, Paris.

Predöhl, A. 1925 Das Standortsproblem in der Wirtschaftstheorie. *Weltwirtshaftliche Archiv* **21**: 294–331.

Ravallion, M. 1979 A note on intraurban wage differentials. *Urban Studies* **16**: 213–15.

Reati, A. 1984 *Rate of profit, business cycles and capital accumulation in UK industry 1959–81*. Commission of the European Communities, Directorate General for Economic and Financial Affairs. Economic Paper No. 35. CEC, Brussels.

Scott, A. J. 1982 Industrial activity in the modern metropolis. *Urban Studies* **19**: 111–41.

Smart, M. W. 1974 Labour market areas: uses and definitions. In Diamond, D., McLoughlin, J. B. (eds) *Progress in planning*. Pergamon Press, Vol. 2(IV).

Sraffa, P. 1960 *Production of commodities by means of commodities*. Cambridge University Press.

Taylor, M. J. 1978 Linkage change and organisational growth: the case of the West Midlands ironfoundry industry. *Econ. Geog.* **54**: 314–36.

Thirlwall, A. P. 1966 Regional unemployment as a cyclical phenomenon. *Scottish Journal of Political Economy* **13**: 205–19.

Townroe, P. M. 1970 Industrial linkage, agglomeration and external economies. *Journ. Town Plann. Inst.* **56**: 18–20.

Townsend, A. R. 1980 Unemployment geography and the new government's regional aid. *Area* **12**: 9–18.

Tyler, P. 1984 *Geographical variations in industrial costs*. Discussion Paper 12, Department of Land Economy, University of Cambridge.

Vipond, J. 1974 City size and unemployment. *Urban Studies* **11**: 39–46

Wachter, M. L. 1974 Primary and secondary labour markets: a critique of the dual approach. *Brookings Papers in Economic Activity* **3**: 637–668.

Walker, S. R. 1977 Linkage structure in an urban economy. *Regional Studies* **11**: 263–73.

Wood, P. A. 1969 Industrial location and linkage. *Area* **2**: 32–9.

CHAPTER 6

Industrial buildings and economic development

Steve Fothergill, Michael Kitson and Sarah Monk

Nearly all theories of industrial location ignore the supply of industrial buildings. The exclusion of property, or indeed any consideration of physical space, from theories of national economic development is even more marked. The implicit assumption has been that industrial floorspace is supplied in the right quantity in the right place at the right time, and that size and design of buildings is as required. In other words, the supply of industrial floorspace simply responds to demand.

These assumptions are rarely made about other factors of production, such as capital or labour (see Ch. 5). However, industrial production needs space in which to take place just as much as it needs to combine labour, materials and finance. If any of these sweeping assumptions about industrial floorspace do not hold, the supply of premises is therefore likely to influence the location and magnitude of economic growth.

In fact, the supply of buildings for industry is probably markedly less responsive to demand than the supply of capital or labour. The highly developed nature of the money market and the mobility of finance between sectors and locations means that few manufacturing firms are likely to be denied access to capital simply because they happen to be in the wrong place. Similarly, labour is geographically mobile. Research shows, for instance, that migration is the main way in which regional labour markets adjust to spatial variations

---

This chapter is based on research financed by the Department of Industry, Department of the Environment and the Economic and Social Research Council. The views expressed are, however, the sole responsibility of the authors.

in the growth and decline of employment (Gudgin et al. 1982). Regions where there is an above average increase in employment receive in-migrants; regions in relative decline lose population.

Industrial floorspace is less mobile. A factory building cannot be moved because the demand for it shifts from its present location to another. Factory buildings last a long time – sometimes a century or more – and increments to the stock and losses from it are mostly small in relation to the total stock. Changes in the spatial distribution of factory buildings thus occur only very slowly. Moreover, the firms which occupy the factory buildings are themselves often immobile, in that the physical removal of production from one building to another is frequently costly and disruptive. A firm which chooses a certain size or type of industrial building because it suits its current and planned requirements may find that, several years later in different circumstances, its buildings have become inappropriate and that there is little it can do about its predicament.

The purpose of this chapter is to examine the interaction between industrial activity and the physical environment in which it occurs. There is growing evidence that in Britain the supply of industrial premises varies from area to area in ways that only partly reflect local demand. There is also evidence that the quality of the industrial building stock leaves a great deal to be desired: much of the present stock was designed and built to accommodate processes and products which have long since been superseded. This is likely to undermine efficiency and have repercussions for the economy as a whole.

It is appropriate to examine the links between industrial buildings and economic development because they are also important for public policy. Unlike many inputs to industry, the supply of industrial buildings is potentially amenable to extensive manipulation by government within the framework of existing legislative powers. Central government already intervenes in the industrial property market through the Department of Industry's factory building arm, English Estates, and through the Scottish, Welsh and Northern Ireland development agencies. In recent years, local authorities have also become involved in the supply of industrial buildings, and the New Town development corporations are important suppliers within their areas. In addition, town and country planning legislation provides wide-ranging controls over the supply of land for industry and the development of industrial property.

The next part of the chapter describes some of the main trends in the supply of industrial floorspace and in its location. This is followed by a more detailed investigation of the quality of the industrial building stock in one part of the country, the Midlands, for which good information is available. The chapter then examines the relationship between employment change and the sort of sites and premises in which firms operate. Finally, the implications for the location of industry and for national economic growth are considered. Much of the evidence comes from our own research and some of the issues are considered at greater length in Fothergill *et al.* (1985).

6.1 Trends in the supply of industrial floorspace

Statistics on industrial floorspace in England and Wales are compiled by the Department of the Environment from data collected by the Inland Revenue in assessing the rateable value of properties. Because of the origin of the figures, they are particularly comprehensive and reliable. Comparable statistics are not compiled for Scotland and Northern Ireland, which must therefore be excluded from the following discussion. As defined, 'industrial' floorspace is essentially that occupied by manufacturing and includes office space attached to factories, but excludes some industrial establishments such as steelworks and chemical plants, where floorspace is difficult to define and measure. It also excludes warehousing, though

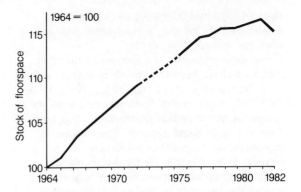

Fig. 6.1 The stock of industrial floorspace in England and Wales 1964–82. *Source:* Department of the Environment

some buildings could of course be used for either industry or warehousing.

Since the mid-1960s there has been a large and sustained reduction in manufacturing employment nationally, and in the recession at the beginning of the 1980s manufacturing output fell well below its level during the preceding decade (see Ch. 1). Such decline should not, however, obscure the substantial increase in manufacturing industry's requirements for floorspace, shown in Fig. 6.1. Between 1964 (the earliest year for which figures are available) and 1981 the stock of industrial floorspace in England and Wales increased by 34.5 m. metres^2 or 16.5 per cent – equal to almost 1 per cent a year. The stock rose every year during this period, though the rate of increase slowed from the mid-1970s onwards and the recession led to small reduction during 1981–82.

The increase in the stock of floorspace has little to do with any tendency for the supply of factory buildings to outstrip the demand for them. Although vacant industrial buildings are now a common feature of most towns and cities, statistics compiled by a leading estate agent, King and Co., show that in April 1982, in the depths of the recession, only just over 4 per cent of the total stock of industrial premises was vacant and on the market (either to let or for sale). Prior to the recession, the share of the stock vacant was much lower, between 1 and 2 per cent, and in August 1979 the figure was as low as 1.2 per cent. In a deep recession, many firms clearly do have spare floorspace within their buildings which is held in anticipation of an economic upturn. Nevertheless, the small share of the total stock which is vacant and on the market, especially prior to the recession, suggests that the long-term growth in the stock of industrial floorspace owes more to manufacturing's growing requirements than any tendency towards over-supply.

The main reason for the increase in the stock appears to have been the growth of output, at least until the recession at the start of the 1980s. Until 1979, output per unit of floorspace remained stable except for minor cyclical fluctuations, or at best showed a small trend increase. The increase in manufacturing output thus led directly to an increase in the demand for factory floorspace. After 1979, output per unit of floorspace fell, though the reduction in the stock of floorspace after 1981, in response to lower output, suggests that the ratio between output and floorspace is likely to return to its previous level. Looking further ahead, there is greater uncertainty about the relationship between manufacturing output and floorspace because the impact of the introduction of micro-processors on the space requirements of industry is not yet clear.

The growth in the stock of industrial floorspace has been unevenly distributed. Table 6.1 shows the net change in six types of area, ranging from London at one extreme to rural areas at the other, between 1967 (the earliest year for which local figures are available) and 1982. The striking feature is the large and consistent urban–rural contrast. In London the stock of floorspace fell by 17 per cent, and in the conurbations by 4 per cent. At the other end of the hierarchy the stock rose by 30 per cent in small towns and 50 per cent in rural areas. This urban–rural contrast in floorspace change is, of course, paralleled by an equally large urban–rural contrast in manufacturing employment change, documented extensively elsewhere (Keeble 1980; Fothergill and Gudgin 1982). It should, however, be noted that the loss of industrial floorspace in London and the conurbations has been much less than the loss of manufacturing jobs. Between 1967 and 1981, for example, the job loss in these areas was 47 per cent and 39 per cent respectively. Similarly, the percentage growth in the stock of industrial floorspace in towns and rural areas has been greater than the growth in manufacturing employment in these places.

For the period 1974–82, shown in Table 6.2, the data allows the net increase in the stock of floorspace to be broken up into components. This reveals that extensions to existing factories make the largest addition to the stock of floorspace. The importance of extensions is frequently overlooked since 'new floorspace' is often confused with 'new

Table 6.1 Industrial floorspace by type of area: England and Wales 1967–82

	Million metres^2		% change
	1967	1982	
London	26.69	22.10	−17.2
Conurbations	65.79	63.11	−4.1
Free-standing cities	37.19	40.76	+9.6
Large towns	26.67	34.18	+15.2
Small towns	44.32	57.55	+29.9
Rural areas	16.46	24.60	+49.5
England and Wales	219.82	242.30	+10.2

Source: Department of the Environment

The areas are groups of local authority districts, defined as follows using 1971 population estimates:

Conurbations:	Manchester, Merseyside, Tyneside, West Yorkshire, West Midlands.
Free-standing cities:	other cities with more than 250,000 people.
Large towns:	towns or cities with between 100,000 and 250,000 people.
Small towns:	districts including at least one town with 35,000–100,000 people, plus coalfield areas.
Rural areas:	districts in which all settlements have fewer than 35,000 people.

Table 6.2 Components of change in industrial floorspace by type of area: England 1974–82 (as % of 1974 stock)

	Stock in 1974 (million metres^2)	Reductions		Additions			Net Change
		Demolitions	Change of use	Extensions	New units	Change of use	
Cities*	127.2	−4.2	−13.3	+7.0	+5.4	+3.5	−1.6
Towns and rural areas	97.8	−3.0	−15.3	+12.8	+11.0	+4.6	+10.1
England	225.0	−3.7	−14.2	+9.5	+7.9	+4.0	+3.6

Source: Department of the Environment
*With more than 250,000 people

factories', and planning policies and debates over public involvement in the industrial property market have often focused exclusively on entirely new factories. In fact, between 1974 and 1982, extensions to existing factories increased the stock of industrial floorspace by 9.5 per cent, compared to a 7.8 per cent increase from new factory units. The main source of loss from the stock of floorspace is change of use, probably to warehousing. The loss of industrial floorspace due to change of use considerably exceeds the gains from changes of use.

Demolitions account for a greater loss of floorspace in cities, but total losses from the stock (the sum of demolitions and changes of use) are actually slightly greater in towns and rural areas than in cities. Additions to the stock of floorspace are thus wholly responsible for the net differences between cities and other areas, and Table 6.2 shows that the location of extensions and new units both favour towns and rural areas. For example, the rate of addition to the stock of floorspace through extensions and new units was three times higher in rural areas than in London. Between 1974 and 1982, a period of depressed economic activity compared to preceding years, extensions and new units together added 30 per cent to the stock of floorspace in rural areas – a major injection of new industrial infrastructure over such a short period.

Three principal agents are involved in the supply of new industrial floorspace: manufacturers who build for their own use, private sector property developers, and the public sector. No figures are available on the relative importance of the two private sector sources of supply, but firms building for their own use undoubtedly make the larger contribution, particularly as most factory extensions are built by their occupiers. The importance of private speculative development appears to vary according to the type and size of premises involved. Specialized premises are not normally built on speculative basis since they are not easily let; nor are large industrial units because there are fewer potential users, and often these users have specialist needs requiring purpose-built premises (Boddy and Barrett 1979).

The involvement of the public sector varies from area to area. English Estates (EE) was established in the 1930s to build advance factories in depressed areas in order to attract new mobile industry. Initially, most of its factories were in North East England, West Cumbria and Merseyside, but its activites became more widespread during the 1960s and early 1970s as the 'assisted areas' were extended to cover more of the country. In the second half of the 1970s, the scale of EE factory building was increased, and in 1981 it took over responsibility for the development and management of factories for the Development Commission, a body with responsibility for rural development, thus extending EE's involvement into several rural parts of southern England.

In August 1983, EE owned 2.7 m. metres^2 of factory floorspace, equivalent to just 1.2 per cent of the stock of industrial floorspace in England, and there were a further 1.2 m. metres^2 of owner-occupied floorspace on EE land. In the assisted areas where EE has been active for a long time, its involvement is greater: in Cleveland, for example, its properties account for 17 per cent of all industrial floorspace, in Durham for 15 per cent and in Tyne and Wear 13 per cent. EE is more important as a source of new industrial floorspace, accounting for 5 per cent of all floorspace in new units in England between 1974 and 1982, and as much as 32 per cent in the Northern region.

Within certain segments of the industrial property market (such as speculatively built new units) EE is still more important and to this must be added further public provision by local authorities and New Town development corporations. The Scottish and Welsh Development Agencies are also active in factory building within their areas. So although in total the provision by the public sector is very small, in some places it is the dominant provider of certain types of property. Whether the public sector ultimately displaces private development in these places, or provides factories which would otherwise not be built, is unclear. It is probably the case, nevertheless, that in many of the areas where EE has been active on a large scale the level of industrial rents is now so low as to deter most private speculative developers.

6.2 The quality of the industrial building stock

There are no nationwide statistics on the quality of Britain's industrial buildings, in terms of their suitability for the businesses that occupy them. CALUS (1979) undertook a study of new speculatively built factory units and concluded that the standardized designs to which these are built are sometimes poorly suited to their occupiers. Our own research looked at the whole of the industrial building stock in selected areas and is probably the largest study of its kind.

In 1982 we surveyed all manufacturing premises in the East Midlands with twenty-five or more workers, with the exception of a few establishments such as brickworks whose premises are highly specialized and atypical. The survey covered 2,447 individual factories, and the information was gathered mostly by a postal questionnaire which achieved a response rate of nearly 80 per cent. The non-respondents were visited by the project team, so that information was finally assembled on the entire population of factories within the scope of the survey in this region. The East Midlands includes three medium-sized cities (Derby, Nottingham and Leicester) plus towns and rural areas but does not include a major conurbation. To enable comparisons across the whole urban hierarchy, details of the industrial building stock in the 'core area' of the Birmingham conurbation were also obtained. These were originally collected in 1979 as part of a different study (JURUE 1980) and the data were adapted to be comparable with that for the East Midlands. After modification, the Birmingham data included 1,292 factories. The exact extent to which the industrial building stock in Birmingham and the East Midlands is typical of the country as a whole is unclear, though the growth and decline of manufacturing employment in different parts of the Midlands has been close to the average for similar areas elsewhere.

Age is an indicator of the quality of premises, in that older building are less likely to be suited to modern industry's needs and more likely to have been designed to house quite different production processes. Older buildings are also more likely to have been extended and altered over time, which may result in a cramped site with poor vehicular access and the haphazard movement of materials, labour and final product. The first part of Table 6.3 shows a marked urban–rural continuum in the age of industrial buildings. Almost two-thirds of the floorspace in Birmingham was built before 1945, and nearly 30 per cent before 1919. Conversely, in rural areas in the East Midlands over two-thirds of the stock of floorspace was built after 1945.

Another indicator of the quality of industrial buildings is whether production is carried out on more than one floor. Multi-storey buildings may have been suited to processes which pre-dated the assembly line, but when modern technology is introduced the need for production to be carried out on several floors may introduce discontinuities in production and may lead to bottlenecks and delays as work-in-progress is moved to the next stage of production on another floor. The figures in the second part of Table 6.3 include all establishments with production on more than one floor and show that roughly three-quarters of the factories in the inner area of a conurbation are

Table 6.3 The quality of the industrial building stock by type of area

	Birmingham inner city	East Midlands		
		Cities	Towns	Rural areas
1. Age				
Percentage of floorspace built in each period.				
pre-1919	29	21	17	13
1919–45	34	27	20	18
post-1945	37	52	63	69
	100	100	100	100
2. Multi-storey premises				
Percentage of factories	74	46	33	25
3. Multi-occupied premises				
Percentage of firms in shared premises	n.a.	13	6	5
4. Room for expansion				
Percentage of jobs in factories where buildings cover less than 50% of site	12	28	26	41
Percentage of jobs in factories with vacant land immediately adjacent to existing site	14	44	50	57

Sources: JURUE; *Industrial Premises Survey*

multi-storey compared with only a quarter in rural areas.

The third part of the table shows that multi-occupied premises comprise a small proportion of the total but again that, in the East Midlands at least, the proportion is higher in cities than in towns and rural areas.

The last part deals with room for expansion. This is important for the simple reason that growth cannot normally be accommodated on sites where there is no room for additional floorspace. Focusing attention on the competitiveness of firms may thus reveal that some are better than others, but unless space is available for additional production a competitive advantage will not necessarily be translated into faster growth of output and employment, at least not in existing factories. Two measures of room for expansion are shown, one dealing with the extent to which existing sites are already built-up and the other with the availability of land immediately adjacent to (but outside) existing sites. In both cases there is a large urban–rural contrast. In Birmingham, in particular, only a small proportion of factories have room for expansion on or adjacent to their existing sites.

In summary, this evidence suggests that in Britain's cities the industrial building stock is substantially worse than in towns and rural areas. On average it is older in cities, it is more likely to be multi-storey and multi-occupied, and there is less room for expansion.

6.3 Physical influences on employment change

The relevance to economic development of these characteristics of the industrial building stock depends on their impact on levels of production and employment. But before looking at this issue it is useful to mention briefly the overall relationship between factory floorspace and jobs.

Between 1964 and 1982, the average number of workers per thousand metres^2 of industrial floorspace in England and Wales fell from 36 to 21, a fall of 40 per cent or nearly 3 per cent a year. During up-turns in the economy, this decline in 'employment density' eased because of greater capacity utilization. During recessions, the decline accelerated as capacity utilization fell. But the long-term trend was unmistakably downwards. The reduction in employment density probably owes a great deal to the changing technical requirements of manufacturing industry. As new machinery is introduced it displaces workers from the shopfloor, labour productivity increases, and employment on any given area of floorspace falls.

The implication of the fall in employment density has generally been overlooked. It means that, on average, in a factory of a given size the number of jobs falls over time. Similarly, in any town or city where the stock of industrial floorspace is static, the level of manufacturing employment falls. Thus if a local authority seeks to maintain its level of manufacturing employment, it is not sufficient simply to maintain the existing stock of factory floorspace in that area. Additions to the stock of floorspace (and the jobs that go with those additions) are constantly needed to offset the loss of jobs on existing floorspace due to falling employment density.

The reduction in the ratio between jobs and land has probably been ever greater than that between jobs and floorspace. Nearly all new factories are built with single-storey layouts for production space; they also tend to be built with more space around them for car parking and circulation of heavy lorries and, where appropriate, for landscaping, outdoor storage and future expansion. Compared with Victorian factory development, modern factory buildings may thus provide little floorspace on a given site. For example, a 1,000 metres^2 site with a four-storey factory covering the whole area would provide 4,000 metres^2 of floorspace. A single-storey factory covering 40 per cent of the site, with the rest used for parking, would provide only 400 metres^2 of floorspace–one-tenth as much–on the same site.

At the level of the establishment, the reduction in employment density is likely to vary considerably, depending on the industry, the technology used, the rate of investment in new machinery and the competitive performance of the firm. The change in employment is also likely to depend on the physical characteristics of sites and premises, which vary in the extent to which they are conducive to efficiency and expansion. Factories on sites that are mostly covered by buildings are less likely to experience employment growth than those on more spacious sites because of the lack of room for expansion. Also, employment in factories in old buildings and multi-storey buildings is less likely to increase because such premises may hinder efficiency and the limited life expectancy of older buildings may deter investment in new extensions on the same site.

To investigate the relationship between employment change and the physical characteristics of sites and premises, we used the data on individual factories in the East Midlands. The analysis looked at the influence of the nature of sites and premises (in terms of age of buildings etc.) occupied by firms in 1968 on employment change in those factories between 1968 and 1982. Factories which opened during this period or which moved between sites were excluded, and those which closed were also omitted because no information could be collected on their sites and premises. Factories with fewer than twenty-five workers and non-respondents to our postal questionnaire were also excluded. In total, the analysis included 990 manufacturing establishments, accounting for about 300,000 jobs or around 60 per cent of total manufacturing employment in the East Midlands.

The information was analysed using multiple regression to measure the strength and direction of the relationships between a number of independent variables (the physical characteristics of sites and premises) and the dependent variable, employment change in individual factories. It is also possible to assess whether the independent variables are statistically significant – in other words, whether they are measuring real relationships rather than random variation in the data. In this instance, the technique was used not to test a fully-fledged theory of factory growth but simply to measure the effect of specific variables, so there remained considerable unexplained variation in the data. In 'cross-section' analysis of this sort, the important results are coefficients on the independent variables which are significant and demonstrate reasonable stability.

Table 6.4 shows the preferred measure of the relationships. A number of points should be noted. Firstly, the size of the establishment and its industry growth are included, in addition to site and premises variables, because these are likely to be important influences on employment change in individual factories. Secondly, whether a factory is single- or multi-storey is not included in the equation because this variable was not statistically significant. In other words, whether a factory is single or multi-storey does not appear to exert an independent influence on growth. Thirdly, the variables included in the equation are all statistically significant. This is indicated by the standard errors, in brackets, which as a rule of thumb should be less than half the value of the coefficient above in order for that variable to be

Table 6.4 Multiple regression analysis of establishment growth

$E = 241.5 - 18.1 \log S - 0.2A - 24.8 \log F + 0.63I$
    (23.9) (4.2) (0.07) (2.9) (0.2)

Where: E = % change in employment 1968–82
       S = % of site covered by buildings in 1968
       A = % of buildings in 1968 which had been built before 1945
       F = employment in 1968
       I = % change in employment 1968–82 in GB in the MLH to which the establishment belongs

significant. Fourthly, the relationship between employment change and size and between employment change and 'site coverage' (the extent to which a site is built-up) are non-linear – hence the use of the 'logarithmic' form of these variables.

The magnitude of the relationships identified in the equation are as follows. For every 10 per cent increase in the proportion of a factory built before 1945, employment growth declines by 2 per cent. The non-linear relationship between employment change and site coverage means that an increase in site coverage from 20 to 30 per cent reduces employment growth on average by 7 per cent, an increase from 40 to 50 per cent by 4 per cent, and an increase from 80 to 90 per cent by 2 per cent. Statistical tests on these results indicate that the relationships identified are robust, and in particular are largely unaffected by correlation between the explanatory variables.

These findings are based on data for only one region, though it covers a large number of factories over a long period. The findings do, however, provide strong evidence of a link between the growth of employment in individual factories and the sorts of sites and premises those factories occupy. Old buildings and heavily built-up sites with little or no room for expansion appear to be associated with above-average job losses.

6.4 Locational consequences

A consequence of the links between sites, premises and employment change is that cities are seriously disadvantaged as locations for manufacturing industry. As noted earlier, enterprises in cities are more likely to operate in older buildings with less room for expansion than their counterparts in towns and rural areas, and these physical characteristics are associated with poor growth.

One possibility is that city firms whose market and competitive performances allow expansion find their growth frustrated, at least on their present sites, by physical constraints. Alternatively, firms with potential for expansion may avoid city locations and opt instead for small towns and rural areas where spacious sites and newer premises are available. Whichever mechanism is at work, however, the net effect is that the physical constraints imposed by sites and premises in cities lead to an urban–rural contrast in manufacturing employment change.

The importance of these influences on urban–rural differences in growth is illustrated by further analysis of the East Midlands factory data. Within this region there have been large urban – rural differences in manufacturing employment change, typical of those in the country as a whole. One analysis we undertook was therefore to compare the employment growth of similar sorts of factories – similar, that is, in terms of the sites and premises they occupy – in cities, towns and rural areas in this region. The conclusion was on the whole that a factory's employmet change is influenced more by the sort of site and premises it occupies than whether it is in a city, town or rural area.

A second analysis confirmed this finding. This involved adding a 'dummy' variable to the equation in Table 6.4 to distinguish between locations in cities, towns and rural areas. If location (in an urban or rural setting) is an important influence on establishment growth, over and above the other variables included in the equation, the dummy variable should have been statistically significant. In fact it was not, and the other variables remained almost unaffected by its inclusion.

This evidence supports the view that urban–rural differences in the growth of employment in existing factories result from urban–rural contrasts in the industrial building stock. In simple terms it means that, on average, the employment of city firms in modern premises on spacious sites grows at the same rate as employment in firms with similar sites and premises in town and rural areas. Conversely, firms in old premises on cramped sites on average experience employment decline wherever they are located. The problem in cities is that a higher proportion of firms operate in old premises on cramped sites.

This discussion has concerned the growth of existing factories, but physical constraints in cities probably also influence the location of entirely new factories. Table 6.5 presents figures we have assembled on the availability of land for new factory building in different types of area in Britain. This is the land which local authorities consider to be 'available' for new industrial development in the short and long term, but excludes land earmarked for specific firms and that within the curtilage of existing factory sites. Land reserved specifically for warehousing is also excluded, though much of the remaining land is available for either manufacturing or warehousing. The table shows that in absolute terms, and especially in relation to the level of manufacturing employment in each area, small towns and rural areas have substantially more land available for new factories than cities. For instance, taking Britain as a whole, rural areas have four times as much available land per employee as the conurbations and over ten times as much as London.

The disadvantage experienced by cities in terms of the quantity of available land – which feeds through to the number and range of sites on offer to firms and developers – is compounded by other factors. A higher proportion of city sites are small, which limits their usefulness, especially to developers seeking to build new trading estates. City land is more expensive than land in surrounding areas. And much available land in inner cities in particular is likely to be recycled from firms that have closed or from public utilities, which may pose expensive problems of reclamation that will not be encountered on greenfield sites in other areas. These factors tend to divert new factory building away from cities.

Table 6.5 Available industrial land by type of area, 1982

	Hectares	Hectares per thousand manufacturing employees
London	746	1.0
Conurbations	4,661	2.8
Free-standing cities	4,213	4.3
Large towns	4,418	6.3
Small towns	9,732	6.9
Rural areas	6,439	10.9
Britain	30,208	4.9

The following counties are excluded: Derbyshire, Nottinghamshire, Hereford and Worcester, Essex, Oxfordshire, Berkshire, Surrey, Buckinghamshire (except Milton Keynes), Gloucestershire, Devon.

6.5 Implications for the national economy

Shortcoming in Britain's industrial building stock have implications for the economy as a whole, as well as the location of jobs within it. We have drawn attention to physical constraints on industrial expansion in cities and their role in generating a shift of jobs to small towns and rural areas. Part of the shift seems to occur within companies as production and employment is diverted from physically-constrained factories in cities to more spacious sites and newer premises in other areas. However, some of the shift probably occurs as urban firms without room for expansion are forced to forego growth and lose business to competitors. This is important because the competitors may be able to increase production because they have the room for expansion, not because of a cost advantage. Moreover, some competitors will be firms in other countries rather than firms in small towns and rural areas in Britain. Insofar as the physical constraints facing firms in Britain's cities undermine their ability to seize market opportunities, they therefore pose problems for the balance of trade and for national economic growth.

Indeed, the inadequacy of a high proportion of factory premises and sites in cities probably constitutes a serious 'supply-side' constraint on the British economy. Britain is one of the most urbanized of all Western industrial countries. Also, because Britain has been a primarily urban industrial society for longer than any other country, it has an unusually large legacy of old factory buildings in densely-developed environments, particularly in inner-city areas. These factories remain important places of production and employment, despite the decline of manufacturing in recent years. Whether they remain adequate for modern production methods or allow firms to expand when market conditions permit is extremely doubtful, and such obstacles to growth are unlikely to be found to the same extent in other industrial countries.

If there is a sustained up-turn in the national economy from the low level of activity of the early 1980s, the problems posed by shortcomings in the industrial building stock will become more pressing. Initially, many firms would be able to expand their output by utilizing spare capacity within their existing premises, but at an early stage in a recovery many can expect to come up against constraints within their buildings. This is especially likely because many firms responded to the recession of the early 1980s by closing whole factories rather than by 'mothballing' capacity in anticipation of a recovery. When physical constraints begin to bite and where there is little possibility of extending existing premises, firms can be expected to look for alternative locations and new premises to accommodate expansion. This poses four problems for the continuation of an economic recovery.

Firstly, the complete relocation of production from one site to a larger building elsewhere is usually disruptive, costly and in some cases impractical. The difficulties associated with relocation are reflected in the very small number of large factories, with a hundred or more workers, that undertake such a move. An alternative is to divert production to a new branch plant, but this can be inefficient because there are often technical reasons (such as economies of scale) why related production activities are best kept together under one roof rather than spread across two or more factories.

Secondly, although the amount of vacant industrial property is at an historically high level, many vacant premises may be ill-suited to accommodate the expansion of growing firms, whose physical requirements may differ from those which closed during the recession. For example, an investigation we undertook in 1982 into vacant industrial premises in one county, Leicestershire, showed that a high proportion of the vacant floorspace was in large, old multi-storey buildings, especially in Leicester city itself. This is precisely the sort of property which estate agents regard as unpopular and difficult to let.

Thirdly, at the same time as an economic up-turn increased manufacturing's demand for floorspace, it would encourage other users to demand more floorspace, and warehousing in particular would compete directly with manufacturing for much of the available space. Fourthly, the construction industry can only respond slowly to an increase in the demand for factory floorspace. As with housing, additions to the stock through new building are small in relation to the total stock, even in peak years of construction activity.

In an economic up-turn the constraints imposed by shortcomings in the supply of industrial building would be felt most acutely in cities, where the constraints are greatest. For this reason a further implication of renewed manufacturing growth nationally would be that in the absence of strong public intervention most of the growth would by-pass cities. New factory floorspace would be diverted to small towns and rural areas where the land is more readily available and more easily developed. Urban and rural trends in manufacturing employment would diverge. Given the present concern of policy to strengthen the industrial base of inner-city areas, the desirability of these spatial trends is highly questionable.

6.6 Conclusions: the need for policy

The large amount of industrial property standing empty in the 1980s suggests to some policy-makers that the supply of industrial buildings is not a problem. The Department of Industry, for example, has been keen to review the efficiency and value of the government's factory building programme in the light of changing national economic circumstances. It has expressed particular concern about the proportion of advanced factories in the assisted areas that are unoccupied.

The evidence presented in this chapter suggests, in contrast, that there is a continuing need for policies to improve and renew Britain's industrial building stock. True, there are a great many vacant premises, but focusing on this fact alone misses the point. The vast majority of industrial floorspace remains in use and much of it is probably poorly suited for the activities that take place on it. The problems are most acute in Britain's cities, where shortcomings in the supply of land and buildings for industry have played a key role in economic decline.

Public intervention in the property market is substantial, through planning controls, the public ownership of land and the role of the public sector as a developer. There is a need, we suggest, for these powers to be brought to bear on the supply of industrial buildings, especially in inner-city areas where the private sector by itself has mostly failed to renew the physical environment in which industry operates. It is especially important that short-term considerations, such as the high margin of spare capacity resulting from economic recession, do not obscure the long-term issue. If industry is to expand, it needs land and buildings to accommodate the expansion. Recession has merely provided a temporary respite to the problem in cities; it has not eased the long-term constraints on urban industry and, as new machinery to the shopfloor continues to reduce employment densities, the capacity of cities to accommodate manufacturing jobs is being continually reduced. If industry is to be competitive, it also needs buildings that meet its requirements and do not impede the smooth movement of materials, products and labour. By reducing the rate of investment in new premises, economic recession has only perpetuated the problems of old, inefficient factory buildings.

References

Boddy, M., Barrett, S. 1979 *Local government and the industrial development process*. School for Advanced Urban Studies working paper No. 6, University of Bristol.

CALUS 1979 *Buildings for industry*. Centre for Advanced Land Use Studies, University of Reading.

Fothergill, S., Gudgin G. 1982 *Unequal growth: urban and regional employment change in the UK*. Heinemann.

Fothergill, S., Kitson, M., Monk, S. 1985 *Urban industrial change*. HMSO.

Gudgin, G., Moore, B., Rhodes, J. 1982 Employment problems in the cities and regions of the UK: prospects for the 1980s. *Cambridge Economic Policy Review* vol. 8, no. 2.

JURUE 1980 *Industrial renewal in the inner city: an assessment of potential and problems*, Department of the Environment inner cities research programme report No. 2.

Keeble, D. 1980 Industrial decline, regional policy and the urban-rural manufacturing shift in the United Kingdom. *Environment and Planning A* **12**: 945–62.

CHAPTER 7

Technological change

John B. Goddard and Alfred Thwaites

7.1 Introduction

This chapter seeks to provide a geographical perspective to the debate about the consequences of technological change (or lack of it) for the British economy. At a national level the continuing loss of competitiveness of British industry has been attributed to numerous causes, but a leading consideration in most analyses has been the failure on the part of firms to introduce new and improved products that have a worldwide market place. The increased rate of innovation made possible by the widespread international diffusion of micro-electronics in the last decade has quickened the rate of technological obsolescene in the products of many parts of British industry, with the resultant loss of markets leading to significant reductions in employment. At the same time, the adoption of new manufacturing processes in established industries as a means of retaining price competitiveness has led to significant job displacements in surviving firms. While technological innovation has in some instances led to the creation of new job opportunities, it is clear that this has not always occurred in the same workplaces or localities where jobs have been displaced. The impact of technological change has therefore been highly differentiated in geographical terms.

Because of the inevitable delay between the invention of a new technology and its full commercial exploitation, it is necessary to consider technological change both regionally and nationally in a long-term historical perspective. Schumpeter, writing in the depressed 1930s, suggested that radical new technologies could themselves create opportunities for investment, profits, growth and employment (Schumpeter 1939). However, such new technologies are essentially a destabilizing influence in the economic system since they tend to occur in a few enterprises and branches of the economy; these innovations subsequently diffuse into a wide range of other industries, giving rise to a 'technological revolution' when this diffusion process transforms the established input/output relationships of industry by creating totally new sectors and destroying others.

Steam power was just such an innovation in the nineteenth century; from its limited role to pump water, steam power subsequently became incorporated into a wide range of industrial products and processes, creating new industries in new locations and undermining old industries and locations. While the basic technology was developed by the end of the eighteenth century, the subsequent refinements that led to the widespread diffusion of the new technology took nearly half a century. Similarly in the twentieth century, the basic technological innovations that made possible the post-1945 boom, such as the internal combustion engine, electric power and basic chemicals, and which also underpinned the growth of the West Midlands and South East

Acknowledgements
This chapter draws heavily on research undertaken in the Centre for Urban and Regional Development Studies, University of Newcastle upon Tyne, over the period 1978–84. Neil Alderman, Tony Edwards, David Gibbs, Peter Nash and Ray Oakey all made significant contributions to the research. Support was provided by the Economic and Social Research Council, the Department of Trade and Industry, the Department of the Environment and the European Commission. All responsibility for the text rests with the authors.

regions had their origins many decades before. Freeman, amongst others, argues that information technology which can act as either a product (e.g. micro-computers), process (e.g. robotics) or managerial information (e.g. telecommunications) alone amongst contemporary innovations has the capability of bringing about a similar radical transformation in the future (Freeman 1984). However, as with earlier technological revolutions, that future upswing in economic activity will initially be confined to only a few geographical locations (Goddard and Thwaites 1980).

Schumpeter and Freeman thus provide a technological interpretation of the long cycles in economic development first described by the Russian economist Kondratiev, although there has been much subsequent debate about the timing of innovation in relation to the long cycles. Mensch has suggested that innovations tend to bunch in the downswing; 'under pressure from high unemployment and under-utilized capital opposition to, and reservations about, untried risky new ideas disappear with a sense that relief might come from anywhere. This situation produces a surge of basic innovation that ends the stagnation gripping the economy in the technological stalemate and ushers in the recovery phase' (Mensch 1979: 162). Mensch identifies innovation clusters around 1764, 1825, 1886 and 1935. Freeman's empirical work reveals a more random distribution of innovation. However, he argues that the diffusion of the basic innovations is more important than their timing and in this respect he has important points to make which are relevant to geographical concerns. A key feature of his analysis is the interaction between the technological, economic and institutional system in bringing about such revolutionary transformations. Following the work of Perez, he argues that in the downswing a mismatch develops between the emerging technical capacity of the new technology and what is actually realizable given existing market and institutional conditions: '. . . such institutional changes include the educational and training system, the industrial relations system, managerial and corporate structures, the prevailing management styles, the capital and financial system, the public and private investment, the legal and political framework' (Freeman 1984: 110).

The features to which Freeman refers have a very clear geographical manifestation. This is particularly the case with respect to skills, management information and public infrastructure. Labour skills are far from mobile, are largely developed in local labour markets, and are very much influenced by the legacy of previous rounds of industrial investment. Areas where skills developed based around heavy engineering may be able to make the transition to electro-mechanical technologies but will need a heavy investment in training to embrace those parts of pure electronics where the emphasis is upon mental rather than physical skills. Moreover, the training capacity in local institutes of higher education is likely to reflect the needs of earlier periods and take a long time to adjust. In terms of management information, the enterprise's scanning of its business environment will be strongly conditioned by existing personal contact networks which are likely, for simple time geographic considerations, to have a strong local orientation. Information reaching enterprises in areas dominated by out-moded technologies consequently may not embrace the latest technological knowledge. Lastly, in the context of the information technology revolution, a lack of demand for telecommunication services arising from a failure to use the ability of this technology to provide access to an international store of technological knowledge may lead to an under-investment in the infrastructure that is likely to be a necessary condition for the next upswing.

Drawing these examples together and following Schmookler (1966), who defined the technological capacity of an economy as 'the accumulated body of knowledge weighted by the number of people who have access to that knowledge', we would suggest that there are significant regional variations in technological capacity or the 'regional ecology' of technological change (Lambooy 1984) within Britain. In the core of this chapter we provide some empirical justifications for this assertion. Before doing so however, some basic definitions are required which relate to the stages by which technological change comes about.

7.2 The definition of technological change

Technological change has its roots in *basic research* in which the quest for knowledge is pursued without any specifically defined economic goal. The research is frequently carried out in institutes of higher education and other publicly-sponsored research institutes. Much of the knowledge emanating from this research eventually becomes public in the academic literature, but developments which in their later stages may have far-reaching industrial consequences may only be revealed through personal contacts between commercial interests and the academic world; these contacts may only bear fruit if there is a sufficient level of technical competence on the commercial side to ensure an effective flow of information between the parties.

The commercial firm generally concentrates upon *applied research* resulting in *innovation* which can be commercially exploited. Firms often need to spend considerable sums in the development of inventions into marketable products, for example the manufacturing and testing of prototypes. The invention becomes an *innovation* once it is commercially marketed or applied. Important economic consequences will emerge at this stage and be reflected in changing demands for capital investment and skills on the part of the innovator and its component suppliers; however, the really significant impacts will only arise with the subsequent diffusion of the innovation. This diffusion may necessitate many *incremental changes* and modifications to original innovations, making them more widely applicable and acceptable. The need for such modifications is likely to emerge from the experience of earlier adopters and may necessitate more developmental research. Widespread diffusion deepens the impact as consumers make choices within the limits of scarce resources to substitute the innovation for existing goods. Ultimately, however, continuing improvements lead to standardization and the eventual abandonment of the innovation.

These stages, which have been referred to as the product life-cycle, may be related to changes in profit levels. In the early stages, innovations can create a market niche within which the firm may have a monopoly position, but although returns per product may be high the small size of the initial market will restrict the size of profits. As the market size increases, this will result in increased profits, but also increased competition from imitators; maintaining a market share may then require competition on price which in turn may neccessitate *process innovations* in the way in which the product is manufactured – for example, replacement of manual by automated methods.

It would be wrong to assume from this discussion that technological innovation is the only way through which firms maintain their rate of profit: market power can be achieved through takeover and merger, extensions of established products into new geographical areas, or relocation of production into lower-cost regions. In each of these activities, technological considerations may be an important consideration. Acquisition may be a means of gaining competence in an emerging area or eliminating the threat of an innovative product. Geographical extensions of markets may necessitate improved intra-company communications and the introduction of *managerial innovations* such as improved telecommunications or computer systems to manage the movement of goods. Relocation of production may also require such managerial innovations and also depend upon process innovations which reduce the skill content required by old technologies in existing locations.

Finally, it should be apparent from the introduction of this chapter that all technical changes are not of an equal weight. *Fundamental innovations* like micro-electronics may require *substantial innovations* in other areas in order to facilitate their widespread diffusion; robots, for example, are combinations of electronics and highly-sophisticated mechanical technology. A considerable track record in incremental technical change within *established sectors* of the economy may therefore be necessary in order to create the appropriate environment within which the fundamental innovations can be accommodated.

Notwithstanding the obvious importance of the so-called high-technology sectors such as electronics and telecommunications for the British economy, the analysis presented in the remainder of this chapter examines technological change in a broader range of manufacturing sectors. The analysis moves through the technological spectrum from the introduction of significant innovations through incremental change to the adoption of process innovations, describing regional differences at the establishment level and then attempting to account for these differences in the terms of the characteristics of the establishments themselves. Finally, the chapter discusses the implications of these findings for employment and regional development.

7.3 Significant innovations

Researchers at the Science Policy Research unit (SPRU), University of Sussex, have recently updated and extended to 1980 an earlier databank of significant innovations introduced into Britain in the post-war period (Townsend *et al.* 1982). The data covers nearly 2,300 innovations occurring in over thirty sectors of manufacturing industry defined at the level of Minimum List Heading (HMSO 1968). While not exhaustive, these sectors represent a broad spectrum of British industry and

Table 7.1 Trends in shares of substantial innovations by area (%)

Years	South East	Non-assisted outside South East	Intermediate Areas	Development Areas
1945–59	36.7	27.5	19.1	17.5
1960–69	31.2	28.4	24.8	14.8
1970–80	34.0	24.9	30.3	10.6
	(33.1)	(26.0)	(28.5)	(11.9)
Total 1945–80	33.7	27.0	25.5	13.7
N =	770	616	583	313
% of manuf. employees*	24.9	30.6	24.6	19.9
% of establishments*	31.0	29.6	23.4	15.9

Sources: *PA1002; Townsend *et al.* 1981
Figures in brackets denote standardization of sectors to produce comparable statistics with earlier time periods

cover a significant proportion of UK manufacturing output and employment.

The innovations included in the databank were identified by some 160 experts who were requested to list those *substantial* ones which had taken place within the identified industrial sectors since 1945. Most innovations were confirmed by more than one expert and subsequent contact with innovating firms provided confirmatory and additional information which allows a regional breakdown of the results. While the majority of innovations included in the databank are products (77–79% of the total), the data also includes some major process (16–17%) and material innovations (3–4%). This data base provides a very useful source from which to derive indications of variations in the location of the first commercial application in the UK of new technology within a wide range of industries. It also permits the identification within those industries of the first application of substantial innovations imported from abroad.

Table 7.1 illustrates the regional trends in innovation between 1945 and 1980. It shows that the South East exhibits a share of significant innovations well above its share of manufacturing employment and more narrowly of manufacturing establishments. In contrast, the Development Areas have a decreasing share of significant innovations and by the last period this is considerably below what might be expected on their basis of employment and number of establishments.

The Sussex study also reveals that the area to benefit most from the transfer of substantial foreign technology into Britain is the South East, which received approximately 217 innovations by this means (Table 7.2). On the other hand, the Development Areas have also benefited from this, largely it would seem at the expense of the non-assisted areas outside the South East. This perhaps reflects the high levels of foreign inward investment to the Development Areas which makes possible technology transfer within corporations. It also suggests that foreign-owned firms do not see any difficulties in transferring and successfully operating new technologies in the Development Areas (Haug *et al.* 1983).

Table 7.2 Percentage of innovations of foreign origin by receiving area

South East		Non-assisted		Intermediate Area		Development Area		Great Britain	
Foreign	Total	Foreign	Total	Foreign	Total	Foreign	Total	Foreign	Total
42.9	33.7	16.2	27.9	23.9	25.5	16.6	13.7	100	100
						N =		504	2282

Source: Townsend *et al.* 1981: Table 8.4, p. 83

7.4 New and improved products

While substantial innovations are perhaps of enormous value to innovating firms and industries and have far-reaching effects upon other industries and consumers, less significant and incremental changes can be very important to the future of individual enterprises and establishments within enterprises. Such changes can lead to product differentiation and the maintenance or enhancement of market shares. It is therefore not only major technological changes which are important to the continued advance of industry in an area, but also crucial updating and differentiation of locally produced goods. In the following paragraphs we provide evidence of regional variations in product innovations (defined as products 'new' to the establishment over the period 1973–77) in Britain.

The data were obtained from surveys of three generally innovative industries well represented in all regions of the country – metal working machine tools, scientific and industrial instruments, and radio and electronic components (the Innovation Survey). A population of 1,368 establishments were identified as manufacturing within these industries. Executives in a random sample of 174 establishments in three regions – the South East (a non-assisted area), the North West (an Intermediate Area) and the Northern region (a Development Area) – were interviewed. A further 633 establishments responded to a national postal survey and, when combined with the interviewed results, provided an overall coverage of approximately 60 per cent of the industries. The 807 respondents gave details of their establishments' characteristics, including levels of employment, corporate status and involvement in R & D, together with evidence of product and process change between the years 1973 and 1977, sources of technical information and estimates of the impact of specified technical change on employment. The basic area of analysis was the economic planning region, but these are also aggregated to four policy areas, viz:

Areas	Economic planning regions
(1) Development	Northern, Scotland, Wales
(2) Intermediate	Yorks & Humberside, North West
(3) Non-assisted	East & West Midlands, South West and East Anglia
(4) South East	South East

Table 7.3 shows that the South East recorded

Table 7.3 Establishment location by status: incidence of product innovation (Percentage of establishments innovating)

Area	Single plant innovation (%)	Group plant innovation (%)	Total innovation (%)
South East	85	91	88
Non-assisted outside the S.E.	74	88	82
Intermediate	76	87	82
Development Area	55	87	73
Great Britain	78	89	84

Source: Innovation Survey

the highest incidence of product innovation while those establishments located in the Development Areas were on average less innovative than their national counterparts. Within these areas, the least innovative set of establishments were operating from within the Northern region and the most innovative plants were operating in the circle of small commuter towns surrounding, but not in, London. Further analysis indicates that the spatial variations in product innovation are small within that set of establishments which form part of a larger group whereas the most noticeable difference is between performances in the independent single plant sector or enterprise 'indigenous' to specific areas. On this evidence, the Development Areas appear therefore to suffer from a local enterprise problem, rather than from the effects of external control.

Although the presence of innovation is important to local economic advance, *where* the development of these innovations occurs may be a better indicator of longer-term potential within a community. Hence, 'in-house' product development may more accurately reflect the innovative and resource base of industrial establishments than the mere introduction of new or improved products. The survey reveals that 83 per cent of establishments claiming a product innovation also claimed that the major development work had been performed on site, with the remaining respondents (109 cases) obtaining their products from 'elsewhere'. This latter group may be termed 'dependent' establishments.

The 'dependent' establishments consisted of the 10 per cent of single plant enterprises and 22 per

Table 7.4 Percentage of establishments with external sources of product innovation by region

Total GB (%)	South East (%)	Non-assisted (%)	Intermediate (%)	Development (%)
17	13	17	15	33

Source: Innovation Survey

Table 7.5 Independent enterprise by on-site development of technology and location

Area	High	Medium	Low	Total
South East				
Number	35	35	46	116
Percentage	30	30	40	
Non-assisted				
Number	3	16	20	39
Percentage	8	41	51	
Intermediate				
Number	5	13	14	32
Percentage	15	41	44	
Development				
Number	1	4	11	16
Percentage	6	25	69	

Source: Innovation Survey

cent of group establishments that obtained their product innovations from some source external to the plant. Establishments located in Development Areas and in particular the North were more inclined to 'import' new products than were establishments located elsewhere (Table 7.4). While the 'dependence' of independent enterprises upon external sources for innovation proved higher in Development Areas than elsewhere, an even more marked difference was observed for group establishments in different locations. One-third of Development Area group plants import technology compared with less than one-fifth of similar plants located in the South East, with the majority of these sources lying in the same corporation.

Of the plants claiming an externally-developed innovation, nearly 50 per cent come from abroad, in particular from the USA; the South East is the greatest beneficiary from this inflow of new products from the US. Within Britain, 60 per cent of external sources noted were in the South East and only 8 per cent in the Development Areas. In addition, of thirty products developed in the South East and transferred to another location for production, 57 per cent were transferred to other South East manufacturing plants and only 18 per cent to Development Area locations: within the UK there is therefore very little interregional transfer of technology.

The preceding analysis has suggested that in quantitative terms, the South East has some advantage in the development and manufacture of new products; it is also the chief beneficiary from the transfer of new products into Britain from abroad. Further examination reveals that these advantages are enhanced when the quality of innovations are considered. To achieve this evaluation, each innovation mentioned by respondents in the survey was categorized with the help of industrial experts as to whether it was 'high', 'medium' or 'low' technology. Table 7.5 describes the distribution of locally-developed product innovations within independent enterprises according to the technological classification. The table clearly suggests that few enterprises indigenous to the Development Areas develop their own high-technology products; when innovation does occur, it is more likely to be low technology.

7.5 Diffusion of process innovations

Firms may be made competitive by the introduction of new manufacturing processes which reduce the costs and/or improve the quality of existing products. These processes are usually embodied in new machinery of various types, machinery which itself may be the result of product innovations on the part of their manufacturers 'off adopter usually purchases the new machinery 'off the shelf' from a sales representative of a supplier and only limited research capacity may be needed in order to incorporate the technology into the customer's own production.

A further perspective on the spatial aspects of technological change in Britain can therefore be obtained by identifying selected products and examining the pattern of take-up amongst potential adopters. Table 7.6 reveals the extent of regional variations in the adoption of five advanced process innovations in 1,234 establishments in nine metal-working industries. The techniques are computerized numerical control of metal cutting, removing and joinery machinery, micro-processors for the control of manufacturing processes, such as assembly, monitoring and inspection, and the use of computers for the coordination of production

Technological change

Table 7.6 Adoption of new technology by assisted area status

		Percentage of respondents in each area having adopted			
	CNC	Computers in commercial use	Computers in manufacturing and design	Micro-processors in manufacturing processes	Micro-processors in products
Development areas	21.3	60.9	28.2	10.1	13.9
Intermediate areas	25.4	70.3	31.1	12.6	20.2
Non-assisted areas	27.0	64.5	29.8	11.6	22.4
South East	24.5	62.6	22.8	11.4	22.7
Great Britain	24.8	64.3	28.3	11.3	20.1
Spatial variation coefficient*	22.9	14.6	29.3	22.1	43.8

Source: Diffusion Survey
* Calculated as: $\frac{\text{maximum \%} - \text{minimum \%}}{\text{mean \%}} \cdot 100$

and design. By way of comparison with the earlier studies, the incorporation of micro-processors into products as a product innovation is also included.

The figures in Table 7.6 refer to the results of a survey of establishments in nine metal-working industries defined at the minimum list heading level of the Standard Industrial Classification and representing 40 per cent of the identified population in 1981 (the Diffusion Survey). It will be at once apparent from the table that the regional variations in rates of process innovation are far less significant than those recorded for project innovation either in the studies reported previously or when the adoption of, say CNC, is compared with the incorporation of micro-processors into products in this survey. Moreover, examining the cumulative proportion of adopters over time suggests that for certain techniques like CNC, the Development Areas led the South East in 1978 in the proportion of firms that adopted the techniques, a finding probably reflecting the influence of regional capital subsidies in bringing forward the purchase of new equipment.

Table 7.7 suggests that international differences in the rate of adoption of new processes are far more significant than those between regions. Based on comparable data for three countries, the table indicates that rates of adoption in Britain of all techniques except the commercial use of computers is dramatically below those recorded in Germany and the US in each of the three sectors surveyed. While regional variations are greater in US than in Britain, this may largely reduce differences in the size of the two countries.

Table 7.7 Adoption rates, by industrial sector and country

Percentage of establishments having adopted	Agricultural machinery			Metal working machine tools			Construction equipment		
	GB	FDR	USA	GB	FDR	USA	GB	FDR	USA
NC machine tools	25.0	41.0	19.5	40.0	56.0	57.0	31.8	47.6	45.8
CNC machine tools	10.0	46.7	23.4	39.6	58.2	55.8	27.3	39.5	44.1
Computers for:									
Commercial use	68.4	87.9	63.0	74.5	83.0	62.5	71.4	83.3	74.6
Design	12.1	26.9	9.7	16.1	19.0	18.3	21.1	33.3	29.8
Manufacturing	18.2	34.5	34.2	29.3	30.3	44.4	36.8	45.9	62.3
Micro-processors/mini-computers in products	11.8	n.a.	11.2	33.0	n.a.	40.0	10.5	n.a.	20.3

Sources: Diffusion Survey; Rees *et al.* 1984; Ewers, H.–J., Kleine, J., International Institute of Management, Berlin

7.6 The influence of research and development (R and D)

While there are many factors at work, we would suggest that regional variation in the commitment to R and D is one of the most important explanations of the low rate of product innovation recorded by industry in the Development Areas. Product innovation is essentially the bringing together of technical information to meet effective market demands. In a rapidly changing world, it is inherently risky to leave the gathering of technical and market information to chance and this accounts for the growth of the numbers of professional R and D workers employed in industry. Employment of R and D workers demonstrates a commitment to technological advance and clearly increases the chance of success in product innovation. It is therefore not surprising that the innovation survey revealed a very strong link between *on-site* R and D and product innovation; indeed, 89 per cent of the establishments recording a product innovation had some R and D effort in the establishment in question. Plants with no R and D on site were a far less innovative group.

Further examination reveals that establishments and, in particular, small independent enterprises located in Development Areas are less likely to carry out R and D activities on-site as compared with similar enterprises located elsewhere and particularly in the South East of England. Again, it is the independent enterprises indigenous to Development Areas which do not appear to pursue technological advance so rigorously or systematically as their southern counterparts. But even within establishments that are part of a large group, there appears to be a greater commitment to R and D in the South East: 50 per cent of such plants in the North employed less than five R and D workers compared with 23 per cent in the North West and 18 per cent in the South East. The regional variations in the mean size of R and D employment per establishment indicated in Table 7.8 suggest important implications for the aggregate level of technological expertise present in the economies of more and less prosperous areas.

A manufacturing establishment can supplement, or to some degree substitute for, its own research effort by using sources of technical information from outside the plant. The dynamic establishment will be expected to exploit these possibilities for technological advance. Innovation survey results suggest a clear difference between manufacturing establishments which are part of a larger group and those which are the sole location of the company in the extent to which external technical contacts take place. As a whole, single-site companies record a lower level of external contact; however, the one-third of establishments which were part of a group recorded 'other locations within the group' as their principal source of external technical information. This intra-group transfer of technical information was of particular importance to manufacturing plants located in the Development Areas. In fact, contacts external to the enterprise as a whole were much less likely in group establishments in Development Areas than in regions like the South East. This may be because of the lack of capability to absorb external information on the part of Development Area plants (i.e. a lack of R and D capability on-site) and/or a scarcity of suitable information sources locally.

In terms of location, the South East region is the primary source of external technical information. In general, plants in Development Areas mentioned relatively few local useful technical contacts. They appear to prefer or are forced to use technical sources at a greater distance than are firms located elsewhere. Although, in general, independent enterprises with one manufacturing site are more locally orientated as regards technical information than are plants which are part of a larger group, this relationship does not hold in the Development Areas, again reflecting the possibly limited supply of suitable technical advice.

7.8 Characteristic of R and D employment by establishment location

Region	Number	Mean	Median	Maximum
North	40	9.7	4	60
North West	51	14.9	4	120
South East	55	21.2	5	300
Total	146			

Source: Innovation Survey (interviews)

7.7 Other influences on products and process innovation

The results demonstrated earlier in this chapter suggest that there are a number of factors in addition to R and D capacity on-site, such as

Table 7.9 Logit analysis of product innovation

	Parameter	Estimate	S.E.
% GM	Single plant independent with low or zero R and D in metal-working machine tools located in the South East	1.23	0.29
RND (2)	Larger R and D effort	1.47	0.29
COMP (2)	Multiplant enterprise <10 plants	0.16	0.24
COMP (3)	Multiplant enterprise >10 plants	1.55	0.41
SECT (2)	Scientific instruments	0.57	0.29
SECT (3)	Radio and electronic components	−0.34	0.28
LOCN (2)	Other non-assisted areas	−0.47	0.28
LOCN (3)	Intermediate Areas	−0.47	0.33
LOCN (4)	Development Areas	−1.01	0.30

Source: Alderman *et al*. 1983

ownership status and enterprise size, that influence the incidence of product innovation. Furthermore, there may be differences between sectors in the rate of technical change and the mix of sectors may vary from region to region. An important question, therefore, is whether regional differences remain after having taken account of such factors.

An attempt can be made to answer this question using logit analysis in which a set of categorical variables reflecting size of enterprise, R and D effort, sector and location are used to predict the likelihood of product or process innovation in an establishment. In Table 7.9 the probability of product innovation is compared with a bench-mark establishment (a single-plant independent firm in a metal-working machine tool sector located in the South East and with low R and D effort). The parameter estimates of the model suggest that an increase in R and D effort, an increase in enterprise size and a shift to production in the scientific instruments sector are the factors most likely to increase the likelihood of innovation. While production in the radio and electronic components sector is likely to depress the likelihood of innovation, a location in the Development Areas is an even more negative influence. In other words, after allowing for other factors, peripheral location does seem to be a significant adverse influence on product innovation.

In the case of process innovation, logit analysis fails to identify any significant regional effects. Moreover, in the case of all technologies in the Diffusion Study, except the incorporation of microprocessors in products, plant size rather than enterprise size or R and D intensity are the most significant influences, suggesting that it is production not enterprise economies of scale that are the most important consideration relating to the introduction of new manufacturing processes.

7.8 Urban and regional contrast

The urban/rural shift in manufacturing production in Britain has been the subject of much debate (e.g. Fothergill and Gudgin 1982) and for certain industries such as pharmaceuticals there is evidence of a non-urban preference on the part of R and D functions (Howells 1984). In addition, detached research laboratories in the public sector, although being concentrated in the South East of England, are in the main located in smaller towns and rural areas (Buswell and Lewis 1970). But what of the on-site R and D capacity which the surveys reported in this chapter suggest are a major determinant of product innovation? Does the incidence of such facilities vary between urban and rural areas and are such variations greater than those between the North and the South of the country?

Table 7.10 attempts to answer this question with respect to the 1893 establishments responding to both the Innovation and Diffusion Surveys. Establishments have been assigned to functional regions and the regions specified according to size (conurbations, cities, towns, rural areas), functional status (dominant, sub-dominant and free-standing), economic structure (manufacturing, commercial, services) and regional location (North, South) (Coombes *et al*. 1982). The table reveals that the proportion of establishments with on-site R and D ranges from 88 per cent in cities subordinate to Greater London to 48 per cent in rural areas in the North. In the case of the conurbations, there is clear evidence of a lower level of R and D activity in the metropolitan centres as compared with surrounding towns and cities, but with the proportion of establishments with on-site R and D being significantly lower in the Northern conurbations and their surrounding towns than is the case of constituent components of the London metropolitan region. While there are no real differences in the case of free-standing cities, the North/South contrast between free-standing towns of all types is quite significant. For example, only 54 per cent of the establishments in towns classified

Table 7.10 On-site R and D facilities and urban status (percentage of establishments in each type of area with on-site R and D)

	%	No.
London		
Dominant metropolitan centres	78	205
Sub-dominant cities	88	67
Sub-dominant towns	83	132
Outer conurbations		
Dominant metropolitan centres	58	223
Sub-dominant cities	68	145
Sub-dominant towns	69	72
Other metropolitan centres		
Dominant metropolitan centres	76	97
Sub-dominant towns	68	76
Free-standing cities		
South	76	196
North	75	85
Manufacturing towns		
South	82	61
North	54	70
Service towns		
South	78	100
North	69	28
Commercial towns		
South	81	76
North	69	75
Rural areas		
South	76	50
North	48	44

Source: Innovation and Diffusion Surveys

as manufacturing in the North have on-site R and D compared with 82 per cent of establishments in similar towns in the South. The contrast between rural areas in the North and South is equally marked. So, as far as R and D effort is concerned, the preference for smaller towns is only a phenomenon in the London metropolitan region and in the regions based on the provincial conurbations; elsewhere, R and D is more significant in metropolitan centres while in the case of smaller towns of similar type there are distinct North/South contrasts. Indeed, the overall picture is one where these North/South differences would appear to be more significant than those between cities of different size.

7.9 Technological change and employment

What do our findings so far imply for jobs in

Table 7.11 The mean employment effect of product and process innovation in the first year by region and plant organizational status

	Product innovation		Process innovation	
	North	South East	North	South East
Single	+0.25	+1.25	−0.50	+0.20
Multi	+6.00	+7.45	−6.53	−0.92
Total	+4.30	+4.92	−3.81	−0.49

Source: Innovation Survey interviews (N= 123)

different regions? We would anticipate that a failure to introduce new products within the peripheral regions of Britain will indirectly result in a loss of markets and jobs. At the same time, the introduction of new manufacturing processes could directly result in labour displacement through new working practices. Our Diffusion Survey lent support to this suggestion with, for example, over 90 per cent of the establishments introducing CNC recording a decrease in average job time, a third a decline in setting-up time of machines and a further third a reduction in inspection time.

In order to provide some indication of the employment changes associated with the introduction of new and improved products, the respondents to the Innovation Survey were asked to estimate the number of jobs created or lost as a direct result of the changes in the year following the innovation. These estimates are compared in Table 7.11 for single- and multisite firms in the North and South East regions. The table reveals that firms in the North gained fewer jobs from product innovation than did firms in the South East whilst in the case of process innovation the most significant job losses occurred in multi-site firms in the North.

The overall regional effects of these changes can be estimated by grossing up the sample to give an indication of the *possible* job gains and losses in the three sectors covered by the survey. Table 7.12 suggests that when account is taken of the different sizes of the three sectors in the two regions, the North has gained relatively fewer jobs from product innovation than the South East, but lost more jobs through process innovation. Combining the two effects gives a marginal increase of employment in one year of 0.2 per cent of total employment in these industries. On the other hand, the South East gains 3,600 jobs or 3.1 per cent of total employment. However, these changes

Table 7.12 Estimated total annual regional impact of product and process innovation

	North	South East
Total employment change (annual average 1978–81)	−700	−4,000
(Percentage)	(−5.2)	(−3.4)
Product innovation	+160	+3,900
(Percentage)	(+1.1)	(+3.3)
Process innovation	−130	−290
(Percentage)	(−0.9)	(−0.25)
Total innovation effect	+24	+3,600
Percentage	(+0.2)	(+3.1)

Sources: Innovation Survey and Department of Employment

need to be seen in the context of the *actual* trends in employment in the three sectors in the two regions. Table 7.12 shows that in the three years after the survey the North experienced an actual annual rate of employment loss of 5.2 per cent compared with a 3.4 per cent loss in the South East. These losses may be attributed to a wide range of factors; however, we may speculate that had firms in the South East also failed to innovate to the same degree as those in the North, then the former region may have experienced an equivalent rate of job loss.

7.10 Conclusions

The analysis that has been presented in the body of this chapter has indicated important differences in patterns of technological innovations in Britain. The single most important cause of these differences would appear to be regional variations in an innovative capacity as reflected in the distribution of R and D activity; the peripheral regions are generally dependent on the core area for sources of technological knowledge and do not seem to provide a supporting environment for innovation especially amongst the smaller and medium-sized enterprises which form the focal point of many contemporary regional industrial development initiatives.

In these circumstances the question arises as to whether public policy can enchance the technological capacity of lagging regions resulting in increased local employment opportunities. On the basis of case studies of firms in central Scotland, South Wales and Berkshire, Sayer and Morgan question whether it is realistic to expect 'that the success of Berkshire can be produced in other regions on a major scale . . . simultaneously'; they suggest the success of this area is

> 'the product of the location preferences of entrepreneurs which has built up a critical mass of high skilled personnel for which there is an international shortage . . . access to good communication, especially Heathrow . . . a core of government research establishments, a rural working environment capable of sustaining elite lifestyles and a marked absence of trade union traditions . . . what distinguishes the area from central Scotland or South Wales is its élite occupational structure and intense decision-making network of activities which have the potential for spawning (and sustaining) new firm formation to an extent not readily apparent elsewhere' (Sayer and Morgan 1985).

It is clear that the advantages of the South East have been produced not only by the private sector but also by the uncoordinated actions of the state – for example, as regards the location of research facilities, the award of defence contracts, the modernization of communications and strict planning controls which have preserved the quality of the residential environment. Lagging regions may take comfort in the hope that a better managed state in which regional concerns are high on the agenda might be able to reproduce some of these conditions elsewhere. They may also be encouraged by the fact that Berkshire has emerged from a relatively backward agricultural region in less than twenty years, achieving its success in part from the migration of a highly mobile élite. Finally, Morgan and Sayer's analysis confirms the importance to a successful region of social networks in developing the technological capacity of an area. The extent to which such networks can be created in lagging regions by the effort of the public sector animateurs remains an open question.

Returning to the theoretical level, the 'success' of the South East and 'the failure' of the North in industrial innovation could be seen as a inevitable consequence of the instability introduced into the market economy by technological change. We have demonstrated elsewhere how advances in telecommunications technology which form the basic infrastructure for the information technology revolution are further enhancing this comparative advantage of the South East (Goddard *et al.* 1985).

But notwithstanding the 'inevitability' of these trends, we would question whether such regional disparities are in the best long-term national, let alone regional, interests. In the long-term, even in the market economy, the highest rates of return can be expected when new technologies diffuse throughout the economy; delays in the process of diffusion may mean that many opportunities are foregone. In our view, the future adoption of information technology in Britain will depend on many incremental product innovations in all corners of British manufacturing industry. A significant regionally-based effort to improve knowledge and skills and a uniform provision of telecommunications, infrastructure and services is clearly called for if the geographical variations recorded in this chapter are to be overcome sooner rather than later; later may be too late with Britain having no manufacturing capacity to update.

References

Alderman, N., Goddard, J. B., Thwaites, A. T. 1983 *Regional and urban perspectives on industrial innovation: applications of logit and cluster analysis to survey data*. Discussion Paper No. 42, Centre for Urban and Regional Development Studies, Newcastle upon Tyne.

Buswell, R. J., Lewis, E. W. 1970 The Geographical Distribution of Industrial Research and Development in the UK. *Regional Studies* **4**: 297–306.

Coombes, M. G., Dixon, J. S., Goddard, J. B., Openshaw, S., Taylor, P. J. 1982 Functional Regions for the Population Census of Great Britain. In Herbert, D. T., Johnston, R. J. (eds) *Geography and the Urban Environment*, 5, Progress in research applications. Wiley, 63–112.

Fothergill, S., Gudgin, G. 1982 *Unequal Growth: Urban and Regional Change in the UK*. Heinemann.

Freeman, C., 1984 Keynes or Kondratiev? How can we get back to full employment. In Marstrand, P. (ed.) *New Technology and the Future of Work and Skills*. Frances Pinter.

Goddard, J. B., Gillespie, A. E., Robinson, J. F. F., Thwaites, A. T. 1985 New Information Technology and Urban and Regional Development: In Oakey, R., Thwaites, A. T. (eds) *Technological Change and Regional Development*. Frances Pinter.

Goddard, J. B., Thwaites, A. T. 1980 *Technological Change and the Inner City*. Working Paper No. 4, The Inner City in Context, Economic and Social Research Council, London.

Haug, P., Hood, M., Young, S. 1983 R & D Intensity in the Affiliates of US-Owned Manufacturing in Scotland. *Regional Studies* **17**: 383–92.

HMSO 1968 *Standard Industrial Classification*. HMSO.

Howells, J. R. L. 1984 The Location of R & D: Some Observations and Evidence from Britain. *Regional Studies* **18**: 13–30.

Keeble, D. E. 1980 The Industrial Decline, Regional Policy and the Urban/Rural Shift in the UK. *Environment and Planning A* **12**: 945–62.

Lambooy, J. G. 1984 The Regional Ecology of Technological Change In Lambooy, J. G. (ed.) *New Spatial Dynamics and Economic Crisis*. Finn Publishers, Helsinki.

Mensch, G. 1979 *Stalemate in Technology*. New York.

Rees, J., Briggs, R., Oakey, R. P. 1984 The adoption of new technology in the American machinery industry. *Regional Studies* **18**: 439–504.

Sayer, A., Morgan, K. 1985 The Electronics Industry and Regional Development in Britain. In Amin, A., Goddard, J. B. (eds) *Technological Change, Industrial Restructuring and Regional Development*. Allen and Unwin.

Schumpeter, J. A. 1939 *Business Cycles*. Mcgraw Hill, New York.

Schmookler, J. A. 1966 *Invention and Economic Growth*. MIT Press, Cambridge.

Townsend, J. F., Henwood, F., Thomas, G. S., Pavitt, K. L., Wyatt, S. M. 1981 *Innovations in Britain since 1945*. Occas. Paper No. 16 Science Policy Research Unit, University of Sussex.

Townsend, J. F., Henwood, F., Thomas, G. S., Pavitt, K. L., Wyatt, S. M. 1982 *Science and Technology Indicators for the UK: Innovations in Britain since 1935*. Report prepared for SERC/SSRC Joint Committee, University of Sussex.

CHAPTER 8

Industrial change, linkages and regional development

J. Neill Marshall

8.1 Introduction

This chapter examines the contribution of behavioural studies of industrial linkage towards an understanding of regional development and industrial change in the UK. It considers material and service linkage studies in the context of trends in the UK economy and assesses the implications of structural changes for linkage analysis.

Linkage studies occupied a central position in industrial location research in the early 1970s, but they are now of more limited interest. We argue that the value of studies of the impact of material linkages on regional growth has been limited by recession and that this has contributed to their decline from pre-eminence. To counter this, an approach towards understanding industrial decline and the adaptability of enterprises to industrial change is described which places linkage studies in a wider frame of reference.

Sectors of the economy have expanded during the recent recession and may provide the springboard for some recovery in employment (e.g. the manufacture of information technology or office equipment, and the provision of certain services). Only a small number of linkage studies have been conducted in service industries, and this chapter therefore examines the role of linkages in business service firm organization, and the implications of this for spatial variations in the growth of services.

Linkages are not analysed for any intrinsic importance, nor as dimensions of organizational structure, but as a means of understanding the interdependence of business enterprise. It is argued that large or dominant firms play a significant role in networks of firm linkages, and that a study of such linkage relationships will assist our understanding of the impact of industrial change on firms and regions.

8.2 The industrial economy

The industrial economy is not static, and material linkages in particular define the channels whereby changes in demand feed through to organizations. Such linkages determine the limits to organizational action, as well as the changes to which firms must respond. Information and service links are also crucial in forming corporate perceptions of industrial change, and the nature of responsive investment or locational decisions. Analysis of linkages needs, therefore, to be set in a broad economic and organizational context.

In general terms, the context for linkage analysis is the decline in the competitiveness of manufacturing industry, which is reflected in a poor export performance and an increase in import penetration. This situation has been compounded since the 1973 oil crisis by a slowdown in the rate of growth of world trade, and since 1979 by a restrictive government fiscal stance. The result measured in employment terms has been a significant decline in the size of manufacturing industry. Total employment in Britain reached a peak in 1966, and approximately 4.5 m. jobs have been lost since, the bulk of them in manufacturing.

The service sector has to some extent cushioned this decline in employment and, as the country emerges from the recent recession, business and consumer services have provided substantial employment growth. There has also, of course, been a wide variety of performance within manufacturing. Metals, mechanical engineering and clothing and textile industries have borne the brunt of the recession, while industries such as pharmaceuticals, radio and electronic components, computers and aerospace have expanded, at least in output terms.

Given these sectoral variations in performance, it is to be expected that regional responses to national industrial change should differ. As the recession intensified, less-favoured regions (e.g. Northern Ireland, the Northern region, Yorkshire and Humberside, the North West and, to which can be added, the West Midlands) experienced major job losses. In the South East and East Anglia the recession has been less intense (Martin 1982), and such regions are emerging more strongly from the crisis (Regional Studies Association 1984). Industrial structure, and in particular the decline in vehicles and metal industries, have been important in the demise of the West Midlands, and textiles in job losses in Yorkshire and Humberside. In contrast, the South East has benefited from the relative buoyancy of marketable services. However, there remain important regional specific responses to the recession. The analysis of firm linkage patterns and dependencies in this chapter aims to contribute to an understanding of both structural and locational processes.

8.3 Problems in linkage analysis

It is difficult to evaluate the impact of linkages on regional economies in the UK. Behavioural industrial studies cover a limited number of industries and regions; more importantly, there are a number of technical and conceptual problems in such work.

Despite the magnitude of recent changes in the UK economy, linkage studies have been largely static and cross-sectional in their approach (McDermott and Taylor 1982). Some studies have provided general insights into linkage change based on secondary-source analysis (Moore 1973; Steed 1970). More importantly, Moseley and Townroe (1973) have considered linkage changes following industrial movement, and Le Heron and Schmidt (1976) have examined linkage changes in the USA associated with technological developments. However, these two studies concentrate on a limited organizational case and a special sectoral–regional context, respectively. Taylor (1978) and Taylor and Thrift (1982a, 1982b) have developed an impressive data base on linkage patterns for several points in time. This work sheds doubt on conclusions derived from static studies. For example, it might be inferred from cross-sectional studies that as the employment size of plants increases, their linkage patterns would become less local, but Taylor (1978) shows that a diversity of linkage adjustments are associated with changes in plant employment. However, Taylor and Thrift's work remains at an exploratory stage. The establishment variables are limited by the original questionnaire survey and, as a result, the analysis assumes that firm characteristics will typify individual plants. There is insufficient evidence to identify the impact of the particular industrial context on the analysis and, therefore, the processes underlying linkage change have to be inferred. There remains, then, a need for further studies of linkages in a dynamic context and, in the absence of such work, any conclusions must be limited.

Further difficulties associated with the measurement of industrial linkage stem from the fact that few studies have collected information on the monetary value of linkage flows, which is important if regional multipliers are to be estimated. On the other hand, a wide range of linkage measures have been used. Some studies have specified particular linkages, for example subcontracting or equipment purchases, others have analysed only the main customer and supplier links, while others again have estimated the proportion of total linkage in certain locations. All studies face the difficulty of quantifying the internal linkages of the organization, which do not always have a clear value when a component of an enterprise is not a profit centre. Nor do all studies account for wholesalers in the same manner when measuring the location of inputs and outputs.

Comparable data on linkages is the product of complex internal as well as external and locational factors. Despite a substantial tradition of linkage research, only a preliminary understanding of the organizational determinants of linkage patterns has been obtained. Much detailed information has been collected on establishment linkage patterns. However, less is known about the organizational decisions and characteristics which affect linkages. Major surveys of company linkages are very time-consuming and, as a consequence, samples tend to be small. This constrains statistical analysis and it is difficult, therefore, to control successfully for the range of factors which influence linkages. In addition, many of the organizational variables used in the analysis are rather simplistic. For example, from an analysis of key linkage studies, McDermott and Taylor (1982) show that organizational structure (i.e. the nature of organizational tasks and responsibilities) has only been inferred from aggregate measures of organization such as employment size. Of the twelve studies examined, only Hoare (1978) uses a measure of internal organization, and this consists simply of the presence or absence of particular departments. Studies also compare dissimilar organizational types: single-site firms are grouped with head offices and even branches, while national and multinational companies are categorized together. The implications of this are that inter- and intra-organizational linkage patterns are compared, as well as organization units which perform very different functions; which makes the conclusions of such work very difficult to interpret.

8.4 Traditional approaches to linkage analysis

Having placed linkage studies in the context of national economic trends, and bearing in mind the methodological problems which affect them, this section considers traditional approaches towards the study of linkage. It is impossible to consider these in detail, so two major fields of interest will be studied, linkages as agglomeration economies and the impact on linkage patterns of external ownership and control in industry. Both these fields of research have aimed to understand the role of material linkage in promoting regional growth, which reflects the fact that such studies developed in an era of economic expansion.

Agglomeration economies and local material linkage

Agglomeration economies associated with local material linkage have been argued to offer cost savings which favour the continued growth of industrial complexes. These economies are seen to derive from cheaper, speedier and more easily monitored local supplier and customer links. Thus, smaller firms, each carrying out a specialized component of production, would 'swarm' together and achieve similar advantages to that obtained by departments in larger organizations. However, studies searching for such agglomeration economies have discovered that linkages could be stretched over considerable distances (Keeble 1969; Townroe 1970), and this seemed to indicate that local external economies were insignificant. While smaller firms did have links with the local area, the importance of this was minimized by the fact that they did not consider proximity to customers or suppliers significant. Nor did firms unequivocally attempt to reduce the costs of their supplier links, the quality of the product and the reliability of the supplier being as important. Finally, when viewed in an international context, it was simply incorrect to suggest that local supplier links would lower costs (Gilmore 1974).

The most extensive investigation of agglomeration economies in the UK examined the iron foundry industry, a sector noted for its local linkages (Taylor 1973). Using a simple interaction model to predict the pattern of firm linkage, West Midlands firms were found to have 2.1 times and East Lancashire firms 6.3 times more local sales than predicted. While this intense local linkage seemed to suggest a search for agglomeration economies, West Midlands firms did not take advantage of the expanding motor vehicle market as expected, and East Lancashire firms did not obtain any advantages from local linkage which improved their survival rate. In fact, as Taylor and Wood (1973) show, firms with local links were small, privately owned and technically unsophisticated. They argue that local linkage must reflect parochialism, and the advantages which accrue must be behavoural in character, e.g. the reduction of uncertainty by retaining traditional customers and suppliers.

Insufficient is known on the present importance of transport costs in industrial location, but their significance was probably enhanced during the rise in petrol prices in the 1970s. Bearing in mind this qualification, which is particularly pertinent in cases where bulky or unstandardized linkages are concerned, it seems safe to assume that agglomeration economies associated with local material linkage have little to do with the recent pattern of regional industrial change. This conclusion has effectively removed material linkage studies from an important position on the industrial location research agenda. It has been recognized, however, that some areas benefit from the local linkages of their enterprises in terms of increased multiplier effects, and research into the regional pattern of material purchases has continued.

Ownership, control and material linkage

The impact of ownership and control on regional

growth and local multipliers has been the second main theme in linkage studies. Ownership refers to the amount of share capital held by an organization, with a distinction usually drawn between independent organizations which own their share capital and subsidiaries that have a substantial proportion, if not all, of their share capital owned by another firm. Control refers to the degree of decision-making freedom delegated to the individual parts of an organization by headquarters.

It is argued that a high level of external ownership and control of industry reduces local linkage and limits regional growth in peripheral regions such as the North, Scotland and Northern Ireland. Why should this be so? Economies of scale derived from 'in-house' production are argued to reduce the value of local linkage to large firms. Plants which are part of a larger organization may also be more aware of supply opportunities available in other regions.

Implicit in this argument is the fact that some externally-owned or headquartered plants will not carry out the full range of business functions and will have less than full autonomy in decision-taking. This will be associated with inputs and possibly outputs being obtained from and supplied to their own organization. There may also be little responsibility to search for local suppliers as well as little reason to develop local involvement. In other words, external ownership *per se* need not necessarily reduce local linkage: linkage patterns depend on the nature of the organizational structure.

This argument can be supported by an examination of major recent linkage studies in the UK. Table 8.1 presents the propensity to import (that is the proportion of sales in the local region minus the proportion of imports) for plants which have head offices outside the local region in four regional economies.

Clearly, plants in non-local firms in peripheral areas have a considerable excess of imports into the local economy relative to their exports. However, this could simply reflect their concentration in industries which are linked to local customers. Since the studies focus on either one industry or stratify the sample by industry, it is possible to compare indigenous versus externally headquartered industry in terms of their imports

Table 8.1 Impact of organization variables on linkages in provincial regions

Organizational variable	Ratio of local linkage in categories of establishment							
	Inputs				Outputs			
	Scot(L)	N.Ir.	Scot(M)	North	Scot(L)	Scot(M)	North	W.Mid.
Employment size	1.5	*	*	1.5	1.3	*	1.5	*
Ownership	1.3	3.5†	1.4	1.2	1.4†	2.7†	1.3	1.1	*
Operations tech.	n.a.	n.a.	–	2.0	n.a.	n.a.	*	–	*
Autonomy	n.a.	2.7	–	2.5	n.a.	2.7	*	–	n.a.
Propensity of non-local firms to import (%)	46.5	7.6	14.1	5.6				

Sources: Scot (L) – Lever (1974); N.Ir. – Hoare (1978); Scot(M) – McDermott (1976); North – Marshall (1979a); W. Mid. – Taylor and Wood (1973)

Key
Employment size	Local linkage in small plants divided by large plants (small 1–99 employees in North, 50 employees average in Scot(L)).
Ownership	Local linkage in independent plants divided by non-regionally headquartered plants
Operations technology	Local linkage in plants with continuous technology divided by non-continuous technology plants
Autonomy	Local linkage in plants within multisite organizations with low autonomy divided by high autonomy plants (individual author's definition)

* Details not presented in paper, but an important factor
† Ownership relationship statistically significant
n.a. Not applicable

and exports. It is not possible to identify single-site companies, branches and headquarters in the local sector, or national and multinational firms in the external sector, in all studies. Excluding Northern Ireland, there is a reasonable degree of consistency in the results, with a slight tendency for non-locally headquartered industry to have fewer local links. Independent local companies have between 1.1 and 1.4 times more local outputs and 1.2 and 1.4 times more local inputs than the external sector. These differences are described as significant in the case of sales in Lever's (1974) study in Scotland, though here this only reflects a loss of local multipliers if such firms could supply intermediate markets where goods are currently imported.[1]

Most studies indicate the importance of organizational factors as well as ownership. Employment size is significant in all cases apart from sales in Lever's study, and production technology is a meaningful discriminator in those studies which incorporate this variable. Here, large plants and more continuous production technologies are associated with little local linkage because the linkage requirements of such plants in terms of their quality and quantity cannot always be met locally. The degree of autonomy is incorporated explicitly in two studies and both find a significant relationship. The concept is also implicit in McDermott's (1976) variable measuring the integration of the marketing function into the rest of the firm's organization. Greater autonomy at the site investigated is associated with local linkage on the supply side because it means that establishments are given more opportunity to use non-company sources for raw materials and components. On the selling side, where a plant is set up to explicitly develop a local market, this will also require greater freedom in decision-taking.

The results with regard to the relationship of autonomy with linkage conflict in the Northern region and Northern Ireland, with the former study arguing that autonomy is a more important predictor than ownership, while the latter argues the reverse. It is likely here that these differences reflect variations in the degree of integration into the company in the two samples. In the Northern region, there is considerable variation in the extent of plant integration into the company and in levels of autonomy. In contrast, in Northern Ireland most plants appear to be tightly controlled from elsewhere and strongly linked into their organization.

The results of an analysis of external ownership and control's impact on linkages have been more conclusive than that for agglomeration economies, but the generalizations from such research are relatively limited. Studies identify a dichotomy between locally-linked small and independent plants with jobbing production technology, and externally-owned plants which may have non-local links because of little autonomy, larger employment size and more continuous production technologies. The inference is, therefore, that peripheral region growth can be constrained in areas with a high level of external ownership and control, due to the lower levels of local material linkage multipliers.

However, these studies ignore the process of industrial change, the fact that linkages are the mechanism whereby *changes* in demand are channelled through to the local economy, and how the role of linkages can be transformed by changes in the national economy. The industries and locations studied for signs of agglomeraton economies encouraging growth have been characterized by de-industrialization during the 1970s (Blackaby 1979). Generalizations based on a former period of economic growth seem inappropriate in the context of industrial contraction. Using evidence from the external control debate, it would appear that regions dominated by branch plants should be at an advantage in a recession because closures would have relatively low local multipliers. In fact, matters are unlikely to be that simple as the following analysis shows.

8.5 Linkages, dependence and industrial change

If linkage studies are to take greater cognizance of their industrial context and also changes in the national economy, it is necessary that they should be placed in a broader framework. Taylor and Thrift (1982a; 1983) outline such an approach which focuses on 'power networks' of linkage associations between firms (see also Fredricksson and Lindmark 1979). Power is defined in terms of the ability of enterprises to control the resources which are necessary for successful business operation. A segmented economy is described consisting of dominant amd dominated firms in linkage networks. Despite recent work (Taylor and Kissling 1983), the categorization of organizations into leaders, intermediaries and laggards, and their relationships, require further analysis. Nevertheless, large organizations are seen to play a central role in transmitting industrial demand through to other firms via their subcontracting arrangements. This is more realistic than traditional approaches because for most firms the aggregate economy is not an abstract entity but rather a group of other organizations. One of the lessons of the recession is that large firms play an important role in the product markets of other firms (Townsend 1983).

Such a framework is capable of producing a greater understanding of the mechanism of industrial change and the role in this of establishment linkages. The effect of industrial changes on linked firms will reflect the share of each establishment's turnover associated with particular products and markets and the position in these of dominant customers. Establishments which are product, market or customer dependent will be more affected by changes in demand. In addition, there is likely to be a behavioural dimension to this dependency, with the degree of dependence of firms/establishments on one another reflecting their perception of the availability of other potential customers/suppliers, and the degree to which linkages are regarded as essential to the firm's development (Rabey 1976).

From this perspective, industrial decline can be interpreted as the natural corollary of expansion. During growth phases in the economy or individual sectors, 'leading' or 'dominant' firms will exert a considerable influence on linked suppliers. As industries grow, a specialist supply capacity is developed when backward integration by the dominant firm is inadequate for its needs. As the rate of growth in the market of the dominant firm slows, these suppliers will be affected by the changing nature of their relationship with the dominant firm as expressed through changes in linkage volumes, or the length and type of contract. In a decline phase, suppliers will need to adapt, where necessary break their links with the dominant firm and restructure into new markets. The ability of firms to respond in this manner will determine the adaptability of the regional economy.

The remainder of this chapter develops this approach to linkages as dependency relations in the context of the recession in manufacturing and the growth of service industry in the UK.

8.6 Industrial decline and material linkage patterns

Few regions are heavily dependent on a narrow range of economic activity, and the linkage of many firms with any one region is small in proportionate terms. Nevertheless, in a general recession, declining forms of economic activity can have an important impact on regions through inter- and intrasectoral linkages (e.g. shipbuilding in North East England and textiles in Yorkshire). In such a situation, the questions posed of researchers are concerned with the role of linkages in industrial decline; as policy-makers and the general public wish to know how many jobs will be lost linked to closure or a reduction in capacity, where the job losses will occur, and what are the prospects for the local economy.

But how can changes in demand in individual broad industries, as they are transmitted over space, be studied and what organizational factors determine firm responses to change and, hence, affect the spatial pattern of decline? Clearly it is impossible to study this process using aggregate statistics. Most major industries or firms have suppliers in a range of sectors. Even where the industries of supplier firms are obvious, it is not possible to simply multiply reductions in production capacity from customers to employment loss in supplier industries because of variations in labour productivity. The only way to obtain reliable information is to identify key enterprises and obtain supplier lists, and interview firms to identify their dependence on particular customers. Even here there are complications because a given loss of output may put one supplier out of business with a loss of jobs in activities unrelated to the declining industry, whereas another firm may be able to switch production to other relatively buoyant markets.

Shipbuilding

A preliminary understanding of the process of industrial decline and linkage adjustment can be obtained by considering organizational response to industrial change in a survey of twenty-eight small shipbuilding suppliers, whose markets contracted when the oil crisis and the subsequent decline in the growth of world trade reduced the overall demand for ships during the 1970s (Gibb et al. 1977a; 1977b; 1977c). These suppliers displayed a high level of customer and market dependence, with twelve out of twenty-five firms having 40 per cent of their turnover accounted for by their five main customers, and twenty out of twenty-eight firms providing output for one market segment. In addition, twenty-two of the firms had made no major changes in their technology, organizational structure or staffing arrangements in response to market trends. Indeed, a picture emerges of firms experiencing changes in their selling activities which are not strategically or consciously planned for or controlled, and organizations appeared wholly reactive to external forces.

Those firms which were most market or customer dependent were more likely to feel customers were dependent on them. They also had a poorly developed marketing function and considered production strengths such as price, flexibility and quality to be key competitive factors. In other words, such firms internalized the major requirements of their key customers. The research was also able to identify those organizational factors most likely to restrict the adaptability of such firms to changes in customer relations. Less adaptable firms tended to have a mixture of the following characteristics: (i) they were old, established family firms; (ii) they were founded in shipbuilding activities; (iii) they were managed by relatives of the founder and staffed by managers with lengthy service; and (iv) they carried out specialist production activities.

Little evidence is available in this study on the role of large organizations in industrial decline and their affect on linkages and negative multipliers. This is an important omission because in many sectors component suppliers are large firms, and almost 60 per cent of the total population of suppliers in the shipbuilding study were branch plants. This omission can be accommodated by an investigation of the motor vehicle components industry where several of the components firms are multinational enterprises.

Motor vehicle components suppliers

Shipbuilding is characterized by long-standing customer–supplier relations, and has a history of stagnation and decline. Motor vehicle components manufacturers have, on the other hand, maintained a strong export performance despite increased competitive pressure (Jones 1981; Dunnett 1980).

Notwithstanding this success, a number of factors have encouraged industrial restructuring. The European car market has grown slowly during the 1970s, mirroring the downturn in developed economies as a whole. European manufacturers have also come under increasing competitive pressure from Japanese producers. The oil crisis and energy conservation measures have produced significant changes in vehicle design and, together with changes in production technology, have increased the investment required of manufacturers for model development and production. The major manufacturers, in an attempt to achieve greater economies of scale in production, have reduced their model ranges, introduced greater vertical integration and rationalized their supplier relations. In addition, the poor performance of the UK national motor industry, and the increase in imported vehicles, has led to a growth in imports of foreign components into the country. Component firms have responded by rationalizing less profitable activities, re-equipping and internationalizing their production operations to improve their access to growth markets and to manufacturers who often purchase local components. Specialist component firms have also increased their cooperation with individual manufacturers to improve vehicle design and performance (Scott-Ward 1978).

The main industries dependent on motor vehicles are metal manufacturing, electrical engineering, rubber, plastics and general engineering. Information available for components suppliers permits an analysis of the location of the industry. A study of the suppliers to a major UK motor vehicle manufacturer showed that the South East accounted for 26.7 per cent of supplies and the West Midlands 25.5 per cent in 1976/77 (Stoney 1982). However, while the South East has a relatively diversified economy, the West Midlands is heavily dependent on the fortunes of the motor vehicle 'industry' (one estimate suggests that perhaps 30 per cent of employment there is directly and indirectly dependent on motor car manufacture (Miller 1983)). These figures, which are the average for the Economic Planning Region, understate the degree of dependence of local areas. Table 8.2 presents the location of the main component suppliers to the motor vehicle industry and shows that seven travel-to-work areas account for 50 per cent of the plants in these firms. Birmingham stands out as having six times as many plants as any other travel-to-work area. Considering relative employment, some other locations are also heavily dependent: Coventry, though it has fewer component suppliers, has its top three firms BL, Talbot and GEC all directly linked to motor vehicles (Jepson 1983).

But what evidence is there concerning the process of enterprise adaption and change in the vehicle component sector? One study conducted in 1975, before the major recession in the motor vehicle sector, identified 820 suppliers related to motor vehicles in the West Midlands (Crompton *et*

Table 8.2 The location of major motor vehicle component manufacturers, 1977/79*

Location by planning region† and travel-to-work area	Sites
Birmingham	14
Dudley/Sandwell	10
Coventry	5
Wolverhampton	4
West Midlands	39
Greater London	6
South East	13
Manchester	6
North West	13
Pontypool	4
Wales	8
Yorkshire and Humberside	8
South West	5
East Anglia	4
East Midlands	3
Northern region	3
Scotland	2
Total	98

* Major components suppliers defined by turnover, with at least 25 per cent of their business in the motor components sector (Scott-Ward 1978). Only sites linked to motor vehicles included (Society of Motor Manufacturers and Traders and Kelly and Kompas Directories).
† Travel-to-work areas listed with four sites or more

al. 1976). A survey of fifty-one firms showed that in the motor car components sector there is a high level of market dependence with a little over half of the firms having 50 per cent of their employment dependent on the motor car market. Therefore, changes in demand can have a dramatic effect on linked firms. The problems faced by shipbuilding firms in adapting to industrial change were also found in motor components, because firms geared their production in terms of volume and quality control towards the needs of the motor car sector. Firms were lacking in expertise to develop some new markets, particularly abroad, and in any case many related engineering and metal-working markets were also depressed. Those plants which were part of a larger firm were additionally constrained in their diversification strategies by potential overlap with products produced elsewhere in the group. On the other hand, small firms were characterized by the most divergent response to changes in demand, with some not being able to diversify due to insufficient financial resources and an inflexible management structure, while others were able to adapt well because, unlike large firms, they did not have substantial fixed capital.

Implications

These studies and other work on industrial decline suggest that, despite the problems of responding to market changes, there is a great deal of volatility in linkage patterns associated with a decline in demand (Taylor and Thrift 1982; Steed 1973). Firms search for new markets as an industry declines, and changes in production processes, designed to improve competitiveness, produce new component or material requirements, which can reduce local dependence. Frequently, industrial decline is associated with strong foreign competition, which encourages firms to restructure to gain access to growth markets abroad, which again reduces internal multipliers.

This suggests that there are likely to be conflicts inherent in the process of industrial change at the regional scale. Enterprise success and the preservation of regional employment depend upon local firms employing a successful diversification strategy in response to a decline in market demand. However, at the same time there is evidence that such a strategy can reduce local linkage multipliers and produce problems for linked suppliers.

These case studies support the argument that the way industrial change feeds through the economy will reflect organizational dependency relationships. They suggest some firms will find adaptive reactions to industrial change difficult, either because their production is specialized or because they lack the experience to diversify into other markets. There are likely, therefore, to be marked variations in the adaptability of organizations to industrial change. Industrial decline is also associated with considerable ownership and organizational change. Horizontal and vertical integration via acquisition is a common response since these strategies bring, on the one hand, advantages of production flexibility and lessen market dependence and, on the other, reduce costs by cutting out intermediaries and allowing short cuts in the production process. This implies that the study of linkages should be placed in a broad economic and organizational context.

From a geographical point of view, an important

question will be whether there are locational as well as organizational determinants of enterprise adaption. This question has been debated in the context of industrial location as a whole (Wood 1978; Massey and Meegan 1982). However, specifically from a linkage perspective, it calls for further research into the impact of the local milieu on enterprises. Spatial variations in the access to business information is one aspect of the local milieu which is likely to play an important role in the enterprise adaption process (Goddard 1980). There is evidence that business communication varies over space, and that the cost of business travel may act as a constraint on locations which are remote from centres of specialist information (James *et al.* 1979). As yet, the link between successful business operation and patterns of managerial communication has not been sufficiently clearly articulated (Marshall 1982a); nevertheless, research in this area remains more relevant than studies of the role of material linkages in economic growth.

8.7 Corporate organization and business service linkages

Background

Business services are an important source of information and advice. Unlike consumer services, they are not tied to local levels of population and income, thus spatial variations in service employment reflect their location. Despite some decentralization of service employment, many of the benefits of the growth in the UK service sector have been concentrated in the south of the country. This reinforces the dependence of provincial economies on the core for information and advice, and limits the access of provincial areas to alternative sources of employment. In the market sector, the pattern of linkage dependencies between organizations plays an important role in explaining and supporting this pattern of development. To demonstrate this, this section examines the organization of the business service sector in provincial areas using directory sources, and the results of surveys of manufacturing and business service linkages.

The analysis is based on an interview survey of forty offices conducted in the Northern region in 1977. Major business service industries in the regional economy were covered, including insurance, banking and finance, advertising, management consultants, and R and D. A further study examining the nature of the demand for and the supply of business services was carried out in 1979/80. This involved a postal survey of 357 establishments in key manufacturing industries in the West Midlands, Yorkshire and Humberside, and the North West Economic Planning Regions, and an interview and postal survey of 378 offices in higher-order business services in the Birmingham, Manchester and Leeds conurbations (see Marshall 1979b; 1982a; 1982b; 1983). The service studies sample differing industries and are not representative of the regional population; the three-region study is additionally hampered by a low response rate. Thus the evidence from these investigations is used here only to indicate the processes at work, and no attempt is made to estimate the actual pattern of manufacturing demand and service industry supply.

Traditionally, research has focused on the role of manufacturing in explaining the pattern of business service employment. Manufacturing demand is argued to lead locally-linked and economically-dependent services. Studies of manufacturing have demonstrated that in the same manner as their material linkage, some branch plants in provincial locations obtain relatively few of their business service inputs locally, and this constrains the regional service sector. More recently, it has been argued that the relationship between manufacturing and the business service sector is more complex. Not all business services are locally linked; a number are tradeable at a regional and national level (e.g. consultancy, industrial design, advertising and computing). Further, an analysis shows that organizations in insurance, banking and finance and miscellaneous service activities, and not manufacturing industries, are the main customers for many business service firms (Marshall 1983). Thirdly, there are spatial variations in manufacturing demand for services which reflect the nature of business service supply (Marshall 1982a). Thus, branch manufacturing plants in business service centres tend to internalize services less within their organizations than branches elsewhere. Multisite organizations appear to exercise choice over the location of their service purchases and evaluate the availability of services at particular sites.

In fact the nature of the supply of business services, determined by the corporate organization of service and manufacturing firms, has had an important impact on the nature of the demand for

Table 8.3 The spatial distribution of business service offices*
(a) All firms, 1980/81

Status of office	Gt. London S.E. Eng				Rest of UK			
	M C	Mkt	Adv	Comp†	M C	Mkt	Adv	Comp†
Multisite firm								
Main office (percentage)	13.1	4.7	7.1	10.8	3.1	0.8	6.5	5.6
Subordinate office (percentage)	7.7	3.5	5.0	18.7	48.2	5.6	25.5	49.8
Single-site firm (percentage)	79.2	91.8	87.9	70.5	48.7	93.9	68.0	44.6

† This information was provided by S. Green, Department of Industry, South-East region.
N: MC (374); Mkt (584); Adv (1,402) Comp (1,490)

(b) Major business service firms§

Status of office	Gt. London† S.E. Eng			Rest of UK		
	Acc	Adv	Comp	Acc	Adv	Comp
Multisite firm						
Main office (percentage)	26.8	41.9	17.5	3.7	6.3	15.1
Subordinate office (percentage)	67.0	51.1	17.5	93.0	85.7	72.7
Single-site firm (percentage)	6.2	7.0	64.9	3.3	7.9	12.1

N: Acc (649); Adv (106); Comp (90)
* Sources: *Accountancy Age*, *Campaign* and *Computer Users Year Book*, Institute of Chartered Accountants, *Advertisers Annual*, British Research and Data, *Institute of Management Consultants Year Book*, Management Consultants Association, Market Research Society, *Computer Services Association Directory*.
§ Major firms defined as those with more than 100 employees in computer services, and by turnover for accountancy and advertising agencies

services. Business service firms in the provincial conurbation study are dependent on a small number of customers, with the three main business clients accounting for on average 47.3 per cent of the output of each office. The data are not sufficient to identify the precise nature of interorganizational dependencies. However, the position of the customers of service firms in a 'dual' economy has important implications for business service firms in the provinces.

The growth of large enterprises in the UK has not been confined to manufacturing; there are parallels in the business service sector (e.g. banking, insurance and finance). Even in services where small or single-site offices are the norm, external control is important in the provinces. Table 8.3 presents data on the corporate organization of selected business services derived from directory sources. This shows that for all services apart from marketing, multisite firms are a significant proportion of larger enterprises. Even in marketing, there are data from the Market Research Society which suggests that 39.5 per cent of its member organizations are owned by another firm, so even here external control is important. Table 8.3 also shows that, as expected, London and the South East has larger proportions of headquarters and single-site firms, while the provinces rely on branch offices. It is likely that this will have an important impact on service industry in this location.

The provincial office market

The study of business service offices in provincial conurbations confirms the importance of multiregional firms, with 53.2 per cent of employment in firms headquartered outside the local planning region. In terms of manufacturing industry markets in the North West, Yorkshire and

Table 8.4 Type of service organization used by manufacturing establishments*

Type of service organization	Manufacturing establishments (%)					
	Nos. sites		Ownership single-site firms		Status multi-site firms	
	Single	Multi	Indep	External	HQ	Branch
Regional company (Offices only within 50 miles)	66.7	55.8	61.8	42.6	46.6	34.8
National company	33.3	44.2	38.2	57.4	53.4	65.2
Nos. of establishments	300		178		104	

* A Chi-square test is significant at the 5 per cent level

Humberside and the West Midlands, service firms headquartered in Greater London account for 51.1 per cent of the main suppliers. National branch offices have been largely set up in the last twenty years, while local firms are somewhat older. Over the period 1976–80, employment in branches of national service firms grew by 44.2 per cent, compared to a 27.3 per cent increase in local firms. So a substantial part of the market once controlled by local business service firms is now controlled from outside.

It is difficult for local firms to break into the market controlled by national service companies. The dominance of national service firms based in the South East is associated with the corporate organization of industry and associated linkages. Table 8.4 shows those types of manufacturing plant which are most likely to use national service companies. Establishments which are part of multisite firms, and in particular branch plants, and within single-site companies those which are externally owned, are most likely to use national suppliers of business services. So there is a link between multisite enterprises in manufacturing and services.

Multisite manufacturing companies largely headquartered in London and the South East organize their service contracts with business service firms headquartered locally who service provincial manufacturing establishments in the regions via their branch network. This pattern of local linkage for South East-based manufacturing firms is reflected in the provincial region study by the fact that these firms have most service suppliers with a head office in London and the South East (79.5 per cent of total suppliers). To provide a wide geographic coverage, branch business service offices each deal with a particular market area and pass business for other areas to the local office with the appropriate responsibility. Since head office plays an important role in generating business, branch offices are more likely to receive business from other offices than are headquarters. Thus, 90.1 per cent of business service branches in the provinces receive business from other offices while 60.3 per cent of head offices do so.

Not surprisingly, then, branch business service offices have relatively little freedom in decision-making (see Marshall 1982b). Only in the case of activities concerned with control of the work on site do more than half of the branch offices make the main decisions. In contrast, particularly financial matters are decided elsewhere in the vast majority of cases. It is also noteworthy that branch service offices in the Northern region have much less autonomy in decision-taking than in the three-region study. In part, this reflects industrial differences in that insurance company and bank branch offices, which are only represented in the Northern region sample, have relatively little freedom in decision-taking. However, even in advertising and computer services, offices in the North have less freedom, which may indicate spatial differences in the organization of business service firms. Key regional centres such as Manchester and Leeds may be given more responsibility to generate and deal with business in more peripheral areas, while peripheral region offices such as those in the Northern region may be in many cases subordinate to offices in more established centres.

There is evidence to support the view that Manchester and Leeds offices do act as centres from which to serve less central areas. Branch service offices in the provincial conurbation study have a number of main customers outside the immediate local area, in contrast to local head offices. Some 39.7 per cent of branch office main commercial customers are from outside the local Economic Planning Region. This pattern, however, differs from location to location. Manchester, for example, serves the North of England, particularly to the west of the Pennines. Offices in Leeds have fewer non-local links and Leeds is a less developed regional centre, though it is expanding as a location from which to serve the North East. Birmingham offices have few non-local links and Birmingham is in fact frequently served by offices from London.

Demand for services

Given the limited level of autonomy in some business service branch offices, it is not surprising to find that local service purchases are low. The Northern region study permits an investigation of the use of services by business service firms. Independent service firms in the Northern region obtain relatively few services such as market research, accountancy and computing activities from their own organization, in contrast to branch offices where no personnel or market research is obtained from specialist suppliers, and other parts of their own organization are the main sources of supply. Table 8.5 suggests that as a consequence there is a substantial leakage of service demand outside the Northern region, with 69.4 per cent of the major suppliers of services to branch offices being outside it, in contrast to 76.6 per cent of independent firm suppliers being in the North. Despite the fact that it is not possible to differentiate inter- and intra-organisational linkages, this suggests that, rather like manufacturing industry, external control within business services will have reduced the demand for local specialist supplies of service activity.

8.8 Conclusions on linkages and regional development

This chapter has argued for a dynamic approach towards linkage analysis which allows for changes in the national economy. It suggests that firm linkage networks are a means of accommodating changes in demand and understanding the role of linkages in the regional economy. Such an approach has been operationalized in the context of sectoral decline in manufacturing and in the context of growth in business service industries. The analysis has been limited because the linkage studies are cross-sectional and employ different methodologies. Yet in manufacturing it was possible to identify meaningful variations in organizational responses to industrial decline. The growth of multisite organizations was also shown to create a dual economy in the business service sector. This confirms the view that linkage studies should be considered as one element in a broader study of the industrial economy, rather than as a separate field of study in their own right.

It is suggested that regional economies have become more open and sectors of the economy increasingly interdependent as manufacturing has contracted. In such circumstances, the competitiveness of the regional economy in international trade and the adaptability of its enterprises are likely to have an important bearing on its economic development prospects. Provincial region economies with high levels of external control have not taken full advantage of the growth in industrial activity during the 1960s and early 1970s due to the limited local material multipliers of certain branch plants. In addition, during a downturn in the national economy, sectors of indigenous industry are not well equipped to respond due to managerial shortcomings and insufficient expertise to diversify. This has intensified the impact of negative local linkage multipliers in the recession.

As government commitment to regional policy has declined, private and public sector agencies have sought to capitalize on local employment

Table 8.5 Location of service suppliers to business service industries in the Northern region by ownership status*

Ownership status	Northern region (%)	Non-Northern region (%)
Independent local company	76.6	23.4
Subsidiary (with a head office outside the Northern region)	50.0	50.0
Branch (with a head office outside the Northern region)	30.6	69.4
Nos. of establishments = 36		

* A Chi-square test is significant at the 5 per cent level

initiatives. Conventional regional policy has also shifted more towards encouraging existing industry in the regions. The conclusions of this chapter have some bearing on such approaches. The analysis of linked networks of firms responding to industrial decline provides a methodology appropriate to, and underlines the need for, a monitoring and rescue service in the local economy. This would offer the opportunity to save any firm at risk due to fluctuations in demand but which potentially has a sound long-term future. The evidence which suggests that the constraints on firm adjustment to recession are ingrained in managerial structures indicates the need for an active approach towards industrial development which takes policy to the factory gate. It also indicates that an industrial policy needs to be supported by educational and training programmes, and incentives aimed at removing managerial bottlenecks to enterprise adjustment such as a lack of strategic planning, a reluctance to accept outside advice or an inability to attract staff with appropriate skills.

The chapter's examination of business service firms suggests that the service sector does not present a panacea for regional employment creation. It also reminds us that initiatives targeted at small and medium-sized firms may have a limited impact where their performance depends on large enterprises elsewhere in the economy. There is a distinct corporate hierarchy in business service industry with the main offices of particularly large firms concentrated in the South East, while provincial areas are increasingly coming to rely on branch offices. The prospects for the local service sector outside London look bleak. The study identified two distinct sectors: local service industry with regional markets, and national service organizations serving a UK market. As large organizations have come to dominate the UK economy, the market for local service industry has grown less quickly than that of their national counterparts. It is particularly pertinent, therefore, to assess the impact of the expansion of national service organizations on provincial areas.

There appears to be no direct detrimental impact of the extension of external control in business services on levels of employment because this is a growth sector. There are, though, signs that service employment growth in locations like the Northern region may be limited indirectly by the organization of business service firms and their tendency to serve such areas from Manchester and Leeds. Many firms use existing branch offices to provide a presence in peripheral regions and, when business is obtained, they 'plug in' the resources of the company from elsewhere. In addition, the growth of specialist suppliers of services linked to national business service organizations is constrained by the tendency of the latter to purchase outside the local area.

1. Lever's study finds a significant variation in the pattern of supplier linkages between indigenous and externally-owned firms, but there is little difference in the proportion of linkage in Scotland.

References

Blackaby, F. (ed.) 1979 *De-Industrialisation*. Economic Policy Papers II, National Institute for Economic and Social Research. Heinemann.

Crompton, D., Barlow, A. T., Downing, S. 1976 *Component suppliers to the car industry*. Department of Industry, West Midlands Regional Office.

Dunnett, P. S. 1980 *The Decline of the British Motor Industry*. Croom Helm.

Fredricksson, C. G., Lindmark, L. G. 1979 From Firms to Systems of Firms: A study of interregional dependence in a dynamic society, pp. 155–187, In Hamilton, F. E. I., Linge, G. J. R. (eds) *Spatial Analysis Industry and the Industrial Environment, Vol. 1, Industrial Systems*. Wiley, 155–287.

Gibb, A. A., Rabey, G. F., Quince, T. A. 1977a *Survey of suppliers to the shipbuilding industry: organisational dependency*. Research Report No. 9, Small Business Centre, Durham University Business School, University of Durham.

Gibb, A. A., Rabey, G. F., Quince, T. A. 1977b *Survey of suppliers to the shipbuilding industry: organisational environments*. Research Report No. 10, Small Business Centre, Durham University Business School, University of Durham.

Gibb, A. A., Rabey, G. F., Quince, T. A., 1977c *Survey of suppliers to the shipbuilding industry: organisational change*. Research Report No. 11, Small Business Centre, Durham University Business School, University of Durham.

Gilmore, J. M. 1974 External economies of scale, inter-industrial linkages and decision-making in manufacturing. In Hamilton, F. E. I. (ed.) *Spatial*

Perspectives on Industrial Organisation and Decision-making. Wiley, 335–62.

Goddard, J. B. 1980 Technological change in a spatial context. *Futures* **12**: 90–105.

Hoare, A. G. 1978 Industrial linkages and the dual economy: the case of Northern Ireland. *Regional Studies* **12**: 167–80.

James, V. Z., Marshall, J. N., Waters, N. S. 1979 *Telecommunications and office location*. Report to the Department of Environment, Centre for Urban and Regional Development Studies, University of Newcastle upon Tyne.

Jepson, D. 1983 The Coventry local economy, the motor vehicles industry and some implications for public policy. *Regional Studies* **17** (1): 56–9.

Jones, D. T. 1981 *Maturity and crisis in the European car industry: structural change and public policy, industrial adjustment and policy*: **I**. Paper No. 8, Sussex European Research Centre, University of Sussex.

Keeble, D. 1969 Local industrial linkage and manufacturing growth in Outer London. *Town Planning Review* **15**: 163–88.

Le Heron, R. B., Schmidt, C. G. 1976 An exploratory analysis of linkage change within two regional industries. *Regional Studies* **10**: 465–78.

Lever, W. F. 1974 Manufacturing linkages and the search for suppliers and markets. In Hamilton, F. E. I. (ed.) *Spatial Perspectives on Industrial Organisation and Decision-making*. Wiley, 309–34.

Marshall, J. N. 1979a Ownership, organisation and industrial linkage: a case study in the Northern Region of England. *Regional Studies* **13**: 531–57.

Marshall, J. N. 1979b *Organisational and locational factors associated with the performance of manufacturing establishments*. Centre for Urban and Regional Development Studies, Discussion Paper No. 26, University of Newcastle upon Tyne.

Marshall, J. N. 1982a Linkages between manufacturing industry and service suppliers. *Environment and Planning A* **14**: 1523–40.

Marshall, J. N. 1982b *Corporate organisation of the business service sector*. Centre for Urban and Regional Development Studies, Discussion Paper No. 43, University of Newcastle upon Tyne.

Marshall, J. N. 1983 Business service activities in British provincial conurbations. *Environment and Planning A* **15**: 1343–1359.

Martin, R. L. 1982 Job loss and regional incidence of redundancies in the current recession. *Cambridge Journal of Economics* **6**: 375–95.

Massey, D., Meegan, R. 1982 *The anatomy of job loss*. Methuen.

McDermott, P. J. 1976 Organisation, ownership and regional dependence in the Scottish electronics industry. *Regional Studies* **10**: 319–35.

McDermott, P. J., Taylor, M. J. 1982 *Industrial organisation and location*. Cambridge University Press.

Miller, D. 1983 The role of the motor car industry in the West Midlands. *Regional Studies* **17**(1): 53–8.

Moore, C. W. 1973 Industrial linkage development paths: a case study of the development of two industrial complexes in the Budget Sound. *Tijd. Econ. Soc. Geog.* **64**: 93–107.

Moseley, M. G., Townroe, P. M. 1973 Linkage adjustment following industrial movement. *Tijd. Econ. Soc. Geog.* **64**: 137–44.

Northern Region Strategy Team 1977 *Strategic Plan for the Northern Region: Economic Development Policies, Vol. 2*. HMSO

Rabey, G. F. 1976 *Contraction poles: An exploratory study of traditional industry decline within an industrial complex*. Discussion Paper No. 3, Centre for Urban and Regional Development Studies, University of Newcastle upon Tyne.

Regional Studies Association 1984 *Report for inquiry into regional problems in the United Kingdom*. Geo Books.

Scott-Ward, J. 1978 *The automobile component industry of the U.K.; structure, growth and future*. Economic Intelligence Unit, Report No. 58, London.

Steed, G. P. F. 1970 Changing linkages and the internal multiplier of an industrial complex. *Canadian Geographer* **15**(3): 229–42.

Steed, G. P. F. 1973 Internal organisation, firm integration, and locational change: the Northern Ireland linen complex, 1954–1964. *Economic Geography* **47**: 371–83.

Stoney, P. J. M. 1982 *The employment impact of the Merseyside motor vehicle assembly industry*. Mimeo, Department of Economic and Business Administration, University of Liverpool.

Taylor, M. J. 1973 Local linkage, external economies and the iron foundry industry of the West Midlands and East Lancashire conurbations. *Regional Studies* **7**: 387–400.

Taylor, M. J. 1978 Linkage change and organisational growth: the case of the West Midlands iron foundry industry. *Economic Geography* **54**(4): 314–16.

Taylor, M. J., Wood, P. A. 1973 Industrial linkage and local agglomeration in the West Midlands metal industries. *Transactions of the Institute of British Geographers* **59**: 129–54.

Taylor, M. J., Kissling, C. 1983 Resource dependence, power networks and the airline system of the South Pacific. *Regional Studies* **17**(4): 237–50.

Taylor, M. J., Thrift, N. J. 1982a Industrial linkage and the segmented economy: 1 Some theoretical proposals. *Environment and Planning A* **14**(12): 1601–14.

Taylor, M. J., Thrift, N. J. 1982b Industrial linkage and the segmented economy: 2 An empirical reinterpretation. *Environment and Planning A* **14**: 1615–32.

Taylor, M. J., Thrift, N. J. 1983 Business organisation, segmentation and location. *Regional Studies* **17**(6): 445–66.

Townroe, P. M. 1970 Industrial linkage, agglomeration and external economies. *Journal of the Town Planning Institute* **56**: 18–20.

Townsend, A. R. 1983 *The impact of the recession.* Croom Helm.

Wood, P. A. 1978 Industrial organisation, location and planning. *Regional Studies*, **12**: 143–52.

Introduction to Industrial Enterprise

What has been termed 'the geography of enterprise' has developed since the early 1970s. This approach takes as its basic assumption the belief that the industrial enterprise is a more useful focus of study than, say, the industrial sector. In other words, therefore, there is more likely to be similarity between a multinational, multiplant chemical company and a multinational, multiplant automobile company in terms of their locational and other operating decisions than there is between a multinational, multiplant chemical company and a small indigenously-owned single plant chemical company. The four chapters which follow look at four different types of enterprise. In Chapter 9, Colin Mason looks at the small firm sector. By the late 1970s, this sector looked like 'an idea whose time has come'. The major multiplant companies in Britain were suffering from very severe international competition and often their branch plants in Britain's Development Areas were their most marginal pieces of capacity, and ripe for closure. Analyses of employment change suggested that the small firms, locally owned and managed, were doing much better at retaining employment, and the quality of employment within such firms was much better. A significant number of programmes to enhance the rate of new firm formation were set in train: they accorded with the political thinking of the time in which Conservative/Republican administrations favoured self-help and independence. Concern was subsequently expressed that increasing the number of new firms might merely increase the closure rate as small firms merely compete within the same local market, which is of a fixed size. More recently it has been pointed out that (a) small firms may be innovators and, as such, capable of servicing more than local markets, and (b) the small firm sector is still holding its employment levels better than the large, multiplant corporation sector. However, Mason points out that there are regional disparities in the rates of new firm formation: some economic stuctures appear to be more conducive to creating new entrepreneurs, and those of the typically depressed regions have not proved a fruitful environment for fostering new firm growth.

The multiplant enterprise has become the most significant enterprise in British industry in terms of output or employment and is likely to continue to play the major role in the geography of employment in the 1980s and 1990s. The diminishing power of regional policy may make their role in adjusting the spatial pattern of employment somewhat less than in the 1960s and early 1970s but there is increasing evidence that local authorities, as the implementers of urban policy, are having to think in terms of planning agreements with the major multiple enterprises. Mick Healey and Doug Watts in Chapter 10 describe the changing role of the multiplant enterprise, and reflect how geographers' concerns with the single plant enterprise in its locational choice have had to be modified to take account of decisions which take at least a national context as their perspective.

More and more, geographers have realized that decisions about industrial investment, employment and output can only be understood in the context of the world economy. The car plant in Birmingham can no longer be viewed solely as competing with those in Dagenham or Merseysde but with plants in Detroit, Spain, Brazil or Poland. As Ian Hamilton points out (Ch. 11), governments have been slow to react to the multinationality of the manufacturing sector. Multinational enterprises have emerged which in economic power (and perhaps in political power) rival small countries. The transnational flows of capital, and consequent employment, represent a fascinating area for study and one which challenges simplistic notions of how the world economy operates. Major British companies, often accused of exporting investment and jobs abroad, are on closer analysis found to earn much of their profit there, which on repatriation creates employment back in Britain.

The growth of the public sector in Britain's manufacturing is the product of two elements. Some sectors – power and steel, for example – are seen as so fundamental to the country that it is regarded as entirely appropriate that the government should manage them; in other cases, major companies fall into difficulties on a scale which makes their rescue only feasible with huge injections of public money. This crypto-nationalization has occurred within the automobile, shipbuilding and aerospace industries. By the early 1980s, almost 20 per cent of manufacturing industry employment lay in public ownership and, as Graham Humphrys describes (Ch. 12), this has a profound effect on how investment and employment decisions are taken.

CHAPTER 9

The small firm sector

Colin M. Mason

9.1 Introduction

Small firms have always played an important role in the UK economy, but it is only within the last decade that they have attracted serious attention from government, the media and academics as a potential source of job generation and wealth creation. Indeed, until the early 1970s the large industrial enterprise was viewed by government as offering the best prospects for economic growth. So, having been 'analogous to plain girls and acne-ridden boys at a high school dance . . . present in large numbers but usually . . . unnoticed' (McGuire 1976; 115), small firms are now extremely desirable partners. This swing in the industry policy pendulum has been prompted by three factors (Johnson 1978). Firstly has been the growing disillusionment with the performance of large firms. In particular, the expected economic benefits from the wave of mergers and acquisitions in the 1960s have not emerged and the restructuring and rationalization of many large firms in the face of global recession has led to large-scale redundancies. Secondly, there has been a growing concern about the political and economic implications of the increasing level of concentration in both the economy as a whole and in individual product markets, including the effects on industrial efficiency, industrial relations and, in the case of multinational enterprises, on national sovereignty. Thirdly, the positive virtues which small firms are believed to possess – notably their contributions to employment creation, innovation and job satisfaction – have attracted greater emphasis.

This renewed interest in the economic contribution of the small firm sector has, of course, been reinforced by the election in May 1979 and re-election in June 1983 of a Conservative government pledged to create a thriving private enterprise economy in which the entrepreneur would play a key role. Riddell (1983: 165) describes their objective as the construction of a society which is 'a cross between nineteenth-century Birmingham and contemporary Hong Kong, located in Esher'. However, it is frequently forgotten that the previous Labour administration set in motion attempts to rejuvenate the small firm sector, for example through the setting up of an inquiry under the chairmanship of Sir Harold (now Lord) Lever to look at the problems of small firms, appointing the Committee to Review the Functioning of the Financial Institutions (under the chairmanship of Sir Harold (now Lord) Wilson) which paid particular attention to the difficulties which small firms encounter in obtaining finance, and by introducing some measures in its later budgets to assist small businesses.

This chapter has four objectives: to provide an overview of and explanation for recent trends in the size of the small firm sector in the UK; to evaluate the role of new and small businesses in economic development; to examine their spatial distribution; and, finally, to assess the role of small firms in the uneven economic development of the UK. However, an important preliminary step is to define a 'small firm', and it is to this issue that we turn in the first instance.

9.2 The small firm – a definition

There is little consensus on what constitutes a

small firm and it is probably the case that they are easier to recognize than to define. Indeed, most countries adopt slightly different definitions, while a single definition is rarely appropriate for all purposes or for all sectors of the economy. However, few would dispute the three essential characteristics of 'smallness' proposed by the Bolton Committee which was set up in the late 1960s to inquire into the role of the small firm in the UK economy (Bolton 1971). In this *economic definition*, a small firm is firstly an enterprise with a very small share of its market. Secondly, it is an enterprise that is managed by its owner or part-owners in a personalized way and not through the medium of a formalized management structure. Thirdly, a small firm is independent in the sense that it does not form part of a larger enterprise and its owner-managers are free from outside control in taking their principal decisions.

Much of the support for the small firm sector, especially from an urban and regional development perspective, is on account of this characteristic of independence rather than because of size *per se*. Firms which are owned, controlled and managed by local individuals are seen as an antidote to the growth of concentration and the spatial division of labour within multiplant companies which have combined to convert many areas into 'branch plant economies' (Watts 1981) – regions in which the majority of industry is owned and controlled by enterprises whose decision-making functions are located in another part of the country or even in another country. Decisions on such issues as purchasing, sales, investment and retrenchment which affect the industrial development of a region are therefore likely to be made at distant headquarters locations on commercial and internal organization criteria and with little regard to the local impact. In contrast, independent, owner-managed firms are assumed to exhibit higher levels of local loyalty and commitment and greater response to local needs although, as Walker and Green (1982) point out, firms which are independent in a legal sense may nevertheless be dependent in operational terms, through a reliance on, for example, a narrow range of key customers or suppliers or on the goodwill of banks and creditors.

Because of the lack of relevant information, the Bolton Committee was unable to use its economic definition to examine the size and structure of the small firm sector and was instead obliged to use a *statistical definition* based on three quantitative indicators: number of employees, turnover and

Table 9.1 The Bolton Committee's definition of small firms and revisions by the Wilson Committee

Industry	Statistical definition of small firms adopted by the Bolton Committee (turnover at 1963 prices)	Revised definition to allow for inflation (turnover at 1978 prices)
Manufacturing	200 employees or less	—
Retailing	Turnover of £50,000 p.a. or less	Turnover of £185,000 p.a. or less
Wholesale trades	Turnover of £200,000 p.a. or less	Turnover of £730,000 p.a. or less
Construction	25 employees or less	—
Mining/quarrying	25 employees or less	—
Motor trades	Turnover of £100,000 p.a. or less	Turnover of £365,000 p.a. or less
Miscellaneous services	Turnover of £50,000 p.a. or less	Turnover of £185,000 p.a. or less
Road transport	5 vehicles or less	—
Catering	All excluding multiples and brewery-managed public houses	—

Source: Wilson (1979: Appendix 1)

number of vehicles. Small firms in the manufacturing, construction and mining industries were defined by employment; those in retailing, wholesaling, motor trades and miscellaneous services by turnover, and in road transport by the number of vehicles (Table 9.1). These thresholds, adjusted where appropriate to take account of inflation, have been widely adopted (despite certain disquiet that the 200 employee cut-off for small manufacturing firms was too high) and now represent the semi-official definition of small businesses in the UK. Despite, however, possessing the advantage of clarity, such a quantitative definition is nevertheless arbitrary. The Bolton Committee chose the 200-employee threshold in manufacturing because most enterprises below this size conformed to their criteria of being owner-managed, independent and with a small market niche, but conceded that many manufacturing firms with up to 500 or more employees also possess such characteristics. Hence, the overlap between independence and smallness is not exact. While most small firms are independently owned, not all independent firms are small. Indeed, from an urban and regional development perspective, it is important to examine the processes that have allowed certain firms to grow beyond the threshold size while still remaining independent. However, the use of a quantitative measure risks excluding such small, *growing* enterprises which have expanded beyond the original definition of smallness.

Table 9.2 The distribution of small firms in the UK by sector, 1976

Sector	Numbers employed in small firms ('000)	Proportion of total employment in small firms (%)	Proportion employed in small firms within private sector (%)
Manufacturing	1,549	27.9	22
Distributive trades	1,236	22.2	39
Miscellaneous services	1,101	19.8	43
Construction	732	13.2	49
Professional and scientific services	373	6.7	48
Others	566	10.2	—
Total	5,557	100	32

Source: Adapted from Bannock (1981a: 30)

On the basis of the Bolton Committee's statistical definition, Bannock (1981a) estimated that the small firm sector employed over 5.5 m workers, or nearly one-third of the total labour force, in the mid-1970s (Table 9.2). The manufacturing sector accounted for the largest share of this employment, although small manufacturing firms comprised a relatively minor share of total employment in the sector – just 22 per cent. Distributive trades (wholesaling and retailing) and miscellaneous services (including pubs, cafes, shoe repairers, garages, cinemas and dry cleaners) each accounted for approximately 20 per cent of total employment in small firms. However, small firms were much more significant in these sectors than in manufacturing, accounting for 39 per cent and 43 per cent respectively of their total employment. Small firms are an even more significant presence in construction and professional scientific services (doctors, lawyers, vets, secretarial colleges etc.) where they provided nearly half of all employment although comprising only 13 per cent and 7 per cent respectively of employment in the small firm sector.

9.3 Recent trends in the small firm sector

The Bolton Committee highlighted both the absolute and relative decline in the size of the small firm sector in the UK over the period from the mid-1930s to the early 1960s (the latest data available to the Committee). They indicated that the decline had been particularly steep in the manufacturing sector (Table 9.3a), although retailing and construction had also experienced similar contractions in both the absolute and relative size of their small firm sectors. The Wilson Committee (1979), which updated the Bolton Committee's data on the size of the small firm sector to take account of developments between 1963 and 1973, suggested that there had been a slight increase in the number of small manufacturing enterprises and a halt to their relative decline (Table 9.3b). However, the contraction in the number of small retail organizations had continued unabated as had their share of total retailing employment and turnover. Small firms in the construction industry had also declined further in numerical terms, although the proportion of total employment and work which they accounted for remained fairly constant. Trends since the early 1970s have confirmed the revival in the numbers of small manufacturing enterprises. However, the number of employees in small manufacturing firms has contracted since the mid-1970s, not surprisingly in view of the national context of 'de-industrialization', but this decline has been much less than from medium and larger-sized firms. Consequently, the proportion of total manufacturing employment accounted for by small firms has risen since the early 1970s as has their share of total net output (Table 9.3c). In contrast, both the number of independent retailers and their share of total retail sales have continued to decline during the 1970s (Dawson 1983) but, because of the lack of data, developments in the remainder of the service sector are shrouded with uncertainty.

The available data, albeit imperfect, therefore indicate that the decline of the small firm sector has been halted and even hints at its re-emergence,

Table 9.3 Small firms in the manufacturing sector, 1935–79

	Number of small enterprises	Percentage of total	Employment ('000)	Percentage of total employment	Percentage of total net output
(a) 1935–63					
1935	136,000	97	2,078	38	35
1958	66,000	94	1,812	24	20
1963	60,000	94	1,543	20	16
(b) 1963–73					
1963	65,700	94.1	1,627	21.3	18.0
1968	66,100	94.9	1,537	20.8	18.1
1970	70,900	95.2	1,623	21.3	18.5
1971	71,400	95.3	1,565	21.0	17.9
1972	69,000	95.4	1,528	21.5	18.4
1973	74,100	95.7	1,506	20.7	17.1
(c) 1975–81					
1975	83,400	96.3	1,558	21.9	18.1
1976	86,300	96.5	1,576	22.6	18.2
1977	86,800	96.7	1,552	22.5	18.7
1978	87,200	96.8	1,516	22.8	19.3
1979	86,800	96.8	1,498	23.1	19.5
1980	87,400	96.9	1,485	24.3	21.5
1981	87,400	97.2	1,408	25.9	22.0

Sources: (a) Bolton (1971: Tables 5.1, 5.2 and 5.3)
(b) Wilson (1979: Tables 2.1 and 2.2)
(c) Binks and Coyne (1983: Tables 5, 6 and 7); *Annual Census of Production*; *Business Monitor* PA 1002
Note: The accuracy of the data, the methods of compilation and the definition of manufacturing have varied over time.

at least in certain sectors. Moreover, official statistics understate the recovery in the small firm sector in at least two ways. First, the paucity of data on small firms in the service sector means that there is no information on the very area in the economy which offers the greatest scope for small-scale enterprise because of the limited economies of scale, the opportunities provided by new technologies and the general increase in the demand for services in a 'post-industrial society' (Curran and Stanworth 1982a). The growth of the 'black economy' since the early 1970s – economic activities for gain which are deliberately hidden from official notice to avoid payment of taxes and other dues – which to a considerable extent comprises small enterprises and self-employed tradesmen, also serves to understate the revival in small-scale enterprise since, by definition, such activities do not appear in official statistics (Curran and Stanworth 1982a). Indeed, in retrospect it can be argued that the Bolton Committee's conclusions on the decline of the small firm sector were unduly pessimistic because it was reporting precisely at the nadir in the fortunes of the small-scale enterprise in the UK (Curran and Stanworth 1982b).

Table 9.4 Proportion of manufacturing employment in small establishments: international comparisons

Country	(Year)	Percentage of employment in small establishments	
		(a) 20–199 employees	(b) 1–199 employees
UK	(1978)	24	30
West Germany	(1979)	30	—
Austria	(1980)	33	35
USA	(1977)	34	39
Sweden	(1978)	37	—
Canada	(1976)	43	48
Norway	(1979)	54	62
Japan	(1978)	54	68
Spain	(1978)	52	64

Source: Ganguly and Povey (1983: Table 3)

But despite these recent trends, the small firm sector in the UK makes a less significant contribution than in most other developed countries (Table 9.4). The proportion of the population which is classified as self-employed (excluding agriculture) is also much lower in the UK than in other developed countries (Ganguly and Povey 1983). Moreover, there is little sign that these differences have narrowed since the mid-1960s when the Bolton Committee first drew attention to this disparity.

9.4 The revival of the small firm sector: alternative explanations

The revival of the small firm in the UK – as demonstrated by the increased share of manufacturing output and employment accounted for by small businesses – need not necessarily be due to an expansion in the number of new enterprises or a decline in the number of small firms that go out of business, but might instead simply reflect the deteriorating performance of large firms. Indeed, there is quite considerable support for the latter explanation. Many large firms, particularly in 'mature' sectors (e.g. vehicles, shipbuilding, steel), have been forced to cut capacity and employment since the early 1970s in the face of the depressed condition of world trade and the superior competitiveness in export markets, and even in the domestic arena, of both low-wage, high-productivity producers from newly industrializing countries and the better designed products of firms in advanced economies. Of course, some large UK-owned companies have remained competitive, but in many cases this has been achieved at the expense of running down their UK operations in favour of overseas production.

There is also evidence to support the alternative view – that numbers of new businesses have increased in recent years. Firstly, the number of new businesses registering as limited companies has increased annually since 1974 (Table 9.5), and when averaged over five-year periods the number of new companies registered in the second half of the 1970s has been higher than for any equivalent period since the Second World War (Storey 1982a). However, this data source only provides a partial guide to trends in new firm formation. On the one hand, by including new wholly-owned subsidiary companies which are registered by established companies and also businesses which are registered but never trade, it overcounts the number of new enterprises. Even more serious, however, is the undercounting which arises from the fact that by no means every firm registers as a limited company, opting instead for partnership or sole trader status. Moreover, the propensity of firms to register varies substantially by sector: whereas over 80 per cent of small manufacturing and wholesale firms and over 70 per cent of small construction firms are limited companies, the equivalent proportions in many parts of the service sector are much lower, at around one-third or less (Bolton 1971). A further complication is that the year in which a business registers as a limited company may be an unreliable guide to its date of start-up because many new businesses begin life in unincorporated form and only subsequently adopt limited liability status (Mason 1983). In addition, by altering the relative attractiveness of different business forms, changes in company law and taxation could produce spurious increases in the number of company registrations (Johnson 1978). In short, by no means every new enterprise will appear in company registration statistics, while not every business which does register is necessarily new, small or independent.

However, the increase in new firm formation suggested by company registration data is largely corroborated by VAT statistics, where new firms are defined as additions to the VAT register. This is also an imprecise definition because it includes both genuinely new businesses and those which had been trading for some time without previously being liable to VAT, for example, because their turnover was below the VAT threshold, and also some wholly-owned subsidiaries. Moreover, it excludes sectors which are exempt from VAT, notably finance, property and professional services. In addition, annual trends can be distorted by changes in the turnover value for VAT exemption. Bearing these problems in mind, the data indicate that the annual number of new registrations has shown a fluctuating pattern since 1974, reaching a

Table 9.5 New company registrations in Great Britain, 1971–84

Year	Number of registrations	Year	Number of registrations
1971	39,445	1978	62,679
1972	54,456	1979	65,058
1973	67,349	1980	68,256
1974	42,496	1981	71,270
1975	45,090	1982	85,656
1976	55,345	1983	94,640
1977	54,520	1984	96,039

Sources: Binks and Coyne (1983: 39); *British Business* (various issues)

Table 9.6 New registrations on the VAT register, 1974–82

Year	Number of registrations	Percentage of stock
1974	147,878	11.3
1975	161,224	11.8
1976	168,847	12.1
1977	157,094	11.2
1978	149,393	10.7
1979	171,292	12.1
1980	157,096	10.9
1981	149,229	10.3
1982	138,715	9.4

Source: Ganguly (1983: Fig. 1)

peak in 1979 and declining since then (Table 9.6), although data based on an alternative definition of a birth suggest that the number of new firms has increased each year from 1980 to 1983 inclusive (Ganguly 1984). A sectoral breakdown indicates that new firms in the period 1980 to 1983 inclusive have been most numerous in retailing (20 per cent), construction (15 per cent) and other services (15 per cent), whereas the manufacturing sector accounted for just 10 per cent of the total (Ganguly 1984).

Other data sources confirm that there has been an increase in new firm formation since the early 1970s. The number of business names registered, averaged over five-year periods, has increased steadily since the Second World War (Storey 1982a) and has risen continuously since 1974 after a drop in 1973–74 (Bannock 1981b), although here again some overcounting inevitably occurs because the registration of a business name does not necessarily indicate that a new economic activity immediately takes place. Data for the East Midlands similarly indicate that new manufacturing firm formation has been higher during the 1970s than in the previous two decades, although the *rate* of new firm formation is not as high as that achieved in the years immediately following the end of the Second World War when the wholly exceptional circumstances of large numbers of demobbed servicemen and pent-up demand existed (Fothergill and Gudgin 1982). However, Gould and Keeble (1984) find no evidence of an increase in the number of new manufacturing firms formed during the 1970s in East Anglia.

With this one exception, these various data sources confirm that new firm formation has increased since the early 1970s. This in turn lends support to the view that the revival of the small firm sector has been achieved, at least in part, by a rising trend in new business formations and is not simply due to the poor performance of the large firm sector. But accompanying this upward trend in new firm formations has been an increase in closures. Annual numbers of company liquidations increased between 1970 and 1976 and particularly steeply from 1980 to 1982, and have continued to rise in 1983 and 1984. However, the number and the proportion of companies deregistering for VAT between 1974 and 1982 both failed to display a consistent upward pattern (Ganguly 1983). Unfortunately, neither data source separately identifies the number of small businesses which closed; consequently it remains unclear the extent to which trends in the size of the small firm sector have been influenced by closures.

Although the number of new businesses formed in recent years has increased, this does not imply that the small firm population is necessarily particularly youthful. Indeed, whereas most new firms are small, the majority of small firms are not new. A survey conducted for the Bolton Committee (which, of course, predated the recent increase in new firm formations) noted that the median age of small manufacturing firms was 22 years, and that only one-quarter could be regarded as recent start-ups being less than 10 years old. Small firms in the motor and retail trades had similar median ages to that in manufacturing, in wholesaling it was 29 years and in the construction industry the average case was 69 years old. A more recent survey (undertaken in 1978) by Watkins (1982) confirms that the majority of small firms have been established for many years. From a sample of 231 small manufacturing firms, fewer than one in ten were less than 10 years old, but over half were post-Second World War creations. There was also a significant proportion, around 20 per cent, that were formed prior to 1900. However, it is conceivable that because such surveys rely for their sampling frame on trade directories and chambers of commerce and Industry Training Board membership lists, they will be biased towards the larger and longer established firms and therefore understate the proportion of young businesses. This view is supported by data made available to the author by a government agency with whom *all* manufacturing firms must register and which (in theory at least) is therefore comprehensive; this indicates that nearly one-third of independent manufacturing firms in south Hampshire in 1979 were post-1971 creations.

9.5 Some factors in the revival of the small firm sector

The expansion in the number of small businesses and the increased share of economic activity which is now accounted for by the small firm sector – a result, at least in part, of the upward trend in new firm formation since the early 1970s – is the product of a wide variety of factors. Undoubtedly, one such factor has been the rise in unemployment during the 1970s and early 1980s and the associated perceived greater insecurity of those still holding a job. Many redundant workers have been 'pushed' into self-employment because of a lack of alternative employment opportunities although, more positively, unemployment has in some cases provided individuals with both the stimulus and the opportunity (redundancy payments providing the launch capital) to set up a business and thereby fulfil a long-held but latent ambition. Regression analysis of time-series data on annual rates of new firm formation and unemployment (usually with a one-year time-lag built into the model) has demonstrated the statistical association between new firm formation and unemployment (Johnson and Darnell 1976; Harrison and Hart 1983).

At a micro-scale, however, there is little evidence of any link between unemployment (or its threat) and new firm formation. On the basis of a review of studies which have examined the subsequent fortunes of workers made redundant, Johnson (1981: 7) concludes that 'it is quite clear that self-employment was not an important avenue for re-employment'. However, the focus of such studies on *manual* workers might serve to overstate the case because other research has shown that redundant non-manual workers have a greater propensity to start a business (Johnson and Rodger 1983).

Micro-scale surveys of the new firm formation process are more equivocal about the relationship between unemployment or redundancy and new business start-ups. For example, in the relatively prosperous area of south Hampshire, a survey of fifty-two manufacturing firms started in the period 1976 to 1980 found only three that were formed because the founder had been made redundant from his previous employment (Mason 1982). It is tempting to relate this low proportion to the below-average unemployment in the area, but a survey of firms started between 1976 and 1979 in Greater Manchester and Merseyside found that only 9 per cent of founders cited either redundancy or the bankruptcy/failure of a previous firm as the reason for starting their own business (Lloyd, 1980). On the other hand, a survey of new manufacturing firms in the Northern region which were started in the early 1970s noted that 35 per cent of founders had been employed in factories which closed either at, or subsequent to, the formation of the new business (Johnson and Cathcart 1979a). Similarly, a survey of the founders of 100 new manufacturing firms in Greater Nottingham which were started in the period 1978 to 1982 found that eighteen were started because the founder had been made redundant, a further twelve because the founder was either unemployed or in insecure employment, while five were prompted by the closure of their own business (Binks and Jennings 1983). Arguably, the link between unemployment and new firm formation might be even greater in the service sector because of the generally lower skills and launch capital required. Certainly, redundancy studies indicate that the majority of workers made redundant who do become self-employed establish their businesses to undertake service activities (Johnson 1981). However, a study of new firms (excluding retailing) in Cleveland that were started between 1971 and 1977 indicates that whereas one-third of the founders in both the manufacturing and construction sectors had been unemployed immediately prior to setting up their new business, the equivalent proportions in both distribution and 'other' services was 29 per cent, and just 9 per cent in financial, professional and scientific services (Storey 1982a).

Rising unemployment is therefore of some significance as a factor in the increase in new business formation, although its influence is by no means consistent. Moreover, the proportion of new businesses which have been started because the founder is unemployed is relatively small, while the proportion of redundant workers who set up a business on their own account is extremely limited.

The processes of industrial restructuring by large firms in response to the recession is also linked in various ways to the increase in new businesses (Shutt and Whittington 1984). First, the search for reduced costs and increased efficiency has prompted some large firms to withdraw from certain markets which are expensive to serve; this includes distant geographical market areas which incur high distribution and servicing costs and particular market segments where the level of demand does not warrant large-scale production. Such withdrawals frequently create limited market gaps which can be profitably filled by small firms (Binks and Coyne 1983). A variant of this trend has been for large companies to divest themselves of subsidiary companies by selling them as going concerns to their management teams with backing from financial institutions. This process, which is termed the management buyout, has generally occurred in one of the following situations: where the subsidiary's activities are peripheral to the company's main business; where it has failed to produce the expected profits; or where the parent company needs to raise cash. The number of management buyouts has increased from thirteen in 1977 to approximately 200 in 1983 (Wright *et al*, 1984). Many large companies are also 'slimming down' in other ways and, in so doing, have created further opportunities for small firms. For example, certain activities which were formerly undertaken 'in house', such as printing and cleaning, are now either subcontracted out to small firms or else are undertaken by former employees working on a self-employed basis. Indeed, many large companies devote substantial resources to encourage their own employees to consider self-employment, although this is frequently in the context of an attempt to reduce the adverse impact at both the individual and local scales of their own contraction. Forms of assistance include training courses on business skills, evaluation of business ideas, provision of resources (e.g. machinery, premises, finance) and orders (Cross 1982; Sargent 1982).

Both the increase in unemployment and the restructuring of large firms are largely a function of the present economic recession. However, a number of other factors which favour small-scale enterprise are a result of longer-term changes in the structure of the economy. The emergence of post-industrial society has involved the decline of large enterprises engaged in the mass production of capital-intensive items and the growth of the service sector. In particular, considerable opportunities for small businesses have emerged in recent years in the expanding knowledge-based activities and professional services of the quarternary sector, and also as a result of demands for new types of personal services (Chappell 1983). Technological change has also increased the opportunities for small firms by reducing the 'minimum efficient scale' of production in many industries. For example, major reductions in the cost of electronics and developments in micro-electronics have made feasible their introduction in more basic equipment, creating a new generation of scaled-down, flexible modern machinery well-suited to small-batch production (Bolland 1983). The development of new materials has also helped to create opportunities for small firms. One illustration of this is the use of plastics instead of steel for many purposes which has provided considerable openings for small injection-moulding and vacuum-forming plastics firms (Bolton 1971). Increased consumer resistance to mass-produced products (at least amongst the affluent middle class) has created an expanding source of demand for small firms which offer tailor-made or regionally-distinctive products. This latter feature is clearly observable in the food and drink industry where the demand for locally-produced and processed foods has increased in recent years. Similarly, consumer dissatisfaction with the quality of beer produced by the large brewing firms, spearheaded by the CAMRA movement, identified a market gap for locally-distinctive, cask-conditioned ales which is being filled by increasing numbers of small-scale breweries (Bolland 1983).

A further significant factor in the revival of small businesses is the emergence (or rediscovery) of alternative forms of business organization, which have been responsible for increasing the numbers of individuals for whom self-employment is now a realistic option. Management buyouts, which were discussed earlier, have been able to tap the latent entrepreneurial drive of managers who were formerly responsible to a distant parent company (Binks and Coyne 1983) and who in all probability would have remained as employees if the opportunity to purchase their own firm had not arisen. The re-emergence of worker cooperatives – enterprises in which the ownership and control of the assets is spread among the people who work in it – has had the same effect. By providing the opportunity for risk-sharing, it is extending the pool of potential entrepreneurs to include industrial workers who might otherwise have been deterred from setting up a business on their own account (Binks and Coyne 1983). Over 900 worker cooperatives were in existence in 1984 compared with only 200 in 1979; most are in the service sector and the vast majority have less than twenty workers. This rapid expansion in the numbers of worker cooperatives can be ascribed to a variety of factors including the desire of those who are unemployed to find alternative work, the increasing interest in industrial democracy, self-management and alternative methods of industrial organization, and the active promotion of the cooperative ideal by various organizations and accompanied by support across the political spectrum (Chaplin 1982; Whyatt 1983).

Franchising is another example of an alternative,

rapidly-growing form of business organization which has simultaneously created small business opportunities and been instrumental in encouraging individuals who might otherwise have remained in employment to become self-employed. Essentially, franchising consists of an organization (the franchisor) with a market-tested business package centred on a product or more usually a service establishing contractual relationships with franchisees (typically aspiring small businessmen) who set up their own *independent* businesses to operate under the franchisor's tradename and market the product or service in accordance with the franchisor's 'blue print'. In return, the franchisee normally pays a royalty to the franchisor. The main advantages to the franchisor are the rapid achievement of national coverage of his product or service with most of the capital put up by the franchisees, and the elimination of many of the motivational and personnel problems which increasingly arise when face-to-face customer contact occurs at a large number of outlets remote from head office. From the franchisee's perspective, the system offers a 'sheltered' form of self-employment with the benefits of trading under a nationally advertised trademark or tradename and access to the franchisor's 'know how' and back-up resources. It has therefore created opportunities for small businesses to be established by individuals who might otherwise have remained as employees throughout their working lives (Curran and Stanworth 1982a). Indeed, surveys of franchisees confirm that many, although by no means all, would not have gone into business for themselves without the backing and business opportunities provided by the franchisor (Hough 1982). It has been estimated that there are currently around 80,000 businesses operating on a franchised basis in the UK (Curran and Stanworth 1982a).

The revival of the small firm sector has also been helped by improvements in the availability of finance for small businesses, at both the start-up phase and subsequently. The government, through its Loan Guarantee Scheme which underwrites 70 per cent (originally 80 per cent) of the value of approved loans to small firms, has attempted to encourage the banks to adopt less conservative lending practices and enabled small business owners to obtain loans without the need to provide their homes as security. However, the banks dispute the claim that the small firm sector has been starved of finance and point to their own range of services to demonstrate their responsiveness to the financial needs of small businesses (Greenhow 1982). The availability of risk capital has also been improved. For example, the existence of tax incentives under the Business Expansion Scheme has attracted back wealthy private individuals to invest in small businesses with growth potential. In addition, the recent development of a venture capital industry in the UK has resulted in the raising of over £300 m from the big financial institutions (e.g. pension funds, insurance companies) in the period 1979-83 inclusive for investing in potentially successful small firms. Similarly, the creation of the Unlisted Securities Market (USM) in November 1980 – a junior Stock Exchange with less onerous admission requirements – has made it easier for rapid-growth small firms to raise capital and for the directors and other investors to realize the value of some of their own shares in the company. Before the advent of the USM, the only way for the owners of such rapid-growth enterprises to realize the value of their shares would have been to sell out to another company.

Finally, both the previous Labour government, towards the end of its period in office, and the present Conservative administration have introduced a wide variety of measures to assist the small firm sector, primarily by removing sources of discrimination against small businesses and by creating a 'climate of enterprise' to encourage more individuals to set up small businesses. Five broad areas have attracted government attention. Firstly, free advice and information is provided via the Department of Trade and Industry's regional network of Small Firms Information Centres and the Small Firms Counselling Service and also by over 300 local enterprise agencies that have been set up with the support of groups of large firms throughout the country. The government has encouraged this development by allowing corporate donations of cash and seconded personnel to enterprise agencies to be tax deductible. Secondly, attempts to reduce the amount of 'red tape' affecting small firms have included a reduction in the requests by government departments for statistical information, the periodic raising of VAT and other tax thresholds, more sympathetic and quicker planning decisions, and the relaxation of parts of the Employment Protection Act for very small firms. Thirdly, tax concessions have been made available to encourage an increase in the stock of small factory units. Fourthly, government-funded enterprise training courses have sought to teach potential business founders the basics of business management so that their new venture might be set up on a more secure footing and to improve the skills of existing small firm owner–managers. Finally, improvements in the availability of finance for small firms have been achieved through tax incentives to private individuals who invest in them (the Business Expansion Scheme), the underwriting of bank loans to small firms that are outside their normal lending criteria (Loan Guarantee Scheme), grants to subsidize the cost of purchasing new machinery (Small Engineering Firms Investment Scheme), and a £40 per week payment for a year to unemployed individuals who set up a new business to offset the loss of their unemployment benefit (Enterprise Allowance Scheme). These measures have been complemented by the efforts of many local authorities to assist small firms, notably by the construction of new factory units and the renovation of older premises as small workshops.

Government support for the small firm sector has undoubtedly been a significant factor in its revival. For example, the guiding principle of the Loan Guarantee Scheme has been 'additionality' – encouraging lending that would not otherwise have taken place because of the high risk or lack of security available to the bank, rather than diverting small firms from existing sources of finance. An assessment of the scheme on behalf of the Department of Trade and Industry suggested that a minimum of 56 per cent of loans and a maximum of 81 per cent satisfied the 'additionality' criteria (Robson Rhodes 1983a). The consultants concluded that it was 'undeniable that the scheme enabled businesses to start which would never have done so otherwise' (Robson Rhodes 1983; 10). The Enterprise Allowance Scheme has similarly encouraged many unemployed individuals who were previously deterred by the loss of unemployment benefit and the uncertain or irregular earnings from self-employment to set up their own business; an evaluation of the pilot version of the scheme concluded that of the firms which set up, about half did so only because of the availability of the allowance (Department of Employment 1984). The Small Firms Counselling Service has also enabled a considerable number of established firms to survive and helped to create many new businesses (Howdle 1979). Arguably, government support for small businesses might have had an even greater impact but for three restrictions. First, much of the assistance has been directed towards manufacturing and manufacturing-related activities (Beesley and Wilson 1982). Many industries which offer the greatest scope for small businesses are therefore excluded from government help. Second, most of the measures have involved direct financial assistance (Beesley and Wilson 1982) but there are other equally serious constraints on small firm development such as marketing and skilled labour shortages which also require to be tackled. Third, much of the assistance has been directed at encouraging new ventures, with relatively little help available for established small firms. But as some observers have noted, the objective of increasing the number of new businesses through direct measures will result in the displacement of many established small firms (Storey 1983a).

In summary, it would appear that while some of the factors behind the revival of the small firm sector are associated with the economic downturn and may therefore only be temporary, other forces are related to more fundamental changes in the structure of the economy and in the attitudes of government, big business and the financial institutions. It is this latter category of factors that are the more significant: not only are they more numerous, but it is likely that their effects will be longer lasting. In turn, this suggests that the revival of the small firm sector is unlikely to be threatened during the remainder of the century.

9.6 The economic impact of small firms

The precise contribution of small firms to both national economic regeneration and urban and regional development is a source of considerable controversy. The popular view, reflected in statements by politicians and by small firm lobby groups, claims that the sector is very significant on account of its important roles in job generation, competition, innovation and industrial diversification (see House of Commons 1983). But, as this section shows, recent research suggests that many of these claims are grossly overstated.

Job generation

The major role of small firms in job generation was originally highlighted by Birch (1979), an American economist, in a widely-quoted (but frequently misinterpreted) study covering both manufacturing and service industries. His main conclusion indicated that two-thirds of the *net* new jobs in the USA between 1969 and 1976 were created by the small firm sector – defined as new businesses and firms with less than twenty

Table 9.7 Manufacturing employment change by size of firm

Area	Time period	Size of firm					
		0–20	21–50	51–100	101–500	501+	Total
		(% of employment change)					
East Midlands*	1968–75	+2.7	+2.3	+1.5	−2.2	−5.9	−1.5
North of England†	1965–78	+8.1	+3.0	−5.7	−4.9	−15.2	−10.7

Sources: * Fothergill and Gudgin (1979); † Storey (1983b)
Note: Openings are placed in their end-year size band but surviving firms and closures are placed in their start-year size band.

employees. Research in the UK, which has been confined to the manufacturing sector and to specific regions and subregions, similarly concludes that most of the net employment growth has occurred in the small firm sector, while employment losses have been concentrated in medium-sized and larger firms (Table 9.7). However, this employment growth has been both modest and insufficient to offset more than a limited proportion of the jobs lost through plant closures in the corporate sector. One illustration of this point is provided by Storey (1982a) who noted that a single retrenchment decision by the British Steel Corporation in 1980 resulted in the loss of 3,000 jobs in Cleveland, whereas surviving new manufacturing firms formed in the period 1965 to 1976 had created less than 2,000 jobs by the latter year, equivalent to just 1.7 per cent of the area's total manufacturing employment. Admittedly, the low rate of new firm formation in Cleveland may make it an extreme case, even amongst depressed regions. Nevertheless, there are no examples of new firms accounting for more than 5 per cent of end-year employment in an area within a period of up to ten years. Indeed, in East Anglia, one of the most prosperous regions in the UK, new firms created during the 1970s accounted for only 4.7 per cent of the region's total manufacturing employment in 1981 (Gould and Keeble 1984).

One of the main reasons for this limited contribution to employment creation is simply that employment growth declines with the age of the firm. Storey (1983b) has noted that most new firms reach their employment peak within six or seven years of their formation and achieve little further growth after that, at least up to year eleven when his analysis ended. Fothergill and Gudgin (1982) provide further support with evidence from Leicestershire where new manufacturing firms formed in the period 1968 to 1975 resulted in the addition of 9,400 jobs by the latter year whereas firms started between 1947 and 1968 only created 3,800 additional jobs during the subsequent seven years. Moreover, firms formed before 1947 actually recorded a decline in employment between 1968 and 1975. It is therefore new firms that have been responsible for most of the job creation within the small firm sector (Storey 1983b), but their employment contribution has been limited by two further factors. Firstly, the typical new firm is extremely small and few expand to become major employers. On the basis of data for Cleveland, Storey (1982a) suggests that the probability of a new firm having 100 or more employees within ten years of start-up is between one-half and three-quarters of 1 per cent. This view is corroborated by Fothergill and Gudgin (1982) who show that in the East Midlands fewer than 1 per cent of post-1968 start-ups employed more than 100 workers by 1975. However, it is this very small group of rapidly-expanding enterprises that account for a disproportionately large share of the total number of jobs created by new firms. Secondly, most new firms have a short life-span. Further evidence from Storey's work in the Northern region shows that at least 30 per cent of new manufacturing firms born in any one year will die by year four (Storey 1983b), with the resultant loss of employment in such firms offsetting the jobs provided in surviving enterprises.

Skill levels

On account of their local ownership and control, small firms provide the entire spectrum of employment functions and thereby enrich the occupational choice within a local labour market. This is in contrast to many externally-owned branch plants which may only provide a truncated range of job types, comprising a large proportion of low-skill assembly functions and few managerial and technical occupations. Small firms can therefore be regarded as playing a valuable role in regional development. However, the manual jobs offered by small firms are generally for skilled and semi-skilled workers. For example, a survey of new manufacturing firms in Greater Manchester and Merseyside found that 60 per cent of all jobs were in these categories whereas only 13 per cent were filled by unskilled workers (Lloyd 1980). It therefore seems unlikely that the small firm sector will be able to achieve a noticeable reduction in what is the largest category of unemployed workers – the unskilled.

The working environment

In contrast to large enterprises which are viewed as being prone to industrial relations problems and low levels of worker satisfaction, small firms are believed to provide a congenial and harmonious working environment which offsets their lower wages and lack of fringe benefits. Here again, recent research suggests that this view has also been overstated. Firstly, even if wage levels are lower in small firms – and there is a lack of unambiguous evidence on this point – it is clear that employees do not regard the intrinsic attractions of working in a small firm as compensating for lower material rewards. Moreover, there is little difference between large and small firm employees in their level of job satisfaction once age and marital status are controlled for (Curran and Stanworth 1981a) and there is no evidence either that the relations between workers are any warmer in small firms than in large firms or that small firm employees have closer social relations with the owner–managers (Curran and Stanworth 1981b). In addition, small firm employees are disadvantaged because of the poorer quality of apprenticeship schemes than in large enterprises and by their greater difficulties in obtaining release to attend outside training courses (Curran and Stanworth 1981c).

Statistics on the strike-proneness of firms of different sizes certainly highlight the low level of industrial conflict in small enterprises. However, this does not necessarily confirm that industrial relations are more harmonious in small firms. Indeed, Scott and Rainnie (1982) argue that the lack of strikes in small firms arises from the difficulty which workers in such firms face in organizing collective action. Conflict in the workplace may therefore be reflected in alternative ways in the small firm sector, notably by increased labour turnover which, as Curran and Stanworth (1981c) show, is much higher in small firms.

It therefore appears that notions concerning an inverse relationship between size of firm and job satisfaction or quality of employment are misleading. While not necessarily offering poorer quality employment than large firms, the claim that small firms provide a superior working environment is not supported by recent research findings.

Industrial diversification

Small and especially new firms are thought to contribute to the diversification of local economies by their ability to seek out new opportunities for the provision of goods and services. In fact, various studies in a number of different regions have shown that new and small firms are concentrated in a narrow range of industries which are characterized by ease of entry, and their distribution is not, in general, related to the growth rate of an industry (Gudgin 1978). For example, in East Anglia, new manufacturing firms formed during the 1970s were numerically most significant in mechanical engineering (22 per cent), paper, printing and publishing (18 per cent), timber and furniture (11 per cent) and metal goods (10 per cent), none of which could be regarded as growth sectors, whereas only 8 per cent were established in the electronics industry (Gould and Keeble 1984). Similarly in south Hampshire, mechanical engineering contained the largest share of new manufacturing firms (21 per cent), followed by the metal goods and timber and furniture industries (18 per cent and 13 per cent respectively) while electronics accounted for only 11 per cent of the total (Mason 1982).

Moreover, there is evidence that industries with a high proportion of new firms also display a high closure rate (Gudgin 1978). This implies that new businesses might simply be replacing other small firms that have gone out of business and are therefore perpetuating the industrial structure of an area rather than acting as a source of diversification. Binks and Jennings (1983) have taken this link one stage further by suggesting that, particularly since the onset of recession and the resultant rise in company failures, many new businesses have benefited from the use of cheap second-hand machinery and low-cost premises released by bankrupt firms. These abnormal cost advantages together with the willingness of their owner–managers to work for limited personal financial return has enabled them to undermine established small firms in the same industry, forcing some into liquidation.

Innovation

New and small firms are also believed to make a major contribution to innovation and to the development of 'sunrise' industries based on new technologies. However, this notion is probably only applicable in the USA where small firms do play an important, although declining, role in innovation. In the UK their role is much less significant but there is some suggestion that the relative contribution of small firms to innovation might have increased since the early 1970s (Rothwell and Zegveld 1982). Nevertheless, at a regional or subregional scale in the UK, the proportion of innovative new or small firms is limited. For example, a survey of new manufacturing firms in the Northern Region found that only 12 per cent were based on a technical innovation (Johnson and Cathcart 1979a). In Greater Manchester and Merseyside, only one in six new manufacturing firm owner–managers claimed to have set up their business on the basis of a new idea, while in south Hampshire the proportion was 23 per cent. However, in both these surveys it was left to the founder to decide whether his company was innovative and it is probable that some were at least a little ambitious in their claims (Lloyd and Mason 1984). Moreover, following through the subsequent fortunes of such innovative firms suggests that in many cases their performance has been characterized by difficulty or failure (Johnson and Cathcart 1979a; Scott 1980). Reasons for this are likely to be associated with the differences in the phasing of research and developmment costs and revenue from product sales, the potentially rapid changes in both demand and the level of competition, the high costs of marketing a new product and the large amount of expenditure on R and D which prevents the simultaneous development of more than one product (Oakey 1984). These factors combine to place most small, innovative enterprises in a highly vulnerable position. However, an innovative firm which is able to overcome these problems does have the ability to expand very rapidly, and even form the basis for medium- to long-term regional economic growth. According to Oakey (1984: 250): 'one new firm in a depressed area with the growth potential evidenced in the past by Fairchild or Texas Instruments could change the image and employment structure of a region within ten years, if correctly nurtured.'

Competition

It is also widely suggested that small firms provide a source of competition to larger firms in their industry, limiting their ability to raise prices or to slip into inefficiency. In addition, small firms are thought to enhance consumer choice. Again, such views are challenged by research findings which indicate that a large part of the small firm sector is in fact complementary to large companies rather than in competition with them. For example, Davies and Kelly (1972), in a survey for the Bolton Committee, found that over half of their sample of small manufacturing firms encountered their main source of competition with other small firms, while only 38 per cent were competing directly against large companies. Moreover, many small firms, especially those engaged in one-off and small-batch production generally on a subcontract basis, are dependent on a handful of large firms for their sales. However, evidence from Lloyd and Dicken (1982) suggests that this characteristic is related to the age of the firm. They show that 60 per cent of new manufacturing firms (up to five years old) were dependent on a small number of key customers, but this proportion fell to 40 per cent amongst young firms (5 to 15 years) and 16 per cent amongst established firms (more than 15 years old).

Sales patterns

Research has also highlighted the reliance of new and small firms on local markets. For example, 41 per cent of new manufacturing firms in the East Midlands primarily served markets within the region, while only 19 per cent engaged in any exporting (Fothergill and Gudgin 1982). Similarly, 60 per cent of new manufacturing firms in south Hampshire contracted upwards of three-quarters of their business within the South East and only 10 per cent engaged in significant exporting activity (Mason 1982). In peripheral regions, dependence on local markets is even more marked. For example, Johnson and Cathcart's (1979a) survey of new manufacturing firms in the Northern region found that half sold three-quarters or more of their output within the region. Similarly, Lloyd and Dicken (1982) noted that over half of the new manufacturing firms which they surveyed in Greater Manchester and Merseyside contracted over half of their sales within their respective conurbations. Moreover, this characteristic was not confined to new firms; young and established businesses in the two conurbations displayed equally high levels of dependence on local markets.

This geographical pattern of sales therefore seriously limits the contribution of small firms to both regional and national economic development. Moreover, it implies that without an increase in demand, new firms may well displace other local small businesses which also serve the local or regional market. On the other hand, contributions to regional economic development are likely to arise from the displacement of regional 'imports' and by undertaking subcontract work for large local firms which aids the latter's efficiency. However, much of this will simply displace the output of competing firms in other parts of the country. Any contribution to the British economy is limited to the minority of small firms which obtain a significant proportion of their sales overseas, although indirect benefits arise from those which successfully compete with imports and from those which supply existing large firms cheaply and efficiently, thereby strengthening the international competitiveness of these companies (Fothergill and Gudgin 1982).

Summary

The clear conclusion to emerge from this review of research findings is that small firms do not fulfil the claims of their enthusiasts, but neither do they necessarily confirm all of the pessimistic views of their critics. Certainly, there is a large measure of agreement that new and small firms create relatively few new jobs, most remain small and a high proportion fail. In addition, small firms make a limited contribution to increased competition and exporting, and their dependence on local market opportunities creates displacement effects. But if a longer-term view is taken, the economic contribution of small firms becomes more significant. For example, Fothergill and Gudgin (1982) estimate that firms founded since 1947 accounted for nearly one-quarter of total manufacturing employment in Leicestershire by 1975. Moreover, the generally pessimistic view of the economic impact of small firms in the aggregate overlooks the very considerable contribution of the small minority which do achieve rapid growth, in terms of their employment creation, export activity and development of new technologies.

9.7 The geography of small firms

The evidence presented in the previous section indicates that the formation and growth of new small businesses does make a significant contribution to local and regional economic development over the longer term, with the main impact accruing from the minority of small enterprises that do achieve substantial and rapid growth. However, the precise contribution of small firms to economic development varies from place to place because of the substantial variations that exist both in the rate at which new businesses are formed and in their subsequent performance. This section therefore examines the role of the small firm sector in the differential urban and regional economic performance in the UK.

Spatial variations in new firm formation

VAT registration statistics represent the most widely used source of information on spatial variations in new firm formation. However, this data source is by no means ideal as it includes as new start-ups, businesses which have been trading

Table 9.8 Regional distribution of new business starts, 1980–83

Region	New firm formation rate* (average of annual rates, 1980–83 inclusive)	Location quotients† (average of annual values 1980–83 inclusive)
South East	13.7	1.15
East Anglia	10.6	0.89
South West	10.6	0.88
West Midlands	12.0	1.00
East Midlands	11.4	0.95
Yorkshire–Humberside	11.5	0.96
North West	10.8	1.03
North	9.3	0.93
Wales	7.9	0.82
Scotland	8.4	0.84
Northern Ireland	5.1	0.64
UK	9.9	—

Source: Derived from Ganguly (1984: Table 4).
* Business starts as a percentage of the regional stock of businesses.
† Regional share of UK starts divided by regional share of UK stock of all businesses.

for some time without previously being liable for VAT (e.g. because their turnover was below the VAT threshold) and some wholly-owned subsidiaries as well as genuine new firms. In addition, some industries are zero-rated for VAT (e.g. food, printed matter, children's clothes) and others are exempt (e.g. insurance, property), although many firms in these sectors voluntarily register and so are included in the statistics.

The most recently published statistics from the VAT register (Ganguly 1984), covering the period 1980 to 1983 inclusive, highlights the South East as having the highest regional rate of new business starts, while the lowest rates were found in some of the peripheral regions – notably the North, Scotland, Wales and Northern Ireland (Table 9.8, column 1). Moreover, the South East had considerably in excess of its 'fair share' of new businesses: its proportion of the national total was well in excess of its share of the UK stock of all businesses. In contrast, with the exception of the North West and the West Midlands, every other region has smaller shares of the national total of start-ups than their shares of the UK stock of all businesses (Table 9.8, column 2).

The Department of Trade and Industry's data on 'enterprises new to manufacturing' provides a further source of information on regional variations in new firm formation. As its name implies, the scope of this data is limited to the manufacturing sector. It also suffers from a number of other disadvantages (see Lloyd and Mason 1984), notably its restriction to firms with eleven or more employees (the cut-off is twenty employees in Greater London and fifty in the West Midlands Metropolitan County), the inclusion of manufacturing establishments set up by previously non-manufacturing firms, and the omission of some firms that satisfy the criteria for inclusion. This data source also points to the generally southern bias in new firm formation, with the highest rates during the 1966 to 1977 period found in East Anglia and the South East while lowest rates were recorded by (in ascending order) the North West, South West, West Midlands, Yorkshire–Humberside and the North (Johnson 1983).

Another source of information on spatial variations in new firm formation involves a comparison of independently-conducted studies of new firms in specific subregions. This procedure is somewhat problematical, however, because most studies have used different data sources, while definitions of new firms may also vary (Mason 1983) and time periods rarely coincide exactly. Nevertheless, this exercise also confirms that new manufacturing firm formation rates are generally highest in the south and southeast quadrant of Britain and are lowest in northern England and Scotland (Table 9.9).

Spatial variations in new firm formation are also in evidence below the regional scale, with the contrast between cities and rural areas being particularly prominent. Fothergill and Gudgin (1982) point to the higher rates of new manufacturing firm formation in rural areas and small towns in the East Midlands compared to the

Table 9.9 Regional comparisons of new manufacturing firm formation

Area	Time period	Number of surviving new firms	Percentage of end-year employment in new firms	Standardized firm formation rate*
Cambridgeshire[1]	1971–81	313	5.2	0.57
East Midlands[2]	1968–75	1,650	4.2	0.42
South Hampshire[3]	1971–81	333	3.5	0.34
Norfolk[1]	1971–81	208	3.5	0.30
Suffolk[1]	1971–81	182	3.1	0.28
Durham[4]	1965–78	236	4.4	0.25
Coventry[5]	1974–82	220	3.5	0.24
Tyne and Wear[6]	1965–78	486	3.6	0.17
Cleveland[7]	1965–78	165	2.8	0.10
Scotland[8]	1968–77	504	2.2	0.08

Sources: (Amended from Gould and Keeble (1984: Table 3, p. 193)
[1] Gould and Keeble (1984);   [2] Fothergill and Gudgin (1979; 1982);   [3] Mason (1982);   [4] Storey (1982b);   [5] Healey and Clark (1984);   [6] Storey (1982c);   [7] Storey (1982d);   [8] Cross (1981)
* Firm formation rate divided by the number of years in the study

cities (Leicester and Nottingham). Gould and Keeble (1984) identify a similar gradation in birth rates within the largely rural region of East Anglia, with the rate of new firm formation highest in rural areas and lowest in large towns (more than 50,000 population). Superimposed upon these urban–rural variations in new firm formation is a second contrast, between large and small plant dominated areas. Fothergill and Gudgin (1982) show that in large plant dominated towns in the East Midlands – towns with over half of their manufacturing jobs in factories which individually employ 500 or more people – formation rates were little more than one-third as high as those in other parts of the region. They go on to suggest that if this relationship holds in all parts of the country then many, although by no means all, of the areas with the lowest rates of new firm formation will be in peripheral industrial regions such as Teesside, Tyneside, Merseyside and Clydeside.

Explanations for such variations in the propensity of different areas to generate new firms are varied, but Table 9.10 summarizes the main factors (see Lloyd and Mason 1984; Storey 1982a; ch. 10 for a detailed discussion). Firstly, the founders of new firms are most likely to set up their business in the industry in which they previously worked. Consequently, in areas with an industrial structure which is dominated by heavy industry with large capital requirements (e.g. shipbuilding, steel, chemicals) formation rates will be depressed. However, this is by no means the most important factor because the skills of an individual worker in, say, engineering could be applied in other industries. Indeed, Johnson (1983) showed that regional variations in new manufacturing firm formation still existed after controlling for industrial mix. The main differences in formation rates across regions are therefore attributable to differences in formation rates in the *same* industry across regions.

Secondly, Fothergill and Gudgin (1982) and Johnson and Cathcart (1979b) have both shown that small firms are likely to produce more new firm founders than large firms. The reason is that large firms do not provide good training grounds for entrepreneurship – their employees have a limited range of work experience, lack contacts with customers, have greater security of employment, enjoy fringe benefits and lack contact with small business owner–managers who might provide an example to follow. The effect, as noted above, is to depress rates of new firm formation in areas where a high proportion of employment is concentrated in large plants.

There is also evidence that individuals are more likely to establish their own business if they are skilled rather than semiskilled, and managers rather than shopfloor workers (Cross 1981). Indeed, Gould and Keeble (1984) suggest that the residential occupational structure has been the most powerful influence on the pattern of new manufacturing firm formation in East Anglia, with the highest rates found in areas which contain a large proportion of males employed in non-manual occupations. Moreover, the growth of new enterprises is strongly influenced by the background of the founders. Firms started by individuals with a management background, particularly if they have a degree or professional qualification, show the fastest rates of growth (Fothergill and Gudgin 1982). Such founders are likely to have higher aspirations and show greater skills in raising finance, developing products and identifying market opportunities than those with less education and manual backgrounds, and may be more likely to start firms in newer industries with greater growth potential. Indeed, for high-technology activities, a degree in science or engineering is likely to be an essential pre-requisite for at least one of the founders. The proportion of managerial, professional and technical occupations – from whose ranks such

Table 9.10 Factors associated with low and high levels of new firm formation

Factor	Low rate of new firm formation	High rate of new firm formation
Industry structure	Heavy industry	Light industry
Plant size	Large plants	Small firms
Workforce skills	Semi skilled, unskilled	Skilled, managerial
Education	Lack of post-school education	Degree-level; technical/professional qualifications
Access to capital	Low level of owner-occupation	High level of owner-occupation
Size of local/ regional market	Low regional income; external control of industry	High regional income; well-developed indigenous sector

Sources: Amended from Storey (1982a: Table 10.1) and Cross (1981; Table 7.1)

successful new firm founders are mainly drawn – is considerably higher in the South East than in other regions (Storey 1982a; Fothergill and Gudgin 1982), not least because the spatial division of labour within multisite enterprises has led to a concentration of head office and R and D establishments in the South East and the dispersal of branch plant assembly operations to peripheral regions. Therefore, on account of its occupational structure, the South East not only contains the largest pool of potential new firm founders, but it is also likely to have a major share of the fastest-growing new enterprises (Fothergill and Gudgin 1982).

New firm founders in the South East are also likely to have the advantages of superior access to capital. In this respect, the importance of personal savings is very important because most firms rely predominantly or even exclusively on the founder's own resources (Storey 1982a; Lloyd and Mason 1984). Often a founder must offer his house as security against a bank loan, or even take out a second mortgage (see Ch. 5). The number of founders able to raise capital in this way is reduced in regions such as Scotland, the North, and Yorkshire and Humberside where the level of owner-occupancy is low, while the amount of capital which can be raised is a function of house prices which are highest in the South East and lowest in Yorkshire and Humberside, East Midlands and the North West (Storey 1982a). Similarly, access to venture capital may favour businesses in the South East and adjacent regions. No less than thirty-eight of the fifty companies that are members of the British Venture Capital Association are based in the South East region (thirty-five in London alone) and, as Rothwell (1982: 368) notes, 'it seems reasonable to suppose that this geographic concentration in sources of venture finance will have some influence on the real availability of venture capital.'

Finally, because most new businesses serve localized market areas, a further influence on the prospects for new firm formation and growth is the buoyancy of demand in the local or regional economy. Once again it would appear that the South East has a more benign environment than peripheral industrial regions for entrepreneurship. Indices of regional wealth and spending power demonstrate that the size of potential regional consumer demand is highest in the South East and lowest in Northern Ireland, Scotland and Wales (Storey 1982a). Similarly, demand from industrial and commercial sources is likely to be greater in southern England on account of the concentration of relatively prosperous industries – notably electronics, defence and professional and administrative services – and low in peripheral regions because of the decline of traditional heavy industry complexes and the limited local purchasing by their externally-owned branch plant sectors.

Although the role of the small firm sector in economic growth and job creation has been overstated, it is clear from this evidence that the benefits which do arise are distributed highly unevenly throughout the UK space–economy, with the main advantages accruing to what might be termed 'Greater South East England' (Regional Studies Association 1983), comprising the South East region itself plus substantial parts of the adjacent East Anglia and South West regions and some of the East Midlands. It would seem valid to conclude that this has been a significant factor in the consistently superior economic development of this area in recent decades.

The locational distribution of 'successful' small firms

In view of the evidence which was presented earlier that only a very limited proportion of small enterprises achieve rapid growth, generally on the basis of a technical or design innovation combined with skilful marketing, it can be argued that spatial variations in new firm formation is a less significant factor in differential economic development than is the locational distribution of these successful small enterprises. Support for this view is provided by a comparison of new manufacturing firms in the two depressed conurbations of Greater Manchester and Merseyside, and the relatively prosperous area of south Hampshire (Lloyd and Mason 1984). This revealed, somewhat surprisingly, that the typical new firm in each area was essentially similar in terms of size (employment, turnover), market orientation, scale of production, dependence on local market opportunities, reliance on a small number of key customers and a lack of innovation. However, differences were observable at the upper end of the distribution, with a handful of rapid-growth enterprises found in south Hampshire but absent from the North West conurbations.

Defining and identifying 'successful' small firms, whether on growth, innovation or other criteria, creates considerable methodological problems. However, taking a *technological* definition of success, the list of ninety-three firms that were identified in a authoritative study of new

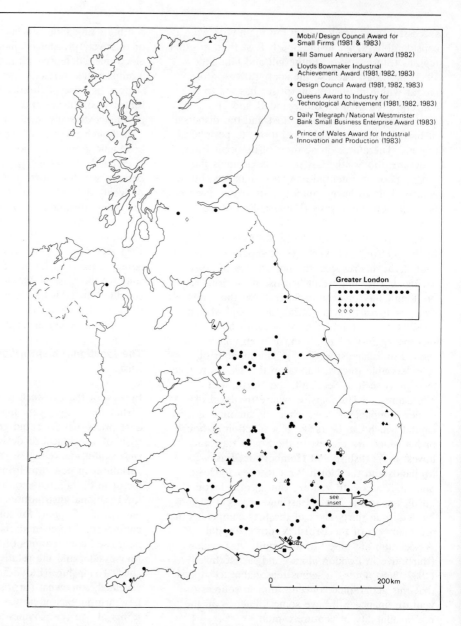

Fig. 9.1 Small firm competition winners (*Source:* Mason 1985)

technology-based firms started since 1950 (Little 1977) and adding locational information (possible in eighty-five cases) reveals the concentration of such firms in southern England, with over half in the South East region (but only 11 per cent in Greater London) and 75 per cent in the South East, South West and East Anglia. An alternative definition – the winners of recent small firm competitions (where the emphasis is generally on technical or design innovation) – similarly highlights the concentration of such firms in southern England (40 per cent in the South East) and their absence in peripheral regions (Fig. 9.1).

Indeed, when the proportion of winners in each region is compared with the regional share of manufacturing firms with less than 200 employees, only the South East, South West and East Anglia have location quotients greater than 1 (East Anglia, with 3.2, had the greatest concentration) while areas with the greatest deficits (each with location quotients of between 0.5 and 0.7) were Northern Ireland, Scotland, Wales and the North. The much higher level of product innovation amongst small firms in the South East compared with their counterparts in peripheral regions has also been highlighted (Thwaites 1982; Oakey *et al.* 1982).

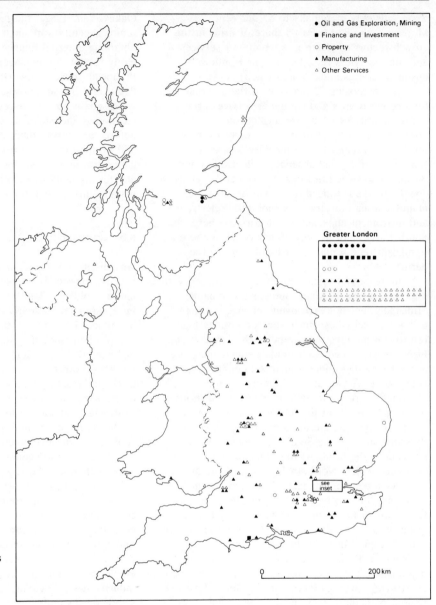

Fig. 9.2 Companies on the Unlisted Securities Market (April 1985) (*Source:* Mason 1985)

Successful small firms might alternatively be defined on the basis of *financial* criteria. One such measure is provided by the Unlisted Securities Market (USM) which was set up in November 1980 to provide smaller, more entrepreneurial businesses with access to a capital market but at a lower cost than is involved in obtaining a full Stock Exchange listing. Companies which gain entry on to the USM are generally profitable and growing, although 'greenfield ventures' – companies without a trading record but with a fully developed project or product which requires financing – are occasionally accepted. The geographical distribution of USM companies at April 1985 (excluding companies registered outside the UK, subsidiaries and companies which previously had a Stock Exchange listing) is heavily weighted towards the South East which contained 59 per cent of the total (Fig. 9.2), with the next largest share accounted for by the South West (8 per cent). Even excluding companies engaged in oil and gas exploration and finance and investment, which tend to have their registered offices in London, fails to dislodge the South East's dominant position: its share falls only marginally to 57 per cent. Moreover, the South East's share of firms in the growth sectors of electronics, computer services and other services (e.g. leisure, distribution, business services and TV and video) is even greater, comprising 72 per cent of the combined total of electronics and computer services firms (a slight majority of which are located outside Greater London) and 67 per cent of firms in other services (the majority of which are located in London). The virtual absence of electronics and computer services firms in the peripheral regions is equally marked. Indeed, only with respect to manufacturing industries, excluding the electronics and instrument engineering categories (where the South East's share is only 28 per cent), is there a significant representation of USM firms in the peripheral areas.

Venture capital organizations are concerned with seeking out fast-growing, innovative companies principally, but not exclusively, engaged in high-technology activities which offer potentially high returns on the invested capital in the longer term. Firms which receive such funding might therefore also be regarded under a financial definition as being successful, or at least potentially so. Information from twenty-four such organizations on the geographical distribution of their portfolios (totalling 369 separate investments) again confirms the dominance of the South East region, with 52 per cent of the total. At the other extreme, Scotland, Wales, Northern Ireland and the North together accounted for just 13 per cent of the companies funded by these organizations, much too low a proportion to be explained simply by the distance of these regions from the concentration of venture capital firms in the South East.

Together, these various sources of data clearly demonstrate the concentration of 'successful' small enterprises – or high fliers – in the South East region, particularly in London itself, notably for computer-related and other service industries, and in the counties to the north, west and south of the capital – Hertfordshire, Buckinghamshire, Surrey and Hampshire, but also including Cambridgeshire in East Anglia. Explanations for this geographical pattern are, in part, associated with the factors which generally promote high rates of new firm formation in 'Greater South East England' – notably the concentration of managerial, professional and technical occupations and the high proportion of the population with degrees and professional qualifications. More specific factors favouring the South East include (Thwaites 1982;

Oakey *et al*. 1982): the technical expertise of many senior executives in small firms; the greater ease in obtaining external finance; the ability to more easily recruit key managerial and technical staff; the locally richer network of technical information flows; and the superior subcontracting opportunities which enables firms, especially in the high-technology industries, both to evaluate products before committing themselves to heavy investment and to minimize their risk and exposure by concentrating on research, design, final assembly and testing while subcontracting out the manufacturing operations.

9.8 Conclusion

Any discussion of the small firm sector in the UK, and this chapter is no exception, is plagued by two major problems. The first is the lack of statistical information, particularly on sectors other than manufacturing; indeed, the position has deteriorated since the Bolton Committee originally highlighted the problem because the government has scaled down its data-collection activities, partly in the interests of economy but also because of its desire to reduce the demands on small businesses to provide statistical information. Second, the bulk of the research on small businesses has been confined to the manufacturing sector. An assessment of the economic contribution of small firms is therefore largely based on the experience in manufacturing industry which might not necessarily be an appropriate guide to their contribution in other sectors. Nevertheless, it is clear that the small firm sector in the UK has undergone a revival since the early 1970s both numerically and in terms of their relative share of employment and output, although not every industry has participated in this trend. Retailing stands out as one activity where both the number of small businesses and their share of total sales have continued to decrease in the face of rising concentration.

To some extent the increase in the relative size of the small firm sector is simply a function of the contraction of the large firm sector in the face of world recession. However, this is by no means the entire explanation; indeed, the *numerical* increase in small businesses can hardly be accounted for in this way. In fact, there has also been a marked upturn in new business formations since the early 1970s and this trend has been a key element in the increase in the size of the small firm sector. Yet the economic significance of the small firm sector, and especially of new businesses, in terms of their contributions to exporting, innovation and job creation and role as seedcorns for future large enterprises remain limited. The main reason for this is not a shortage of small businesses *per se* but a lack of rapid-growth enterprises based on high technology which do have the potential to generate substantial numbers of new jobs and wealth over a relatively short period. Despite the general increase in the size of the small business sector, new or small firms with such characteristics remain in extremely short supply. An assessment of the UK small business sector might therefore legitimately conclude that 'reports of its birth are much exaggerated' (*The Economist* 1983).

It has even been suggested that the revival of the small business sector will be short-lived. This view stems from the belief that the current political enthusiasm for small businesses represents a temporary response to an unusually deep recession and so might not endure indefinitely. Moreover, disillusionment with small firms could set in when it becomes apparent that they have not fulfilled the exaggerated claims of their supporters, leading to a powerful backlash. Another line of argument predicts that when the promised economic upturn emerges, this will weaken the impetus of the small firm sector by reducing the number of new business formations which are motivated by unemployment. In addition, a return to economic growth might be expected to see a recurrence of the corporate sector's earlier habit of expanding via the takeover of 'leader' small firms.

However, there are probably stronger grounds for anticipating that the small firm sector will continue to increase both numerically and in terms of their relative significance because of longer-term economic changes that are unconnected with the recession. Firstly, the seemingly inevitable continuation in the shift from a goods-producing to a service-based economy will create many more opportunities for small-scale businesses, particularly in the leisure field where there is considerable demand for personal services of various kinds. Indeed, there are relatively few economies of scale in the service sector, and in certain circumstances – notably a widespread branch network dealing directly with the public – large size actually has diseconomies arising from difficulties of supervision and staff motivation. This problem has been responsible for the rapid growth of franchising over the last decade (Curran and Stanworth 1982b) and this trend seems set to continue for the foreseeable future. Secondly, at least one leading industrialist has suggested that in the future large companies will 'concentrate on the core activities of their business, relying for everything else on specialized suppliers who would compete for their custom' (Sir Adrian Cadbury, quoted in *Financial Times*, 3 May 1983). This development will also lead to new opportunities for small firms, but seemingly only in a 'satellite' role. Thirdly, the long-wave theory of economic development suggests that during the next decade there will emerge a whole range of new industries based on technologies currently in their infant stage: examples include information technology, biotechnology, energy-related technologies, advanced medical electronics and robotics. In the early stages of new industries such as these, it is likely that small firms will play a key role. Existing and new technology-based small firms will have opportunities for innovation and exploitation while indirect opportunities will be generated for small firms as suppliers of specialized components and subassemblies to the major corporations operating in these industries (Rothwell and Zegveld 1982).

The crucial point from a geographical perspective is that on the basis of recent experience it seems most improbable that an expanding small firm sector will contribute to a narrowing of regional disparities in the UK and might even be instrumental in widening them. As this chapter has highlighted, the capacity of different areas to generate new businesses varies: the benefits from new firm formation and growth have therefore also been distributed unevenly across the space–economy. Even more significant is that the minority of 'high fliers' in the small firm population – enterprises which have created substantial numbers of new jobs, generated sizeable local multiplier effects and achieved overseas sales and substituted for imports – are disproportionately concentrated in the South East and parts of the adjacent regions. This dearth of successful small firms in the peripheral regions was recently underlined by the difficulties encountered by a venture capital fund, set up by the National Enterprise Board to invest in fledgling enterprises using advanced technologies in the assisted areas, in finding suitable investment opportunities. The Californian venture capitalist hired by the NEB to manage the fund went even further, claiming that 'if there are any good entrepreneurs in the North of England I sure as hell haven't seen them', while noting that 'there are some good opportunities in the South East particularly around London' (quoted in *Financial Times*, 6 July 1982). Unless this situation can be changed, for example by a regional innovation policy to encourage the creation of new technology-based firms in the regions, it seems inevitable that the small firm sector will continue to contribute to uneven regional economic development in the UK.

References

Bannock, G. 1981a *The Economics of Small Firms: return from the wilderness*. Basil Blackwell.

Bannock, G. 1981b The clearing banks and small firms. *Lloyds Bank Review*. **142**: 15–25.

Beesley, M., Wilson, P. 1982 Government aid to the small firm since Bolton. In Stanworth, J., Westrip, A., Watkins, D., Lewis, J. (eds) *Perspectives on a Decade of Small Business Research*. Gower, 181–99.

Binks, M., Coyne, J. 1983 *The Birth of Enterprise: an analysis and empirical study of the growth of small firms*. Institute of Economic Affairs, Hobart Paper No. 98.

Binks, M., Jennings, A. 1983 New firms as a source of industrial regeneration. Paper to the 6th National Small Firms Policy and Research Conference, University of Durham.

Birch, D. L. 1979 *The Job Generation Process*. MIT Program on Neighbourhood and Regional Change, Cambridge, Massachusetts.

Bolland, A. 1983 Technology, economic change and small firms. *Lloyds Bank Review*. **147**: 42–56.

Bolton, J. 1971 *Small Firms: Report of the Commission of Inquiry on Small Firms*. HMSO, Cmnd 4811.

Chaplin, P. 1982 Co-operatives in contemporary Britain. In Stanworth, J., Westrip, A., Watkins D., Lewis, J. (eds) *Perspectives on a Decade of Small Business Research*. Gower, 85–109.

Chappell, H. 1983 The mauve economy. *New Society* 28 July, 123–4.

Cross, M. 1981 *New Firm Formation and Regional Development*. Gower.

Cross, M. 1982 The entrepreneurial base of the large manufacturing company. In Stanworth, J., Westrip, A., Watkins, D., Lewis, J. (eds) *Perspectives on a Decade of Small Business Research*. Gower, 143–56.

Curran, J., Stanworth, J. 1981a A new look at job satisfaction in the small firm. *Human Relations*. **34**: 343–65.

Curran, J., Stanworth, J. 1981b The social dynamics of the small manufacturing enterprise. *Journal of Management Studies* **18**: 141–58.

Curran, J., Stanworth, J. 1981c Size of workplace and attitudes to industrial relations in the printing and electronics industries. *British Journal of Industrial Relations* **19**: 14–25.

Curran, J., Stanworth, J. 1982a Bolton ten years on: a research inventory and critical review. In Stanworth, J., Westrip, A., Watkins, D., Lewis, J. (eds) *Perspectives on a Decade of Small Business Research*. Gower, 3–27.

Curran, J., Stanworth, J. 1982b The small firm in Britain – past, present and future. *European Small Business Journal* **1**: 16–25.

Davies, J. R., Kelly, M. 1972 *Small Firms in the Manufacturing Sector*. Committee of Inquiry on Small Firms: Research Report No. 3. HMSO.

Dawson, J. 1983 Planning for local shops. *The Planner* **69**: 18–19

Department of Employment (1984) Evaluation of the pilot Enterprise Allowance Scheme. *Employment Gazette* **92**: 374–7.

Economist (1983) British small business. 23 July: 66–73.

Fothergill, S., Gudgin, G. 1979 The job generation process in Britain. *Centre For Environmental Studies, Research Series No. 32*.

Fothergill, S., Gudgin, G. 1982 *Unequal Growth: urban and regional employment growth in the UK*. Heinemann.

Ganguly, P. 1983 Lifespan analysis of businesses in the UK 1973–1982. *British Business* 12 August: 838–45.

Ganguly, P. 1984 Business starts and stops: regional analyses by turnover size and sector 1980–83. *British Business* 2 November: 350–3.

Ganguly, P., Povey, D. 1983 Small firms survey: the international scene. *British Business* 19 November: 486–91.

Gould, A., Keeble, D. 1984 New firms and rural industrialization in East Anglia. *Regional Studies* **18**: 189–201.

Greenhow, F. 1982 The banks and Bolton ten years on. In Stanworth, J., Westrip, A., Watkins, D., Lewis, J. (eds) *Perspectives on a Decade of Small Business Research*. Gower, 31–43.

Gudgin, G. 1978 *Industrial Location Processes and Regional Employment Growth*. Saxon House.

Harrison, R. T., Hart, M. 1983 Factors influencing new business formation: a case study of Northern Ireland. *Environment and Planning A* **15**: 1395–1412.

Healey, M., Clark, D. 1984 Industrial decline and government response in the West Midlands: the case of Coventry. *Regional Studies* **18**: 303–18.

Hough, J. 1982 Franchising – an avenue for entry into small business. In Stanworth, J., Westrip. A., Watkins, D., Lewis, J. (eds) *Perspectives on a Decade of Small Business Research*. Gower, 63–83.

House of Commons 1983 Small businesses. *Parliamentary Debates* **47**: 1100–61 (4 November).

Howdle, J. 1979 *An Evaluation of the Small Firms Counselling Service in the South West Region*. Department of Industry South West Regional Office, Bristol.

Johnson, P. 1978 Policies towards small firms: time for caution? *Lloyds Bank Review* **129**: 1–11.
Johnson, P. 1981 Unemployment and self-employment: a survey. *Industrial Relations Journal* **12**: 5–15.
Johnson, P. 1983 New manufacturing firms in the UK regions. *Scottish Journal of Political Economy* **30**: 75–9.
Johnson, P. S., Cathcart, D. G. 1979a New manufacturing firms and regional development: some evidence from the Northern Region. *Regional Studies* **13**: 269–80.
Johnson, P. S., Cathcart, D. G. 1979b The founders of new manufacturing firms: a note on the size of their 'incubator' plants. *Journal of Industrial Economics* **28**: 219–24.
Johnson, P. S., Darnell, A. 1976 New firm formation in Great Britain. University of Durham, Department of Economics, Working Paper No. 5.
Johnson, P. S., Rodger, J. 1983 From redundancy to self-employment. *Employment Gazette* **91**: 260–4.

Little, A. D. Ltd 1977 *New Technology-Based Firms in the United Kingdom and the Federal Republic of West Germany*. Anglo-German Foundation.
Lloyd, P. E. 1980 New manufacturing enterprises in Greater Manchester and Merseyside. *University of Manchester, North West Industry Research Unit. Working Paper No. 10*.
Lloyd, P. E., Dicken, P. 1982 *Industrial Change: Local Manufacturing Firms in Manchester and Merseyside*. Department of the Environment Inner Cities Directorate.
Lloyd, P. E., Mason, C. M. 1984 Spatial variations in new firm formation in the United Kingdom: comparative evidence from Merseyside, Greater Manchester and South Hampshire. *Regional Studies* **18**: 207–20.

McGuire, J. 1976 The small enterprise in economics and organization theory. *Journal of Contemporary Business* **5**: 115–38.
Mason, C. M. 1982 New manufacturing firms in South Hampshire: survey results. University of Southampton, Department of Geography, Discussion Paper No. 13.
Mason, C. M. 1983 Some definitional difficulties in new firms research. *Area* **15**: 53–60.
Mason, C. M. 1985 The geography of 'successful' small firms in the United Kingdom. *Environment and Planning A* **17**: 1499–1513.

Oakey, R. P. 1984 Innovation and regional growth in small high technology firms: evidence from Britain and the USA. *Regional Studies* **18**: 237–51.
Oakey, R. P., Thwaites, A. T., Nash, P. A. 1982 Technological change and regional development: some evidence on regional variations in product and process innovation. *Environment and Planning A* **14**: 1073–86.

Regional Studies Association (1983) *Report of an Inquiry into Regional Problems in the United Kingdom*. Geo Books.
Riddell, P. 1983 *The Thatcher Government*. Martin Robertson.
Robson Rhodes 1983a *Small Business Loan Guarantee Scheme: commentary on a telephone survey of borrowers*. Department of Industry.
Robson Rhodes 1983b *An analysis of some early claims under the Small Business Loan Guarantee Scheme*. Department of Industry.
Rothwell, R. 1982 The role of technology in industrial change: implications for regional policy. *Regional Studies* **16**: 361–9.
Rothwell, R., Zegveld, W. 1982 *Innovation and the Small and Medium-Sized Firm*. Frances Pinter.

Sargent, V. 1982 Large firm assistance to small firms. In Watkins, D., Stanworth J., Westrip, A. (eds) *Stimulating Small Firms*. Gower, 67–89.
Scott, M. 1980 Independence and the flight from large scale: some sociological factors in the founding process. In Gibb, A., Webb, T. (eds) *Policy Issues in Small Business Research*. Saxon House, 15–33.
Scott, M., Rainnie, A. 1982 Beyond Bolton: industrial relations in the small firm. In Stanworth, J., Westrip, A., Watkins, D., Lewis, J. (eds) *Perspectives on a Decade of Small Business Research*. Gower, 159–78.
Shutt, J., Whittington, R. 1984 Large firm strategies and the rise of small units: the illusion of small firm job generation. North West Industry Research Unit, School of Geography, University of Manchester, Working Paper No. 15.
Storey, D. J. 1982a *Entrepreneurship and the New Firm*. Croom Helm.
Storey, D. J. 1982b *Manufacturing Employment Change in County Durham Since 1965*. Durham County Council.
Storey, D. J. 1982c *Manufacturing Employment Change in Tyne and Wear Since 1965*. Tyne and Wear County Council.
Storey, D. J. 1982d *Manufacturing Employment Change in Cleveland Since 1965*. Cleveland County Council.
Storey, D. J. 1983a Regional policy in a recession. *National Westminster Bank Quarterly Review* November: 39–47.
Storey, D. J. 1983b Job accounts and firm size. *Area* **15**: 231–7.

Thwaites, A. T. 1982 Some evidence of regional variations in the introduction and diffusion of industrial products and processes within British manufacturing industry. *Regional Studies* **16**: 371–81.

Walker, S., Green, H. 1982 The role of the small firm in the process of economic regeneration: the evidence from Leeds. In Collins, L. (ed) *Industrial Decline and*

Regeneration: proceedings of the 1981 Anglo-Canadian Symposium. University of Edinburgh, Department of Geography and Centre for Canadian Studies, 85–108.

Watkins, D. 1982 Management development and the owner-manager. In Webb, T., Quince, T., Watkins, D. (eds) *Small Business Research*. Gower, 193–215.

Watts, H. D. 1981 *The Branch Plant Economy: a study of external control*. Longman.

Whyatt, A. 1983 Developments in the structure and organization of the co-operative movement: some policy considerations. *Regional Studies* 17: 273–6.

Wilson, H. 1979 *The Financing of Small Firms: interim report of the Committee to review the functioning of the financial institutions*. HMSO, Cmnd 7503.

Wright, M., Coyne, J., Lockley, H. 1984 Management buyouts and trade unions: dispelling the myths. *Industrial Relations Journal* 15(3): 45–52.

CHAPTER 10

The multiplant enterprise

Michael Healey and H. Douglas Watts

10.1 Introduction

Multiplant enterprises are the most significant type of organization in manufacturing industry in the UK whether this is measured in terms of output or employment. In 1980, the top 100 private sector enterprises in terms of net output accounted for 41 per cent of manufacturing output and 36 per cent of manufacturing employment. Multiplant operation is also characteristic of many smaller firms. Between 3,500 and 4,000 private sector manufacturing enterprises operated more than one plant in the UK in 1980. These enterprises accounted for about one-quarter of the plants and about three-quarters of employment in manufacturing. If anything, these data underplay the role of the multiplant enterprises because they do not reflect their control over the many small firms which supply them with goods and services (Taylor and Thrift 1982).

In discussing multiplant enterprises, the term *plant* refers to the whole of the premises under the same management at a particular address at which manufacturing activities take place. This is broadly equivalent to what the Business Statistics Office (BSO) calls a *local manufacturing unit*. The term *establishment* is used here interchangeably with plant, although in official statistics the BSO sometimes has to use data on establishments which refer to activities at more than one plant (Healey 1983a: 28). A *firm* is regarded as one or more productive plants operated under the same trading name; while an *enterprise* consists of one or more firms under common ownership or control. A *multiplant enterprise* thus refers to an enterprise operating more than one plant.

No official data are published in the UK on the extent of multiplant operation, but some indication of how its relative importance has changed over time is given by the average number of plants operated per enterprise. Prais (1976: 60–6) suggests there was a gradual evolution in multiplant working from the end of the last century, with a rapid acceleration in the 1960s. In 1958 the 100 largest enterprises operated an average of twenty-seven plants each, but by 1972 this had risen markedly to seventy-two (Table 10.1). Since 1972

Table 10.1 Extent of multiplant operation in enterprises in the UK, 1958–79.

	Plants per enterprise	
	The 100 largest employers	All enterprises
1958	27	n.a.
1963	40	1.29
1968	52	1.33
1972	72	1.50
1976	n.a.	1.32
1979	n.a.	1.32

Sources: Prais 1976: 62; *Report on the Census of Production*, various years.

Acknowledgements
The authors are indebted to Cadbury Schweppes for the provision of employment data and to Howard Stafford who commented upon an early draft of the Chapter.

the extent of multiplant operation seems to have fallen. The Census of Production data records 1.32 local manufacturing units per enterprise in both 1976 and 1979 compared with 1.50 in 1972. This fall may in part be due to increases in the birth of new firms (Chs. 1 and 9) and the closure of plants by large companies, but may also reflect changes in the way the number of local units employing ten or fewer employees are estimated. Nevertheless, in 1976 the 232 enterprises with between 1,000 and 1,500 employees operated an average of five plants each, and the twenty-four enterprises with between 20,000 and 50,000 employees operated an average of about fifty plants each (Macey 1982: 35).

Not surprisingly, the extent of multiplant operation varies between industries (Table 10.2). Unfortunately, the most recent data is only available for the average number of *establishments*, not *plants*, per enterprise. Nevertheless, they show a marked variation between industries. The mean number of establishments per enterprise for all manufacturing in 1980 was 1.20. The highest figure occurred in the manufacture on non-metallic minerals products (1.37) and the production of man-made fibres (1.29). The lowest figures were in timber and wooden furniture industries (1.08) and other manufacturing industries (1.04).

The official data on multiplant enterprises can be supplemented by survey evidence which identifies the importance of multiplant operation in particular areas. These surveys further emphasize the importance of multiplant concerns in the British economy. For example, they accounted for 94 per cent of manufacturing employment in Cleveland in 1965 (Cleveland County Council 1982: 21); 87 per cent of manufacturing employment in Coventry in 1982 (Healey and Clark 1984a: 50); 80 per cent and 63 per cent of manual employment in manufacturing in 1975 in Inner Merseyside and Inner Manchester, respectively (Dicken and Lloyd 1978: 188, 193); 74 per cent in the Warwick district in 1983 (Healey 1984a); and 71 per cent of employment in the West Yorkshire wool textile industry (Hardill 1982: 21–2).

Not only are multiplant enterprises an important element in national and regional economies, but they are distinctive from single-plant concerns. The justification for recognizing multiplant enterprises

Table 10.2 Multiplant operation in the UK by industrial class, 1980

Class[1]		Establishments per enterprise[2]
24	Manufacture of non-metallic mineral products	1.37
26	Production of man-made fibres	1.29
22	Metal manufacturing	1.25
43	Textile industry	1.25
41/42	Food, drink tobacco manufacturing industries	1.24
25	Chemical industry	1.23
	All manufacturing	1.20
34	Electrical and electronic engineering	1.18
32	Mechanical engineering	1.15
47	Manufacture of paper and paper products, printing and publishing	1.15
48	Processing of rubber and plastics	1.15
35	Manufacture of motor vehicles and parts thereof	1.13
44	Manufacture of leather and leather goods	1.12
31	Manufacture of metal goods	1.11
33	Manufacture of office machinery and data processing	1.11
36	Manufacture of other transport equipment	1.11
21	Extraction and preparation of metalliferous ores	1.09
37	Instrument engineering	1.09
45	Footwear and clothing industries	1.09
46	Timber and wooden furniture industries	1.08
49	Other manufacturing industries	1.04

Source: Report on the Census of Production 1980: PA1002, Table 13.
[1]The figures for the individual classes only include the establishments classified to those classes; many of the larger enterprises have establishments allocated to more than one class.
[2]These figures understate the number of *plants* per enterprise since some *establishments* consist of more than one plant.

as a distinct group may be argued at both plant and enterprise levels. At the former level, production units operated by multiplant enterprises are part of a corporate system and their status within that system may be a more important factor in determining survival and prosperity than the profitability of the particular plant. On the one hand, a plant making a loss may be subsidized by more profitable plants in the enterprise. On the other hand, a profitable plant may be closed as part of a rationalization scheme to concentrate production at another location. Neither of these strategies is available to the single-plant enterprise. Dicken (1976: 404) says 'it is indisputable that the behaviour and performance of any single unit of a multi-plant enterprise has to be seen within the context of that enterprise as a whole, and that decisions affecting the unit's activities may well be made on non-local criteria'.

A further distinction is that the plants of a multiplant enterprise may be dependent on materials and services from other plants in the enterprise and may send all, or the majority, of their production to other group plants and not have to sell in the open market. Such plants are likely to be less influenced by their local environment than single-plant enterprises. Indeed, any external economies which single-plant enterprises derive from locating in an industrial agglomeration may be relatively unimportant to the plants of multiplant concerns. This allows multiplant enterprises a greater potential flexibility in locational choice.

The same point about locational flexibility applies at the enterprise level. In any investment decision, a multiplant enterprise has the choice not only between *in situ* expansion or a new location but also, should it choose the former, the choice of which of its plants to expand. Perhaps even more significantly, a multiplant enterprise has the potential both to reorganize its product locations between plants and to select particular plants for closure, choices which are denied the single-plant enterprise.

It is because of the economic significance and distinctiveness of the multiplant enterprise (Hayter and Watts 1983) that it is important to assess its contribution to changes in industrial location in the UK and the processes by which these changes are brought about. Towards this end, the remainder of this chapter is structured around four main questions:

1. What are the processes by which multiplant enterprises change the location of their activities?
2. What is the nature of spatial growth in multiplant enterprises and what factors affect it?
3. How do they rationalize their activities and what are the constraints on this process?
4. What are the consequences of these actions for urban and regional development?

Unfortunately, despite a major growth of interest in the last twenty years in the process of locational change in manufacturing industry, the specific contribution of multiplant enterprises to the process in often not distinguished. Consequently, recourse often has to be made to data sets which refer either predominately to multiplant enterprises, such as large corporations, or to activities in which they are involved such as the establishment of branch plants.

This chapter concentrates upon examining the locational behaviour of private sector multiplant enterprises within manufacturing industry in the UK. The behaviour of nationalized concerns is taken up in Chapter 12, while the impacts of the overseas investments of domestic companies on their UK activities are discussed in Chapter 11 which deals with multinational companies. Multiplant enterprises are for the most part treated in this chapter as a single category, although it is accepted that for some purposes it may be desirable to treat the small- and medium-sized multiplant concerns as a separate group (Healey 1983b: 339–40).

10.2 The geography of multiplant enterprises

'Each type of business organisation has a particular establishment structure that is laid out over space' (Taylor and Thrift 1983: 458) and, in the case of multiplant enterprises, these structures take on distinctive forms. The enterprises not only vary markedly in size; they also vary in the extent to which they straddle local, regional and, indeed, national boundaries. Studies of manufacturing activity in specific regions inevitably take only a partial view of enterprises which operate multiregional systems.

Within the UK, no data are available to permit a description of the distribution of employment in multiplant enterprises, although a special tabulation of data relating to the largest thirteen multiplant enterprises in 1971 indicated that the distribution of their employment was very similar to that of manufacturing as a whole (Watts 1980: 44). At a

The multiplant enterprise

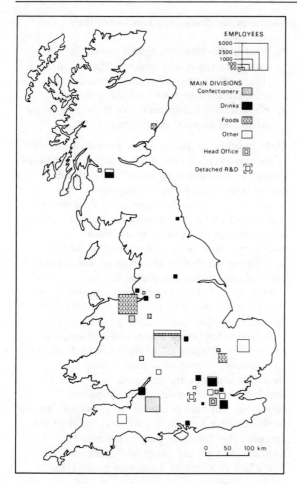

Fig. 10.1 The multiplant enterprise: Cadbury Schweppes, 1984 (sites with 50 or more employees). *Source:* Cadbury Schweppes plc

more disaggregated level, maps have been produced of the distribution of employment in individual companies, but these provide little indication of the different activities at different sites. An example of such a map is shown in Fig. 10.1 which illustrates the distribution of employment in Cadbury Schweppes in early 1984.

Cadbury Schweppes is about 27th in size by turnover amongst enterprises operating primarily in manufacturing industry and it employs 22,900 workers in the UK. These represent 60 per cent of its world-wide labour force and help to stress the extent to which the larger multiplant firms are embedded into a world-wide corporate production system. Figure 10.1 illustrates clearly its multiregional, multiplant system. The firm is known primarily for its chocolate and drink activities and it is, like many multiplant firms,

organized on a multidivisional basis. The principal UK divisions engaged in manufacturing are confectionery, drinks, tea and foods, and health and hygiene. The product range is wide, varying from Cadbury's Milk Tray chocolate to disinfectants, from Chivers Old English Marmalade to aerosols, and from Schweppes Tonic Water to Parazone bleach. Like many major UK multiplant enterprises, it has a distinct core region in which much of its employment is concentrated, in this case the West Midlands. A particularly striking feature of the spatial arrangement of the enterprise's activities is that some R and D activities (in Reading) are detached from production sites, as are its head office activities which are located in central London.

This purely descriptive account of a corporate system helps to recognize that within such systems it is possible to identify *locational hierarchies*. Conventionally, the hierarchy may be traced from the head office staff through R and D staff to skilled production workers and finally to unskilled production workers at the base of the hierarchy. Within a multiplant enterprise, there will be a tendency for the different types of job to be found in different areas, hence the term locational hierarchy.

There is strong empirical evidence to support some but not all the aspects of the locational hierarchy. The hierarchy emerges most clearly in the way that white-collar staff are concentrated in particular regions and some of the evidence is shown in Fig. 10.2. In Fig. 10.2a; the concentration of the head offices of the 100 largest (by turnover) manufacturing enterprises stands out most vividly: no less than 74 per cent were located in the South East in 1982 and those still outside it reflect mainly concentrations of particular industrial activities such as brewing and whisky in Scotland (Watts 1972; Goddard and Smith 1978).

R and D activities display a similar tendency to concentrate in the South East (Fig. 10.2b), although the concentration is less marked with only 47 per cent of the research establishments in industrial firms in the private sector. Some sectors do show a higher concentration of R and D activities than this, and Howells (1984) records that 68 per cent of the main R and D centres of multisite firms in the pharmaceutical industry are in the South East. Within the South East, R and D units tend to be located in the less urbanized counties; a marked contrast to the head offices where the majority are located in central London.

Aggregate data on the distribution of different

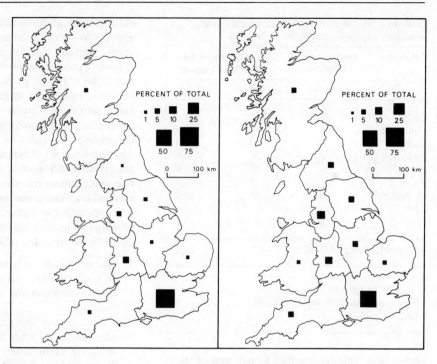

Fig. 10.2 The locational hierarchy:
(a) head offices of the 100 largest private sector manufacturing enterprises, UK, 1982;
(b) research establishments in industrial firms in the private sector in Great Britain, 1980.
Sources:
(a) *Times 1000*, 1982–3 (1982);
(b) Department of Trade and Industry (1983)

types of production labour are not available and so it is difficult to examine the assertion that peripheral locations within the multiplant system are characterized by a preponderance of unskilled production workers in contrast to the high proportion of skilled workers in the key parts of the system. Some limited data from GEC Telecommunications plants appear to support this view although female employees are used as a surrogate measure of unskilled workers (Table 10.3). Until there has been a systematic study of the distribution of different types of shopfloor jobs within multiplant enterprises, assertions about systematic spatial variations in the distribution of different types of production workers need to be viewed with some caution.

Table 10.3 Distribution of unskilled workers within a corporate system: GEC Telecommunications, 1976

Location	Female employees as a percentage of production workforce
Coventry	55
Glenrothes	62
West Hartlepool	63
Kirkcaldy	82
Newton Aycliffe	86
Middlesborough	94
Treforest	95

Source: Meegan (1982: 35)

10.3 Patterns of changes

While the contemporary geography of multiplant enterprises is of interest, most concern is usually expressed about the way in which these companies modify employment opportunities in different places. These modifications are often constrained by the existing structure, as will be shown later, but the theme of locational change has become of increasing interest as the numbers employed in UK manufacturing have continued to fall. Table 10.4 records the losses in employment between 1979 and 1983 among the largest nine private sector multiplant enterprises. Both Courtaulds and GKN shed almost half their employees. In contrast, Ford and GEC shed almost one-tenth of their jobs. These are, of course, net losses and almost certainly underestimate the number of jobs lost, since plant closures and contractions in some parts of the organization may have been offset to some extent by the acquisition of plants formerly operated by other enterprises, the opening of branch plants and the taking on of labour in other parts of the organization.

The impact of large enterprises on job losses at a more local scale is well illustrated by the case of Coventry. The largest fifteen enterprises in the city in 1982 employed only half of those working for the companies comprising the top fifteen in 1974, a net loss of 46,500 jobs (Healey and Clark

Table 10.4 Net employment loss in selected large multiplant enterprises: 1978–79 to 1982–83

Enterprise	Employment, 1978–79	Net employment loss, 1978–79 to 1982–83	
	('000)	('000)	(%)
GKN*	69	−32	−46.3
Courtaulds	103	−47	−45.6
Lucas	70	−21	−30.0
Imperial	90	−26	−28.8
ICI	93	−26	−27.9
Thorn†	101	−27	−26.7
Unilever*	69	−20	−22.5
GEC	155	−18	−11.6
Ford*	74	−8	−10.8
All manufacturing 1978–82§			−18.9

Source: Company reports
* 1977–78 to 1981–82
† Four years only 1979–80 to 1982–83
§ 1982 estimated

1985: 1355). This represented 88 per cent of the net decline in manufacturing employment in the city over the period. By far the largest cutback was made by BL who reduced their Coventry workforce by 21,000 or 71 per cent. Of the thirteen manufacturing establishments operated by BL in Coventry in 1974, only six were still run by the company in 1982 (Healey and Clark 1984b). However, most of the reduction in the labour force of the largest companies was accomplished by them shedding labour in their existing plants rather than by closing factories. The top fifteen enterprises in 1982 had lost 39,000 jobs in the thirty-one plants in which employment contracted in the previous eight years, while only 6,000 jobs went as a result of them closing thirteen plants. These data suggest that to understand the process which has changed a boom city into a depressed industrial area in less than twenty years, it is essential to examine the investment and disinvestment behaviour of a few large companies and the position of their Coventry facilities within their wider corporate structures and strategies (Healey 1985).

The extent to which job losses reflect planning across the whole of a multiplant enterprise's operations is difficult to judge and Townsend (1983: 80) suggests that master plans are more characteristic of public sector enterprises. In some companies (e.g. Lucas and ICI) employment decline tends to be handled at divisional level whereas in organizations dominated by small plants, highly decentralized decision-making appears to be the rule. Courtaulds, for example, claim that they have no master closure list.

In an examination of the geography of these employment losses, particular interest has focused upon the role of the larger multiplant enterprises, and relatively few studies have concentrated upon the medium-sized multiplant company (an exception is the study by Healey (1983b) which is described below). Studies of the recent recession are growing but most of the readily available material relates primarily to the late 1960s and early 1970s. These studies range over a number of firms operating in rather different environments, although in many cases the studies relate to only one division of a diversified firm:

1. Output growth and stable employment. The brewing industry and its major firms such as Bass, Whitbread and Allied Breweries (Watts 1980).
2. Output growth and declining employment. The scientific instruments and electronics sectors, including examination of firms like ICL and Plessey (Massey and Meegan 1979).
3. Stable output and declining employment. The brick and paper and board industries, including the London Brick Company, Reed International and Bowater (Massey and Meegan 1982).
4. Output decline and declining employment. The electrical engineering industry, including supertension cables and heavy electrical machinery and firms such as BICC (Massey and Meegan 1979).

Over time, the case studies have moved from a concern with changes in production sites towards an interest, first, in the distribution of employment across those sites and then, secondly, in the types of employment on them. The main interest in the last of these stages has been in the evolution of locational hierarchies.

In beginning to understand changes in the geography of multiplant enterprises, the first task is to describe the way in which they change the location of their activities. The spatial evolution of multiplant enterprises consists of two closely-related elements:

1. Spatial growth: corporate changes associated with an increase in the number of plants operated.
2. Locational adjustment: corporate changes associated with static or decreasing numbers of plants.

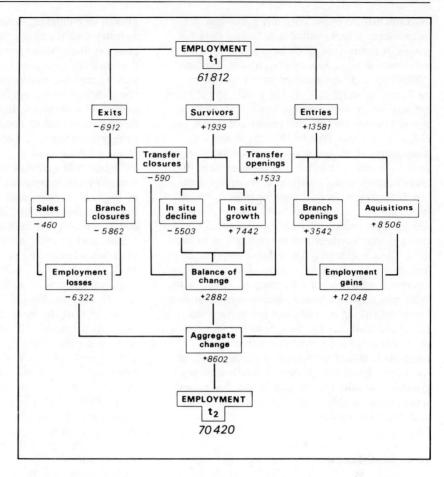

Fig. 10.3 Corporate components of change in an accounting framework: multiplant enterprises in the textile and clothing industries, 1967–72. *Source:* Based on Healey (1983b)

A number of firms have displayed periods of spatial growth followed by locational adjustment, notably in the UK brewing industry, but in the majority of cases the two elements of spatial evolution are so closely inter-related that it is almost impossible to separate them out (Healey 1984b). What is required is an approach which not only recognizes the many different ways in which corporate systems can change, but one which also recognizes how those different processes are related to one another. It is for this reason that there is an increasing interest in the components of change in multiplant companies.

Figure 10.3 indicates the main components of change and presents them in an accounting framework which illustrates the mechanisms which lead to an alteration in an enterprise's employment. There are changes in the number employed in plants owned throughout the time period (*in situ* change) to which can be added employment in new plants entering the corporate system (entries) and from which should be subtracted the employment lost in plants which leave the corporate system (exits). Among the exits, job losses are of two types: plant closures and sales of plants. Entries consist of new branches and acquisitions of plants from other firms. Of particular interest are transfers of plants from one location to another when the employment at the new location may be less than that which existed at the old one. This is part of what Massey and Meegan (1979) term 'in transit loss'. There may also, of course, be 'in transit gains'. Jobs may also be gained or lost 'in transit' when products are switched between existing plants, but these more subtle changes are not picked up in the accounting framework, although the changes may be reflected in *in situ* expansions and contractions.

While it is relatively easy to identify the components of change, the assembly of data to assess the relative importance of individual components to particular companies or groups of companies is a difficult task. Amongst the largest multiplant enterprises, studies of individual concerns may be useful but among medium-sized companies it is more valuable to collate data for groups of enterprises which have some characteristic (e.g., industrial sector) in common.

This approach was adopted with a sample of medium-sized enterprises in the textile and clothing industry which were examined over a six-year period (Healey 1983b). The results are shown in Figure 10.2. From 1967 to 1972 the dominant components of change were acquisitions and *in situ* growth; *in situ* contraction and closures were of approximately equal significance in employment losses. Strikingly, branch plant openings were less important than acquisitions and *in situ* expansions, and the new branches failed to compensate for either the employment lost in contractions *in situ* or that lost in closures. The relative importance of the different components is influenced strongly by the particular period of time examined and, as the 1970s progressed, it became increasingly clear that locational change in multiplant enterprises was influenced markedly by changes within an existing set of plants. It is for this reason that most emphasis is placed upon locational adjustment in the remainder of this chapter. Nevertheless, spatial growth, especially that by acquisition, has played some role in modifying the recent geography of multiplant enterprises.

10.4 Spatial growth

Although the recent past has been characterized by plant closures rather than plant openings, many of the contemporary changes may be influenced by earlier growth patterns, and some multiplant enterprises continue to open new plants even if the general trend is for the number of industrial establishments to fall. Spatial growth is of two types. Internal growth occurs where sites are added to the corporate system by the establishment of branch plants; while external growth takes place where sites are added to the corporate system by the acquisition of other firms and their associated plants. External growth adds plants only to the corporate system, whereas internal growth adds sites to both the corporate and national industrial systems. It is for this reason that most interest has focused on internal growth, and it is an interest enhanced by the possibility of diverting internal growth between areas by urban and regional policy measures.

Since components of change studies suggest that transfers are relatively unimportant in the spatial growth of multiplant firms, the Department of Industry data recording the opening of new branch plants is of particular interest. Some of these branches are established by single-plant enterprises moving into the multiplant category while others are additions to the operations of existing multiplant enterprises. It is not possible to calculate the relative importance of these two categories in the UK, but survey evidence from plants established in the assisted areas between 1972 and 1975 indicates that among branches owned by UK companies, 43 per cent were set up by single-plant firms and 57 per cent by multiplant firms (Herron 1981: 6). These proportions may, of course, be very different in the rest of the UK.

The most recently published data on new branches relate to the period from 1976–80 (Killick 1983). Whereas from 1966 to 1976, 43 per cent of employment in new branch plants was in moves of over 68 km, by 1976–80 the proportion had fallen to 29 per cent. In this latter period, branch plants established less than 19 km from the firm's nearest plant in a similar activity accounted for over half (53 per cent) of new branch plant employment. Some indication of the recent reduction in branch plant formation is provided by comparing the average of 30,320 jobs per annum created in branch plants from 1966–75 (Pounce 1981: 14) with the average of only 10,120 jobs per annum from 1976–80. The fall in the numbers employed in moving and the reduction in the average distance moved means that the interregional growth of the corporate systems through the establishment of branch plants is now of reduced significance.

While there is a wealth of official data on the opening of branch plants (Nunn 1980; Pounce 1981), examination of the spatial growth of multiplant enterprises by acquisitions is less well documented. The evidence that is available indicates that acquisition of firms for market advantage is relatively rare and although a recent study (Watts 1980; 1981) has illustrated the important role of market shares in stimulating the spatial growth of large firms in the brewing industry, the firms studied by Massey and Meegan (1979) are probably more typical of manufacturing industry as a whole. In their study of firms in the electrical engineering industry, acquisitions resulted from a desire to cut capacity and to reduce costs, from a need to achieve economies of scale in production and from a need to increase the market power of particular firms. In these cases the location of the acquired firm was seemingly of little relevance in the acquisition decision. Although the management of firms rarely give location as a factor influencing their choice of which firms to take over (Healey 1979: 153-9), when their decisions are examined in aggregate there is evidence that distance can constrain their takeover decisions. One study of the textile and clothing industry found a significant 'neighbourhood effect' with twice as many plants being acquired in the same regions as the head offices of the acquiring enterprises as would be expected from the distribution of the industries involved (Healey 1983b: 336).

Evidence on the spatial patterns of takeover activity is nevertheless fragmentary and sometimes conflicting. Acquisition patterns are constrained by the distribution of potential acquisition candidates, but whereas examination of acquisition activity in the food, chemicals, textile and clothing industries in 1973-4 showed that acquired firms were reasonably evenly distributed across the planning regions (Leigh and North 1978), a more detailed study of multiplant enterprises in the textile and clothing industry with head offices in the East Midlands, North West England and Yorkshire and Humberside, showed a tendency for acquired plants to be over-represented in the South East, East Anglia, the West Midlands and the South West (Healey 1983b). An analysis of acquiring firms in Leigh and North's national study showed that the acquirers were predominantly based in the South East. The over-representation of acquirers in this region is strikingly illustrated by the fact that while the South East accounted for 21 per cent, 28 per cent, 24 per cent and 4 per cent of employment in food, chemicals, clothing and textiles respectively, it contained 56 per cent, 62 per cent, 65 per cent and 18 per cent of the acquiring firms. What is more, firms based in the South East tended to acquire more firms than those based elsewhere (Leigh and North 1978).

In the long term, the spatial growth of multiplant enterprises by both internal and external methods produces increasingly dispersed manufacturing firms. Such dispersion raises interesting questions about the long-term viability of the different sites within the corporate system as well as more searching questions about the long-term contribution of recently-established branch plants to regional employment totals and the post-acquisition experience of acquired firms. Questions like these are of central importance in studies of locational adjustment, and it is the processes of locational adjustment rather than those of spatial growth which are predominant at the present time.

10.5 Locational adjustment

The essential feature of locational adjustment is that it involves reorganizations made by an enterprise where the number of its plants remains

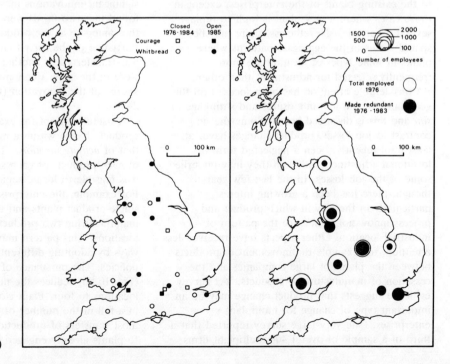

Fig. 10.4 Locational adjustment (a) Courage and Whitbread 1976-84; (b) Metal Box Open Top Division, 1976-82. *Sources:* (a) Corporate data; (b) Based on Peck and Townsend (1984)

the same or is reduced, while their capacities or products are changed. Some examples are shown in Fig. 10.4. Courage (the brewing division of the Imperial Group) reduced its brewing plants from six to three between 1976 and 1984 by closing four plants and opening one. Whitbread reduced its plants from seventeen to eight over the same period by closing ten and opening one, while from 1976 to 1983 the open-top Division of Metal Box reduced its plants from eleven to seven by closing four. In the latter case, divisional employment fell from about 10,000 to around 6.000 employees, which included jobs lost in contraction *in situ* as well as closure (Peck and Townsend 1984).

Locational adjustment thus encompasses many of the components of change identified in Fig. 10.3, including plant closures, sales and *in situ* changes as well as some plant openings. Other forms of adjustment *in situ* include changes in the products and processes, the occupational structure of the labour force, the capital – labour ratio, and the organization of production in individual establishments.

Locational adjustment patterns

Spatial patterns of locational adjustment are more constrained than are those for plant movement or acquisition (section 10.4) because the opportunities available to the decision-makers are clearly limited to the existing plants of the enterprises, except in those cases where a new plant is built to replace several other ones. Nevertheless, there may be spatial biases in the aggregate patterns where plants in some types of location are more frequently selected for adjustment than others.

Considerable attention has been focused on the geography of job loss in closures and shrinkage *in situ* and this is discussed below. Somewhat in contrast to job losses, product changes have, at least until recently, been a neglected form of locational adjustment although they may underlie some of the job losses. In the last few years though, there has been a growing interest, particularly in the way in which product and process innovations influence the pattern of industrial location. Other aspects have received less attention, for example the movement of products between the plants of large companies and the cessation of manufacture of products. Yet survey evidence suggests that product change may be an important type of change for multiplant enterprises. One very large survey reported that a third of a sample of over 1,400 multiplant firms claimed that they had substantially altered their production profiles between 1960 and 1972 by discontinuing older lines of production and substituting new lines as a result of innovation or experiment within the firm, or of better market prospects and improved technology (Hamilton 1978: 163). Another study found that product changes which resulted in an alteration to the minimum list headings in which a plant manufactured were as common as decisions to open branch plants and acquire plants (Healey 1981b: 361). These findings help to account for the observation that 'changing a location is both expensive and time consuming for an industrial company, changing the use made of any given location need not be difficult' (Townroe 1974: 305).

Different types of product change seem to have distinct patterns. One study found that the plants gaining additional products were further from the head office of their enterprise than the plants which lost products (Healey 1983b: 338). The mean distances were 58 and 12 km respectively, suggesting a net outward movement of products within the plants of the enterprises, though the difference was not statistically significant. In contrast, where the additional product is a new one for the enterprise, there is a strong tendency to first manufacture it in the main plant of the enterprise or close by. For example, a study of the location of the first commercial manufacture of significant innovations in multiplant companies found that two-thirds were produced 'in house' at the same site as the product was developed; while thirty-one of the forty-four cases where the product was transferred to another location for manufacture were in the same Economic Planning Region as the source of the innovation (Oakey *et al.* 1980: 242, 248).

A useful concept for examining the effects of product change within a multiplant enterprise is that of *activity locations*. This refers to the location of each product, or each stage of production which it is feasible to locate separately (Healey 1984b). For example, the enterprise in Fig. 10.5a has two single-product plants and two plants each manufacturing two products, making six activity locations. This pattern may be altered in various ways by adopting different spatial rationalization policies. The pursuance of a policy of specialization (Fig. 10.5b) reduces the number of activity locations to four. Plant closure also usually results in a fall in the number of activity locations, as most transfers of production following closures are to plants already engaged in those activities. In the

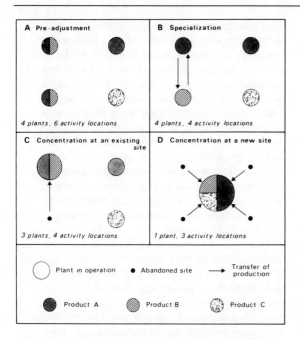

Fig. 10.5 Activity locations in a hypothetical multiplant enterprise. *Source:* Based on Healey (1984b)

two cases illustrated in Fig. 10.4, concentrating production at an existing site (Fig. 10.5c) leads to the number of activity locations being reduced from six to four, while following a policy of concentration at a new site (Fig. 10.5d) reduces the number to three.

Locational adjustment: explanations

In searching for explanations for the different patterns of adjustment, there is a vigorous debate between research workers who emphasize the national and international economic forces as the prime causes of adjustment (Massey and Meegan 1979; 1982) and others who stress the way in which the characteristics of the enterprises and their plants influence change (Healey 1981a; 1981b; 1982). While it would be mistaken to ignore the national and international context in which change takes place, primacy is given here to explanations arising from the response of firms and plants to the environments in which they operate.

In searching for explanations of locational adjustment patterns at enterprise and plant level, it is important to distinguish between production reorganizations resulting from *investment and technical change* where adjustments occur in the context of significant investment, often related to changes in techniques of production, and reorganizations arising from *rationalization* or cuts in a firm's capacity (Massey and Meegan 1982). Minor adjustments may also result from the process of *intensification* by which labour productivity is increased but without major new investment or substantial reorganization of production techniques. It is often difficult to make these distinctions in practice, although they are important conceptually. Here attention is focused on the first two kinds.

The London Brick Company (LBC) provides a simple example of the kind of production reorganization associated with investment and technical change since it is, in essence, a single-product firm (Massey and Meegan 1982). By a series of acquisitions, it had come by the early 1970s to account for the whole of the fletton brick industry. In the face of stagnant demand, it found that it was necessary to reduce costs in order to maintain its profitability. This led to the building of two completely new brickworks at Whittlesey in the Peterborough area. In the larger of the two works, labour productivity was three times its previous level. This investment in new plants led directly to the closure of several other works.

Rationalization, the other type of production reorganization, is used by Massey and Meegan to mean disinvestment, without major reinvestment in plant or machinery or new factory premises. The general reason for its occurrence is, they say, a lack of profitability, though this may be expressed in terms of excess capacity. The rationalization of the UK motor assembly industry since 1982 provides an example, as it was mainly a reflection of the fall in demand in both domestic and export markets. The decline in output to less than half the 1972 figure led to capacity in excess of the requirements for profitable production, and the closure of five of the thirteen motor vehicle assembly plants (Speke, Liverpool; Canley, Coventry; Abingdon; Solihull and Linwood near Glasgow) (Law 1985). Considerable investment has, however, taken place in many of the surviving plants, suggesting that this case does not meet Massey and Meegan's definition of rationalization in full, but this expenditure has had more to do with the introduction of new models and improvements in efficiency than with creating excess capacity. For instance, BL spent over £250 m. on the semi-automated Metro production line at Longbridge.

The descriptions of adjustments to different economic environments help to emphasize the difficulty of isolating the factors influencing the locational adjustment process and of assessing their relative importance, but a useful way of categorizing the factors influencing different types of adjustment is to group them according to the scale at which they most frequently operate (Healey 1984b). It is recognized though that some influences operate at more than one level in the scale hierarchy and the factors at different scales interact together to shape the particular form of adjustment adopted.

At the international and national scales, three factors are particularly important: sectoral trends, enterprise interactions and the socio-political environment. Clearly, the performance of the different sectors in which an enterprise works will be influenced by macro-economic features such as the overall performance of the UK economy, including both its ability to compete in export markets and its ability to withstand foreign competition (through imports) at home.

Within any sector, interactions between individual enterprises at the national scale can influence patterns of change, and the significance of enterprise interaction has been well documented for the brewing industry (Watts 1978; 1980) and can be seen as a factor influencing job loss in, for example, the steel industry. Discussions between British Steel and private steel producers in the early 1980s about reducing excess capacity led to the closure of several private plants including Hadfields (Sheffield), Duport (Llanelli) and Round Oak (Staffordshire), (Townsend and Peck, 1985).

Whereas sectoral trends and enterprise interaction can act as stimuli to locational adjustment, the socio-political environment affects the ease with which it can take place. This is particularly important where job loss is involved. It is often suggested, for instance, that redundancies are easier to implement in times of recession when individual occurrences do not stand out as much as they would in periods of growth. There may even be a band-wagon effect contributing to the 'waves' of redundancy in Britain in 1980–81 (Townsend 1983: 70–1). Socio-political factors can work at a more local scale too as they can affect the decisions of multiplant enterprises as to which of their plants to cut back or close. Some enterprises claim to adopt a socially responsible policy. For example, British American Tobacco claim that their decision to contract their operations in Liverpool rather than transfer all the production to Southampton was because 'the company is very conscious of its responsibilities on Merseyside'.

This was despite the fact that manufacturing costs in Southampton were half those in Liverpool (*Southern Evening Echo*, 7 February 1984). More usually, negative features of the local environment are cited to support the closure or shrinkage *in situ* of particular plants. The existence of a militant labour force has, for example, been cited as a reason contributing to the closure of the major assembly plant at Speke (Law 1985: 8); while the lack of room for expansion, poor access and deterioration in the urban residential environment have been suggested as factors encouraging multiplant firms to evaluate their inner-city plants less favourably than those located elsewhere (Keeble 1978: 106–10).

The various influences discussed so far have all been external to the companies, but the nature of the enterprises themselves and the characteristics of the plants they operate may also affect which enterprises and which plants are involved in locational adjustment. Smith and Taylor (1983: 639) have stressed that adjustments cannot be understood 'without reference to the larger decision-making structure of the firm as a whole, especially in the case of multiplant and multiproduct enterprises'. This is seen particularly clearly in those cases where changes in the organizational structure of individual companies precede locational adjustment. For example, the formation of a workshop division within British Rail gave a national perspective on their railway engineering operations which had previously only been considered on a regional basis. A major rationalization scheme followed where product rationalization was accompanied by investment in selected sites (Watts 1980: 128–9). Despite several case studies which point to the influence of the characteristics of enterprises on their locational behaviour (see, for example, Watts 1980: 235) there have been few systematic attempts to test the relationships found in these studies on larger samples. An exception is a study of small and medium-sized multiplant enterprises in the textile and clothing industry which found relatively few significant relationships between enterprise characteristics and locational behaviour (Healey 1979; 1981b; 1982). For example, although plant closures were found to occur more frequently in larger enterprises and were least frequent in companies following a growth strategy, differences in profitability, previous rates of growth, organizational structure and product nature had no influence on the frequency of closures.

Examinations of the effects of plant

characteristics have been more rewarding. Evidence to support the influence of these is available from both case studies and larger surveys. For example, the decision of the London Brick Company as to which of their plants to close following the opening of the new brickworks at Whittlesey was influenced primarily by their relative efficiencies, but was also affected by the cost of transport of the output to the market, the ease with which a works could be re-manned to meet cyclical increases in demand, and variations in the quality of clay (Massey and Meegan 1982: 175–6).

Healey (1981a), in a study of the influence of ten plant characteristics on eight types of locational adjustment, found general support for the contention that the nature of the plants operated by companies had a constraining influence on their locational behaviour. Four of the attributes of the plants were particularly important influences on locational adjustment – size of plant, space for extension, recently-opened plants and plant status. For example, the main contributor to the significant differences between the status categories was the locational behaviour of the head office plants of the parent companies. They were less frequently closed and more frequently extended than were branch plants or the head office plants of their subsidiary companies. They were also more likely to gain or lose products than other plants.

One result of the increased significance of external growth in the recent past is that many plants experience a change of ownership. The usual argument put forward is that recently acquired plants are particularly likely to be involved in locational adjustment in an attempt to integrate them into the enterprise system. For instance, studies of the brewing (Watts 1977) and iron foundry (Smith and Taylor 1983) industries and of large plants in the Northern region (Smith 1979) found that recently acquired plants were more at risk of being closed than other plants. The extent of the risk seems to vary over time though, with closure being most likely a few years after a merger. One study found that there was a short period in which acquired plants were no more likely to be closed, but then after about five to six years of operation the probability of their closure became higher than that for the other plants (Healey 1982: 46–7). There is also evidence that plant closures are more frequent following horizontal takeovers than others types (Healey 1982: 47; Leigh and North 1978: 24).

Our understanding of the factors influencing locational adjustment is still very fragmentary, but multiplant enterprises do appear to form a distinctive group of organizations which have an identifiable locational behaviour. It remains to examine what effect their behaviour has on urban and regional development.

10.6 Multiplant enterprises and the urban and regional system

Two main elements have been characteristic of the changing distribution of UK manufacturing industry over the last two decades. These elements consist of the urban – rural shift and the new spatial division of labour (see Ch. 1). Identifying the role of the multiplant enterprise in contributing to these changes is not easy and the the evidence is limited.

Urban – rural shifts

Evidence of the role of multiplant enterprises in the urban – rural shift comes from both regional and sectoral studies but these refer, almost inevitably, to only a few regions or sectors. Indeed, the regional evidence relates only to the East Midlands (Fothergill and Gudgin 1982: 81–3).

Table 10.5 Urban–rural contrasts in the East Midlands, 1968–75

	Employment change* as percentage of 1968 employment in each area	
	Multiplant firms	Single-plant firms
Cities	−17.8	−9.5
Larger towns	−13.6	−5.1
Smaller towns	−7.9	−3.8
Rural areas	+2.9	+2.7

Source: Fothergill and Gudgin 1982: 96
* Employment change in existing factories made up of net employment change in survivors minus employment loss through closure.

Table 10.5 examines employment change from 1968 to 1975 in plants belonging to single-plant and multiplant firms within the East Midlands. The multiplant firms show much greater urban – rural contrast in growth, suggesting that they may be diverting expansion away from the cities to existing branches in smaller towns and rural areas. The contrast between urban and rural areas might be even greater if the data also included new branch plants set up in rural areas to undertake work diverted from city plants. In the East Midlands, city-based firms were more likely than those in other areas to create employment in new branch plants within the region.

While the rather restricted regional evidence does indicate a small role for multiplant firms in contributing to the urban – rural shift, the sectoral evidence provides much stronger support for the view that some multiplant enterprises are abandoning certain urban areas. Healey (1983b: 335) was able to detect that closures were more common in large cities in his sample of multiplant textile and clothing enterprises, and in the brewing industry there is evidence that older city plants have a greater propensity to close than those elsewhere (Watts 1980). Perhaps the most detailed study to date and one in which the conclusions can be related directly to an urban – rural framework is Massey and Meegan's (1978; 1979) examination of restructuring in the electrical engineering sector.

The most interesting group of firms were those facing restructuring because of over-capacity and high costs. These included firms manufacturing heavy electrical machinery (especially turbine generators, switchgear and transformers), supertension cables and aerospace equipment. The firms were, prior to reorganization, located primarily in the West Midlands, North West, South East and Northern regions. The financial restructuring of the industries (i.e. the amalgamation of firms into larger organizations) was necessary to coordinate the cuts in capacity and to re-allocate investment to more profitable activities. The cutbacks in capacity resulted in both partial and complete plant closures.

Examination of the intraregional patterns shows that the reorganizations had an important effect on employment in the major cities of Birmingham, Greater London, Liverpool and Manchester. These cities had more than their fair share of job loss: they had 32 per cent of the jobs prior to reorganization but they accounted for 44 per cent of employment loss. This over-representation in the cities could be explained largely by the fact that the choice of plants for closure was based on a desire to increase labour productivity. The plants closed were relatively labour-intensive and such plants were located predominantly in the major cities (Massey and Meegan 1978: 276–9).
Alternative interpretations are that these city plants occupied restricted sites which were unsuitable for the introduction of new techniques or, where new techniques were introduced, they reduced the labour inputs (Fothergill and Gudgin 1982).

The new spatial division of labour

It will be recalled that one of the characteristic features of the larger multiplant enterprise is its locational hierarchy. An outcome of these hierarchies is a new spatial division of labour (Massey 1979). Whereas in the nineteenth century, different regions tended to develop and specialize in different industrial sectors, in the late twentieth century the specialization is by occupation. Nowhere is this seen more clearly than in the concentration of white-collar jobs in the South East.

There is little doubt that multiplant enterprises have played a role in creating the new spatial division of labour, but their contribution to this over-representation is not yet clear. However, Gudgin et al. (1979: 152–6) estimate that multiregional organizations (a useful surrogate for larger multiplant enterprises) account for an excess of about 140,000 white-collar jobs in manufacturing in the South East. Out of this total, approximately 50,000 jobs are in detached head offices (these are mainly the offices referred to in the 'head office' literature) and another 20,000 in corporate R and D establishments. The remainder (another 70,000 jobs) are in activities whose significance had previously been unrecognized and these were the subcorporate head offices associated with branches, divisions or subsidiaries. These findings stress that the emphasis in the literature on corporate head offices ignores an important element in the concentration of white-collar activities in the South East.

The presence of a high proportion of white-collar workers in a particular region can be seen to bring a number of advantages to it. With a declining production workforce, there is a tendency for non-production workers to increase or, at least, decline more slowly and such workers appear to be less sensitive to fluctuations in the business cycle. Second, their high incomes can stimulate demand for services and some of this income is fed back into the regional economy. Third, the concentration of decision-makers (particularly those making major investment and disinvestment decisions) might work to a region's advantage through new investments being located primarily in the head office region simply because of proximity to it. This latter argument has been put particularly forcibly by representatives of regions on the UK periphery who see distance from the main centres of decision-making as a major disadvantage. They stress that as a result of the concentration of control functions of multiplant firms in the South East, a large number of regions are characterized by high levels of external control.

External control is measured conventionally by the number of employees in externally-owned plants. In the Northern region, among plants with over 100 employees, almost 80 per cent were in externally-owned plants in 1973 (Smith 1979: 429), while in Scotland the proportion was 64 per cent in plants of all sizes in 1977 (Cross 1981: 519). However, ownership data provide only a crude indication of the extent of external control.

It may be higher than these figures suggest where plants sell a major part of their output to a firm outside the region and are thus reliant on sales to that firm for survival and it may be lower where subsidiary plants have a high degree of autonomy from their parent firm. In Wales, an examination of externally-controlled plants found that about half were subsidiaries with considerable autonomy (Tomkins and Lovering 1973) while the proportion in the Northern region in this category was just over a third (Marshall 1979). Clearly, the centralization of control functions of multiplant firms in the South East has produced externally-controlled economies elsewhere, although the high percentages cited as an indication of the extent of external control need to be interpreted cautiously.

The extent of external control has developed quite recently. In the Northern region, employment in externally-owned plants rose by 20 percentage points from 1963 to 1973 (Smith 1979: 427). The most important element contributing to the increase in external control was acquisition with a much smaller contribution from the establishment of new branch plants. However, there is now some evidence to suggest that levels of external control may be falling (Healey and Clark 1985: 1356–7) where, in recession, externally-controlled plants are losing labour faster than their indigenous counterparts.

The implications of the emergence of branch plant economies have stimulated considerable research interest in the last decade (Watts 1981) in order to establish whether or not external control has any important implications for a region's development. There has been considerable speculation as to its effects upon employment.

Evidence supporting the view that external control results in the loss of white-collar jobs is strong and is an almost inevitable repercussion of the concentration of white-collar activities in the South East. It is sometimes argued that it also results in the loss of skilled jobs and an increase in the proportion of unskilled ones but, as was shown earlier in this chapter, this is speculation supported only by limited evidence.

There is rather more evidence relating to job stability although it is not clear-cut. Recent evidence, however, suggests externally-controlled plants are particularly vulnerable to closure. Plants established in peripheral areas (usually characterized by high levels of external control) from 1966 to 1975 as a result of interregional moves showed a much higher propensity to close than moves to elsewhere (Henderson 1980: 159). This may reflect, in part, multiplant companies withdrawing from their more peripheral operations, a viewpoint confirmed in Healey's (1983b) study of clothing and textiles where the median distance of closed plants from their head offices was higher than that for the plants remaining in operation. Townsend's (1983: 81) results for the twenty largest companies in the UK suggest that the assisted areas may have continued to fare worse in the late 1970s and early 1980s in that this group of multiplant enterprises showed a distinct tendency to announce a job loss in the assisted areas before announcing any job losses in the rest of the country. It was also noted earlier (p. 22) that acquired plants can be at greater risk of closure than other plants within a corporate system. There are few aggregate data sets relating to the whole of the UK, but data for the iron foundry industry (Smith and Taylor 1983: 653) indicate the plants within multiplant groups were 50 per cent more vulnerable to closure compared with independent plants. A similar trend was noted among acquired plants in the brewing industry (Watts 1980: 220).

While closures of plants of multiplant enterprises are important, perhaps most important to the provision of employment opportunities is the rate at which *in situ* employment change takes place in the externally-owned sector in comparison with the indigenous sector. In the Northern region, a comparison of plants which survived throughout the study period (1963–78) and which did not change status showed that the externally-owned plants which were part of multiregional corporate systems experienced an increase of 17.0 per cent compared with a fall of −4.6 per cent in independent plants which were predominantly single-plant firms. These data relate to only one region and it appears that independent firms showing employment growth were acquired by external firms. Plants acquired between 1963 and 1973 had expanded their employment by 18 per cent prior to acquisition but the number of jobs had fallen by 14 per cent afterwards (Smith 1979).

Studies of regional economies have drawn attention to the role of the larger multiplant firms which are often described as the 'prime-movers' of the regional economy (Lloyd and Reeve 1982). The plants belonging to the multiregional firm may well benefit a region by allowing access to the parent's financial resources, by permitting entry into the parent's markets and by access to any technological or administrative innovations adopted by the parent firm. Against these advantages must be set the feeling that distant control results in an insensitivity to local needs, a point of view expressed forcibly by nationalist politicians in Wales and Scotland and by local leaders in regions such as Tyneside and Teesside. Jobs may be de-skilled, branch plants closed and local entrepreneurial initiative damped down by high levels of external control, but against this must be set the greater competitiveness of the multiregional enterprise in the national economy. Merseyside's loss may be Britain's gain, although increasingly there is a suspicion that the gains are being felt not in the rest of the country but abroad as large multiplant enterprises internationalize their production (for example, Lloyd and Shutt 1983; Ch. 11).

10.7 Conclusion

The multiplant enterprise will continue to play a dominant role in changing the geography of employment opportunities in the late 1980s and 1990s and this chapter has illustrated the characteristics of the locational behaviour of these companies over the last two decades. It should be emphasized that the patterns of change described and the explanations for them depend upon the historical circumstances in which the enterprises were operating at the times the investigations were undertaken. Admittedly their responses were not determined by the environments in which the companies operated, but such environments did limit the range of options open to them.

The current political climate makes it unlikely that new policy initiatives focused on the multiplant firm will emerge from central government to change the geography of employment opportunities. Some local agencies may well intervene in the affairs of some smaller local firms (witness the work of the Welsh and Scottish Development Agencies as well as the initiatives by some local authorities) but the abolition of the Metropolitan counties which have been particularly active in this field and the rate-capping which is applied to some other authorities suggests the degree of intervention over the next few years is unlikely to increase and may well diminish. Even in more prosperous times and under different political masters, the concept of the Planning Agreement failed to get off the ground (Watts 1980: 270–81).

Although it seems unlikely that controls will be introduced in the immediate future on the plans of large corporations to close or make substantial reductions at particular plants which are as stringent as already exist in some other countries (Sweet 1981), there is some evidence that private sector multiplant firms are taking more account of the effects of their employment cutbacks on particular local economies. The case of British American Tobacco was referred to earlier (p. 160) and other examples are not difficult to find. ICI, for example, among other activities, is involved in 'Business Link' intended to promote job generation in small firms in the Runcorn – Widnes area where at one time it employed about 15 per cent of the working population. Similarly, Unilever is involved with 'In Business' designed to aid small firms in the Wirral. It must be said in most cases that the funds made available by the private sector have been small relative to the turnover of the companies involved and the provision of funds may best be seen not so much as acts of social conscience but more as deliberate attempts to ease the run-down of their workforce in particular areas by appearing to help to stimulate new jobs. Such measures, however, are palliatives rather than cures to the problems caused for local economies by corporate reorganizations.

The dominance of multiplant enterprises in the British economy means that there is a need for future research to continue to probe into the ways in which they adjust and change the whole range of their activities. This is particularly important because examining the locational behaviour of multiplant enterprises is one of the keys to understanding the process of urban and regional development.

References

Cleveland County Council 1982 Manufacturing employment change in Cleveland since 1965. Planning Department, Cleveland County Council, Report 21.

Cross, M. 1981 *New firm formation and regional development*. Gower.

Department of Trade and Industry 1983 *Regional industrial policy: some economic issues*. Department of Trade and Industry.

Dicken, P. 1976 The multi-plant business enterprise and geographical space: some issues in the study of external control and regional development. *Regional Studies* **10**: 401–12.

Dicken, P., Lloyd, P. E. 1978 Inner metropolitan industrial change, enterprise structures and policy issues: case studies of Manchester and Merseyside. *Regional Studies* **12**: 181–97.

Fothergill, S., Gudgin, G. 1982 *Unequal growth: urban and regional employment change in the UK*. Heinemann.

Goddard, J. B., Smith, I. J. 1978 Changes in corporate control in the British urban system, 1972–1977. *Environment and Planning A* **10**: 1073–84.

Gudgin, G., Crum, R., Bailey, S. 1979 White collar employment in UK manufacturing industry. In Daniels, P. W. (ed.) *Spatial patterns of office growth and location*. John Wiley, 127–51.

Hamilton, F. E. I. 1978 Aspects of industrial mobility in the British economy. *Regional Studies* **12**: 153–65.

Hardill, I. 1982 Components of employment change in the West Yorkshire woollen textile industry, 1972–1976. Centre for Urban and Regional Development Studies, University of Newcastle upon Tyne, Discussion Paper 44.

Hayter, R., Watts, H. D. 1983 The geography of enterprise: a re-appraisal. *Progress in Human Geography* **7**: 157–81.

Healey, M. J. 1979 Changes in the location of production in multi-plant enterprises with particular reference to the United Kingdom textile and clothing industries, 1967–72. Unpublished PhD thesis, University of Sheffield.

Healey, M. J. 1981a Locational adjustment and the characteristics of manufacturing plants. *Transactions Institute of British Geographers* **6**: 394–412.

Healey, M. J. 1981b Product changes in multi-plant enterprises. *Geoforum* **12**: 359–70.

Healey, M. J. 1982 Plant closures in multi-plant enterprises – the case of a declining industrial sector. *Regional Studies* **16**: 37–51.

Healey, M. J. 1983a The changing data base: an overview. In Healey, M. J. (ed.) *Urban and regional industrial research: the changing UK data base*. Geo Books, 1–29.

Healey, M. J. 1983b Components of locational change in multi-plant enterprises. *Urban Studies* **20**: 327–41.

Healey, M. J. 1984a Industrial change in Warwick District 1974–83, Department of Geography, Coventry Lanchester Polytechnic, Industrial Location Working Paper 6.

Healey, M. J. 1984b Spatial growth and spatial rationalisation in multi-plant enterprises. *Geo Journal* **9**: 133–44.

Healey, M. J. 1985 Industrial decline, industrial structure and large companies. *Geography* **70**: 328–38.

Healey, M. J., Clark, D. 1984a Industry in the inner and outer areas of Coventry: similar or different? *City of Coventry Economic Monitor* **2**(84): 43–56.

Healey, M. J., Clark, D. 1984b Industrial decline and government response in the West Midlands: the case of Coventry. *Regional Studies* **18**: 303–18.

Healey, M. J., Clark, D. 1985 Industrial decline in a local economy: the case of Coventry 1974–82 *Environment and Planning A* **17**: 1351–67.

Henderson, R. A. 1980 An analysis of closures amongst Scottish manufacturing plants between 1966 and 1975. *Scottish Journal of Political Economy* **27**: 152–74.

Herron, F. 1981 The Post Industry Act (1972) industrial movement into, and expansion in, the Assisted Areas of Great Britain: some survey findings. Government Economic Service Working Paper 56.

Howells, J. R. L. 1984 The location of research and development: some observations and evidence from Britain. *Regional Studies* **18**: 13–29

Keeble, D. 1978 Industrial decline in the inner city and conurbation. *Transactions Institute of British Geographers* New Series **3**: 100–14.

Killick, T. 1983 Manufacturing plant openings 1976–80: analysis of transfer and branches. *British Business* **17** (June): 466–8.

Law, C. M. 1985 The geography of industrial rationalization: the British motor car assembly industry 1972–82. Geography **70**: 1–12.

Leigh, R., North, D. J. 1978 Regional aspects of acquisition activity in British Manufacturing industry. *Regional Studies* **12**: 227–45.

Lloyd, P. E., Reeve, D. E. 1982 North West England, 1971–77: a study in industrial decline and economic restructuring. *Regional Studies* **16**: 345–59.

Lloyd, P. E., Shutt, J. 1983 Recession and restructuring in the North West Region: the policy implications of recent events. North West Industry Research Unit, School of Geography, University of Manchester, Working Paper Series 13.

Macey, R. D. 1982 Job generation in British manufacturing industry: employment change by size of establishment and by region. Government Economic Service Working Paper 55.

Marshall, J. N. 1979 Ownership, organisation and industrial linkage: a case study in the Northern Region of England. *Regional Studies* **13**: 531–57.

Massey, D. 1979 In what sense a regional problem? *Regional Studies* **13**: 233–43.

Massey, D., Meegan, R. 1978 Industrial restructuring versus the city. *Urban Studies* **15**: 273–88.

Massey, D., Meegan, R. 1979 The geography of industrial reorganisation: the spatial effects of the restructuring of the electrical engineering sector under the Industrial Reorganisation Corporation. *Progress in Planning* **10**: 155–237.

Massey, D., Meegan, R. 1982 *The anatomy of job loss:*

the how, why and where of employment decline. Methuen.

Meegan, R. 1982 Telecommunications technology and older industrial regions. Centre for Environmental Studies, Paper 7.

Nunn, S. 1980 The opening and closure of manufacturing units in the United Kingdom 1966–75. Government Economic Service Working Paper 36.

Oakey, R. P., Thwaites, A. T., Nash, P. A. 1980 The regional distribution of innovative manufacturing establishments in Britain. *Regional Studies* **14**: 235–53.

Peck, F. W., Townsend, A. R. 1984 Contrasting experience of recession and spatial restructuring: British shipbuilders, Plessey and Metal Box. *Regional Studies* **18**: 319–38.

Pounce, R. J. 1981 *Industrial movement in the United Kingdom 1966–75*. HMSO.

Prais, S. J. 1976 *The evolution of giant firms in Britain*. Cambridge University Press.

Smith, I. J. 1979 The effect of external takeovers on manufacturing employment change in the Northern region between 1963 and 1973. *Regional Studies* **13**: 421–37.

Smith, I. J., Taylor, M. J. 1983 Takeover, closures, and the restructuring of the United Kingdom iron foundry industry. *Environment and Planning A* **15**: 639–61.

Sweet, M. L. 1981 *Industrial location policy for economic revitalisation: national and international perspectives*. Praeger.

Taylor, M. J., Thrift, N. 1982 Industrial linkage and the regional economy: 1. Some theoretical proposals. *Environment and Planning A* **14**: 1601–13.

Taylor, M. J., Thrift, N. 1983 Business organisation, segmentation and location. *Regional Studies* **17**: 445–65.

Tomkins, C., Lovering, J. 1973 *Location, size, ownership and control tables for Welsh industry*. The Welsh Council.

Townroe, P. M. 1974 Post-move stability and the location decision. In Hamilton, F. E. I. (ed.) *Spatial perspectives on industrial organisation and decision-making*. John Wiley, 287–308.

Townsend, A. R. 1983 *The impact of recession*. Croom Helm.

Townsend, A. R., Peck, F. W. 1985 An approach to the analysis of redundancies in the UK (post-1976): some methodological problems and policy implications. In Massey, D., Meegan, R. (eds.) *The politics of method: contrasting studies in industrial geography*. Methuen.

Watts, H. D. 1972 Further observations on regional growth and large corporations. *Area* **4**: 269–73.

Watts, H. D. 1974 Spatial rationalisation in multi-plant enterprises. *Geoforum* **17**: 69–76.

Watts, H. D. 1977 Market areas and spatial rationalization: the British brewing industry after 1945. *Tijdschrift voor Economische en Sociale Geografie* **68**: 224–40.

Watts, H. D. 1978 Inter-organisational relations and the location of industry. *Regional Studies* **12**: 215–25.

Watts, H. D. 1980 *The large industrial enterprise: some spatial perspectives*. Croom Helm.

Watts, H. D. 1981 *The branch plant economy: a study of external control*. Longman.

CHAPTER 11

Multinational enterprises

F. E. Ian Hamilton

11.1 Introduction

> For ... American investment ... some definite, clear-cut and well-defined policy, on the part of the UK would seem desirable.
> (Dunning 1958; 321)
>
> ... very few resources are devoted to understanding MNE corporate strategies, assessing their UK impact and considering how UK policy instruments might be employed in influencing them at national or regional level.
> Department of Trade and Industry, 1983: 327)

These two telling statements by leading British economic analysts of multinational enterprise (MNE) are significant in several respects. They underline the continued inaction or inertia on the part of the UK government in formulating a policy or set of policies towards MNEs; they indicate that a concern about the operations of US-based corporations has been replaced by a broader interest in the activities of all MNEs in the UK. That trend is related to two further implications. The web of MNE involvement in the UK economy has become far more complex and dynamic since the 1950s and is one form of the much increased importance of internationalization or 'globalist' tendencies. Finally, that the study of MNEs that has burgeoned worldwide since 1970 still does not add up, certainly in the UK case, *either* to sufficiently comprehensive data collection and monitoring of recent and contemporary changes in MNE activity *or* to adequately detailed analysis of behaviour to yield policy-aiding predictions of individual or aggregate decisions by such corporations.

The foregoing quotations, however, refer to inward foreign investment only. In 1982 the UK still remained, after the USA, the world's second largest source amongst developed capitalist economies of the stock of direct foreign investment located abroad, although annual outward flows of capital investment from Britain had been surpassed in the early 1970s by flows from West Germany and in the mid-1970s also by those from Japan. Yet the patterns and processes of growth and development of British industrial firms overseas into MNEs are not officially documented; the details must be gleaned laboriously, if this is at all possible, from such sources as company reports and the *Financial Times*. Such a situation is both a cause and an effect of the lack of policy.

11.2 Multinationals and the changing international position of the UK economy

Ever since the latter half of the nineteenth century when entrepreneurs initiated industrialization outside Britain – especially in the US, Germany and Japan – manufacturing in the UK and by UK firms globally have had to endure relative decline in world importance. For almost a century this resulted from slower rates of growth, adjustment and development than in other countries. Not until 1966 did aggregate employment in British industry begin to fall, but since 1978 several manufacturing sectors have suffered an absolute decline in *output* as well. Given the components of the UK industrial system, the operations of both domestic and foreign-based multinationals have been only one set of forces in the changing international role of manufacturing located in Britain. Nevertheless, they constitute very significant forces. In broad terms, one can hypothesize that multinationals have shaped relative and absolute trends in manufacturing in the UK in at least five inter-related ways:

1. The slower growth of UK-owned multinationals as compared with those headquartered in other countries.
2. The shift by UK-based multinationals of proportionately more production, trade and services to locations overseas.
3. A relative decline in the attractiveness of the UK for inflows of foreign direct investment and hence a drop in the rate of development of foreign-owned firms there.
4. Alterations in the sectoral composition of both home and foreign multinationals operating in UK space.
5. Qualitative and quantitative changes in the population of domestic and foreign multinational enterprises and their establishments in the UK.

Focus in this chapter is mostly on the first three trends.

According to Dunning (1982), foreign direct investments were probably as important in the world economy in relative terms before 1914 as in the 1960s and 1970s. This somewhat surprising finding does not contradict his earlier observation that: 'The increase in the contribution of MPEs to world industrial output is one of the most impressive economic features of the last two decades' (Dunning 1971: 19). In fact, the two world wars, economic recessions and widespread nationalism in the 1920s and 1930s, introduction of anti-trust legislation in major industrial nations, and nationalization of foreign assets especially in the USSR, Eastern Europe and East Asia (after 1917 and 1945–54) – all these combined to produce a deep 'sag' in global foreign direct investment between 1914 and 1954. Thereafter, the increase referred to by Dunning became an integral part of the unprecedented world industrial boom of the 1950s and 1960s: it was that which restored the relative importance of foreign direct investment to its pre-1914 levels. Yet multinationals have become ever more overt because of the increasing number of enterprises that have 'gone transnational', the greater concentration of the production, trade and services of certain sectors in the hands of a few giant corporations, the general confinement of their activities to the non-communist world and their recent accelerated interpenetration mainly of the advanced capitalist economies of North America, Western Europe and Japan. Most importantly, however, the establishment or acquisition of *manufacturing* subsidiaries have increasingly substituted for primary activities and portfolio investments in the foreign investment process. Such a fundamental change has been encouraged by market growth in a proliferating number of independent states protected differentially by tariffs in a world of more liberalized trade. It has been further facilitated by technological innovations in transport, communications and production processes which have reinforced the internationalization capabilities of firms in general and larger corporations in particular. But the perception of higher risk has deterred significant growth of activity in the Third World.

11.3 The relative decline of UK multinationals

The apparent similarities in the relative roles of foreign direct investment (fdi) in the world economy before 1914 and after 1960 should not obscure dramatic changes which are especially marked in the UK case. The estimated stock of accumulated fdi rose from about US $14 bn in 1914 to more than US $400 bn today (Table 11.1). Yet the geographic pattern of sourcing has altered greatly: in this respect the UK forms part of the broader relative decline in capital sourcing from Western Europe, a drop from about 77 per cent in 1914 to about 40 per cent in 1978, in the face of rising American and, latterly, Japanese fdi. The UK still remains, after the US, the second largest source of fdi, yet its share has seen the steepest and most continuous decline of any West European country – from almost half in 1914 to less than one-eighth in 1978. Even so, the stock of UK investment abroad in 1978 was more than four times greater than it had been in 1960, although this is to be compared with a fifty-fold increase in fdi from firms headquartered in Japan, a seven-fold rise in that from other EEC-based firms and a five-fold growth in that of US firms in the same period.

These broad trends hint that UK-based multinationals have declined relatively in world importance. Much evidence would have to be collected and sifted before one could prove any such hypothesis. What data are to hand, however, are restricted in their coverage and comparability both through time and at points in time. But they are indicative. Analysis of information presented in the Fortune 500 for the period 1978–82, for

Table 11.1 Estimated stock of direct investment abroad by area or country of origin 1914–78

Area/country	1914		1938		1960		1978	
	US$bn	%	US$bn	%	US$bn	%	US$bn	%
USA	2.7	18.5	7.3	27.7	32.8	52.0	168.1	43.5
UK	6.5	45.5	10.5	39.8	10.8	17.1	45.9	11.9
Rest of EEC	4.6	31.5	6.35	24.1	11.9	18.9	78.4	20.3
Other Europe	—	—	—	—	2.4	3.8	33.8	8.8
Japan	0.02	0.1	0.75	2.8	0.5	0.8	26.8	6.9
Rest of world	0.5	4.4	1.45	5.6	4.7	7.4	33.2	8.6
Total	14.3	100	26.35	100	63.1	100	386.2	100

Source: Dunning 1982

instance, permits a comparison of the world sales rankings of UK multinationals in manufacturing as a whole and in the wide industry groups used by Fortune. Though this corresponds to the years of most severe 'de-industrialization' in Britain in job terms, the data indicate divergent trends within and across industry groups and size groups of multinationals. The investigation covered the world's top 800 multinationals, this number coincidentally being decided by the lowest comparable sales level in the Fortune 500 Directories respectively of US and of non-US industrial firms. In 1982, the 500th non-US firm had sales of US $786.7 m., whereas 343 US corporations enjoyed sales in excess of that figure. The discrepancy of forty-three in numbers of firms is accounted for by 'double' entries in the two tables combined: the non-US tables included some US-owned subsidiaries (e.g. Ford Britain, Ford Germany, Ford Brazil, Ford Mexico etc.), while the US tables listed some subsidiaries of non-US firms (e.g. Lever Brothers, subsidiary of Unilever; BASF Wyandotte etc.).

Several clear, yet complex, trends emerge from such an investigation:

1. Apart from the two petroleum multinationals that retained their identical world sales ranks between 1978 and 1982 – Royal Dutch Shell (Anglo-Dutch owned) and British Petroleum (BP) – an almost equal number of UK-owned multinationals improved their world placings (39) as dropped rank (41). Some of this pattern of change is, of course, accounted for by transfers of subsidiaries or assets between the UK and non-UK multinationals in the Fortune listings. The pending takeover by Sumitomo of Dunlop's French subsidiaries and its northeast England branch, or the 'swapping' of plastics capacities in the UK between ICI and Royal Dutch Shell are recent examples. However, while an aggregate of seven UK multinationals climbed into the top 800 for the first time after 1978 (e.g. Racal Electronics), it is significant that two-fifths of those UK-owned multinationals which dropped rank each plummeted by more than 50 places while only one-quarter of those which raised their rank did so by more than 50 places.

   This would broadly indicate, therefore, a decline in the world status of UK multinationals in aggregate on the basis of sales. On the other hand, it is certain that the leading UK multinationals in the world's top 800 performed globally well or at least not too badly compared to other multinationals, and far better than UK-located manufacturing as a whole. While this implies that facilities located overseas played a major and increasing role in maintaining the position (or cushioning the decline) of UK multinationals, it is also probable that their performance *in the UK itself* was superior to (or at least less poor than) that of smaller, non-transnational British firms.

   This aggregate pattern, however, masks some interesting and diversified rank movements by multinationals, especially when these are considered by sectors. It is to these trends that we must now turn.

2. All UK-owned multinationals in the chemicals, rubber and tyre, textile, vehicle and automotive parts and office equipment industries declined in world sales rank between 1978 and 1982. With the exception of the latter, all operate in the sectors most affected by inflated oil purchase

prices. On the face of it, this might be construed as evidence that UK multinationals may have fared no less badly than their foreign rivals in these sectors. In practice, while this appears to be true of chemicals, with ICI in particular holding up its position well, firms such as Courtaulds (textiles and artificial fibres), British Leyland (BL) vehicles and its component suppliers such as Lucas or Guest Keen and Nettlefolds (GKN) and Dunlop (rubber and tyres) dropped by some 25 to 60 places in their rank amongst world industrial multinationals, and they also slipped badly in rank *within* their sectors. That office equipment manufacturers dropped rank is especially worrying given that this is a growth industry; it suggests poor competitiveness in comparison with foreign firms.

3. Metal manufacturing and metal products, machinery, industrial equipment, papermaking and timber processing were all typified by more UK multinationals dropping rank, both amongst all industrial firms and within their sectors, than firms which raised their rank. Particularly hard hit were corporations in engineering in which four out of five UK multinationals slipped by more than 50 places. Clearly this reflected the failure of a majority of UK multinationals in these fields to maintain market shares and hence to restructure enough or at all in the face of the world recession in capital goods, (and to some extent also consumer goods' demand).

4. More UK multinationals rose than fell in rank in the foods, drink and tobacco and in the electronics sectors. Those in the latter were broadly in step with the world-wide tendency for electronics multinationals to improve their position, though the UK performance was generally below average for the sector. Greatest movement of firms occurred in food, drink and tobacco, industries in which UK-owned corporations hold world leadership (see Tables 11.2 and 11.3): that nine multinationals dropped rank, two by more than 50 places, was more than offset by thirteen which climbed, four of them by more than 50 places in the top 800. Such trends clearly express a continued process of concentration into very large firms headquartered in the UK, partly as a result of takeovers, mergers and acquisitions amongst the twenty-two UK multinationals themselves in these sectors.

5. UK or part-UK firms in only one sector – petroleum refining – showed considerable stability in world ranks, with Royal Dutch Shell in second place (after Exxon Corporation) and BP fifth (after also General Motors Corporation and Mobil) but Ultramar (Table 11.3) gained at the expense of Burmah Oil.

6. UK firms in a surprisingly disparate group of industries all improved their world rankings both in their sectors and in manufacturing as a whole: Clark (leather and shoe), Pilkington

Table 11.2 World rankings of UK large industrial organizations within sectors 1982 (by sales)

Sector	Total	World rankings of UK companies in the top 50 of the sector
		Rankings
Food, drink, tobacco	17	1 9 12 15 18 24 25 27 31 32 34 35 38 41 43 49 50
Petroleum	2	2 5
Metal manufacturing	3	10 11 50
Engineering*	5	21 30 32 35 46
Electrical and electronics	4	14 25 39 43
Metal goods	3	6 12 19
Chemicals†	4	2 6 33 34
Textiles, clothing & footwear‡	4	2 10 19 22
Paper, publishing & printing	4	6 9 35 43
Pharmaceuticals‡	3	12 19 23
Total	49	

Sources: Fortune 500 2 May 1983 and 22 August 1983.
* Vehicles, aerospace, industrial and farm equipment
† Includes rubber, tyre and soaps
‡ Of top 25

Table 11.3 The largest UK industrial organizations 1982

Company	Main industrial field	Sales Total (US $ m)	Sales World rank	Employment World rank	Employment ('000s)	Prod. abroad as a percentage of sales	UK exports as a percentage of UK prod.
Royal Dutch Shell*	Petroleum	83,760	2	21	163		
British Petroleum	Petroleum	51,322	5	24	145		
Unilever*	Food, soaps, etc.	23,120	17	8	283		
BAT Industries	Tobacco	15,478	35	19	178	74	11
Imperial Chemical Industries (ICI)	Chemicals	12,872	50	38	124	46	23
National Coal Board (NCB)	Coal	9,033	72	10	282		
General Electric Company (GEC)	Electrical	8,006	82	18	184	35	22
British Steel Corporation (BSC)	Steel	6,580	101	52	104	4	24
Rio Tinto Zinc (RTZ)	Metals	6,438	103	104	70		
Grand Metropolitan	Food	6,087	113	40	129	22	2
George Weston Holdings	Food	5,696	124	47	114	54	5
British Leyland (BL)	Vehicles	5,374	143	51	105	13	35
Imperial Group	Tobacco, food	5,146	151	57	101	18	1
Dalgety-Spillers	Food	4,685	169	21	332	37	1
Thorn–EMI	Electrical	4,654	171	60	98	33	8
Allied Lyons	Brewing, food	3,807	215	86	76		
British Aerospace (BAe)	Aircraft	3,591	225	79	79	—	56
Courtaulds	Textiles	3,419	238	78	80	33	21
Guest Keen & Nettlefolds (GKN)	Engineering	3,309	247	103	72	43	10
Reed International	Paper	3,245	253	132	57	20	6
BICC	Cables	3,147	269	157	50	47	16
Rank Hovis McDougall (RHM)	Food	2,902	293	163	47	12	1
BOC International	Gases	2,774	306	224	38	72	4
Cadbury Schweppes	Food	2,760	308	224	38	36	5
Bowater	Paper	2,740	309	272	32	54	3
Beecham Group	Pharmaceuticals	2,689	316	268	33	60	8
Burmah Oil	Petroleum	2,688	317	306	25		
Dunlop	Rubber, tyres	2,668	319	129	59	66	9
Ultramar	Petroleum	2,648	322	810	45		
Rolls Royce	Aeroengines	2,612	324	159	49	5	46
Bass	Brewing	2,504	340	101	72	5	1
Hawker Siddeley	Engineering	2,462	350	163	47	42	23
Metal Box	Packaging	2,284	376	159	49	41	6
Lucas Industries	Engineering	2,226	390	109	68	27	26
Rothmans International	Tobacco	2,165	396	317	23	60	15
United Biscuits	Food	2,108	403	203	40	33	3
Union International	Food	2,014	420	652	13		
British Shipbuilders (BS)	Shipbuilding	1,960	426	107	69	—	30
Consolidated Goldfields	Mining	1,917	433	57	101		
Plessey	Electronics	1,840	442	171	43	25	14
Pilkington Brothers	Glass	1,832	443	203	40	62	8
Brooke Bond	Food	1,830	445	102	72		
Babcock International	Engineering	1,753	457	279	30	57	14
Gallaher	Tobacco	1,744	462	299	27		

Table 11.3 (contd)

Company	Main industrial field	Sales		Employment		Prod. abroad as a percentage of sales	UK exports as a percentage of UK prod.
		Total (US $ m)	World rank	World rank	('000s)		
Glaxo Holdings	Pharmaceuticals, food	1,590	508	293	28	43	33
Reckitt & Colman	Food	1,576	576	279	30	68	6
Tube Investments	Engineering	1,552	522	272	32	23	20
Northern Engineering Industries	Electrical	1,517	531	268	33	23	20
Coats Patons	Textiles	1,498	546	165	46	69	9
Distillers	Beverages	1,467	554	493	20	17	97
Rowntree Mackintosh	Food	1,397	590	287	29	41	8
Blue Circle Industries	Building mats.	1,313	602			37	8
ICL	Electronics	1,304	606	287	29	31	16
John Brown	Engineering	1,301	608	723	10	36	23
Whitbread	Beverages	1,296	612	215	39	3	2
Pearson	Printing, ceramics	1,257	626	277	31	23	20
Vickers	Engineering	1,148	668	302	24	27	29
IMI	Metals	1,107	688	302	24	18	26
Turner & Newall	Various	1,087	696	203	40	43	31
Wellcome Foundation	Pharmaceuticals	1,077	701	521	19	65	2

Sources: Fortune 500 2 May 1983 and 22 August 1983; *Labour Research* April 1983.
* Jointly owned between UK and Netherlands capital.

Brothers (glass), Beecham and Wellcome Foundation (pharmaceuticals), Thomas Tilling (building and construction equipment), British Aerospace and Rolls Royce (aerospace). Apart from Thomas Tilling, in which diversification has been an important strategy, this group seems to share one major common advantage, namely a technological lead at global (or at least European) level in new products, the design and quality of products, and in production processes.

Although this latter category is encouraging for the British economy, the overall decline of UK multinationals in relative sales position should give rise for deep concern since – leaving aside the formerly widespread 'overmanning' problem using employment as an indicator – it is linked in no small measure to their sectoral distribution and production system characteristics. As Dunning and Pearce (1981) found for 1978 (and there seems to have been no fundamental change by 1983), more than half of the UK multinationals are in 'low research intensity' sectors, especially food, drink, tobacco, paper and wood products, textiles and clothing, leather and shoes, and building materials. Table 11.3 shows that twenty-seven out of the top sixty UK firms were in this category in 1982. Such a predominance does not augur well for the UK economy when one views the progression, adaptation or restructuring of the 'successful' manufacturing economies of the world, especially Japan, West Germany and the USA, from mainly 'unskilled labour-intensive' through 'capital/raw materials-intensive' and 'capital/machinery-intensive' towards mainly 'knowledge-intensive' industries (Linge and Hamilton 1981). Indeed, the importance of 'low research intensity' multinationals exposes the UK economy in two prime ways.

First, orientation to mainly consumer markets makes the multinationals is these sectors particularly vulnerable under the conditions of deindustrialization in the UK and the steady decline in the world ranking of the UK in per capita Gross National Product. Even goods traditionaly perceived as having inelastic demand, such as tobacco and tobacco products, have recently shown signs of dramatic market contraction as tax increases and anti-smoking campaigns shrink demand in the industrialized countries. Under such pressure on profit margins, giants like BAT Industries (Table 11.3) adopt strategies to transfer more of their production and trade in these products out of the UK and into relatively stable markets abroad and to diversify their operations both at home and overseas.

Second, firms in these industries often face severe competition because their products are at a mature stage in the product life-cycle and are more standardized; comparative advantages have often shifted overseas. Production costs are thus of more critical importance, even though advertising, product differentiation and market imperfections may sustain branded goods in certain market areas. Although UK wages may be significantly lower than elsewhere in much of the developed world, lower productivity still offsets this advantage and weakens the competitiveness of UK multinationals in these industries. This is why firms like Cadbury Schweppes are automating food and confectionery production lines to an even greater extent than they were before. Yet continental European and North American corporations, especially in foods and drinks, are making significant inroads into UK markets with 'additive-free' or 'natural' products, the appeal of which is increasing with rising social concern for health and the presence of US or West European supermarket chains willing or committed to marketing them: it is time that UK firms met this challenge posed by changing consumer preferences.

As is evident from Table 11.3, only fourteen of the top sixty British multinationals operate in the category of 'medium research intensity' industries, i.e. metals, engineering, vehicles and automotive parts, rubber and tyres. Of these, however, three are currently state-owned (British Leyland, British Shipbuilders, British Steel Corporation), although political decisions regarding their 'privatization' are pending. Dunning and Pearce (1981) found that seventeen of the seventy-five UK multinationals in their 1978 sample were in this category; this number had risen to nineteen by 1982. Most tend to be in the lower half of the UK '60' or '80' and the world '800' in terms of sales and most declined in rank.

In 1978, according to Dunning and Pearce (1981), twenty of the seventy-five UK multinationals exhibited 'high research intensity', though in 1982 only thirteen of these were in the top UK sixty. The encouraging aspect about this group, however, is that ten were amongst the top thirty UK companies, with the most strongly represented sectors being petroleum and chemicals (with three of the top five – Royal Dutch Shell, BP and ICI – supported by BOC International, Burmah Oil and Ultramar), aerospace (BAe, Rolls Royce), electronics (GEC, Thorn-EMI) and pharmaceuticals (Beecham). These are joined by a number of smaller firms such as Plessey, ICL, Racal and Delta (in electronics), and Glaxo and Wellcome Foundation (in pharmaceuticals).

In general, however, only a quarter of UK multinationals operate in 'high-research' sectors as compared with roughly one-third of Dunning and Pearce's (1981) world sample of 831 multinationals for 1978 and a similar proportion of the sample of 800 for 1982 discussed in this chapter. Broadly, this indicates a 'backward' structure of the population of UK multinationals. The importance of this does not lie so much in the rates of growth of multinational firms as Buckley, Dunning and Pearce (1981) found 'no discernible impact on growth' of the three-fold division of industrial sectors into those with 'high', 'medium' and 'low' research intensity; indeed, they argue that 'nationality' was a major factor and that UK multinationals did not perform as well as those of other 'nationalities'. However, nationality combines in the UK case with predominance of 'low research intensive' multinationals to account for the weaker position of British multinationals in the world sales league. Although the UK still had the second highest number of firms eighty-two, plus two jointly (Anglo-Dutch owned) after the USA (347) in the top 800 in 1982, their aggregate sales were placed only fourth after those of US, Japanese and West German multinationals. And in fact the share of world sales achieved by UK multinationals declined from 31 per cent in 1962 through 21 per cent in 1977 to a figure little above 17 per cent today.

Much research is still required into the size, growth and performance of UK (and other) multinationals. What evidence there is, however, should be treated as indicative and with caution, particularly in the light of the findings of Smyth *et al.* (1975: 111, 114) that: 'The validity of many empirical studies in . . . industrial organization depends on the extent to which alternative measures of firm size may be interchanged because often such studies . . . use whatever measure is conveniently available . . . If sales is the measure of firm size rather than assets, then concentration is increasing at an even greater rate than is indicated by studies using assets.' These issues become even more pertinent when attempting to measure profitability. As Smyth *et al.* (1975: 113) go on to observe: 'A number of UK studies have found that the rate of profit declines with firm size measured by net assets ... As size measured by sales increases more slowly than size measured by net assets (in the UK), ... profit rates decline faster with firm size is measured by sales rather than if it is measured by net assets'.

It is clear from Table 11.3 that the size range of the top sixty UK corporations in 1982 (and hence of the seventy-five in 1978 and eighty-two in 1982 referred to earlier) is huge in terms of sales and assets. Even if one omits the Anglo-Dutch jointly-owned Royal Dutch Shell and Unilever, the sales in 1982 of BP (the UK's largest 'purely' British corporation) equalled the combined sales of the last thirty-two firms in Table 11.3, i.e. from Rolls Royce to the Wellcome Foundation. If the 'British-owned' sales of Royal Dutch Shell and Unilever are added, then the combined sales at the top six multinationals (i.e. these two corporations plus BP, BAT Industries, ICI and GEC are greater than the combined sales of the remaining top fifty-three firms (excluding the NCB). To that extent, concentration is high amongst UK multinationals, a finding which further corroborates the work of Prais (1976) who found that concentration had increased and accelerated in recent decades so that by 1970 'nearly half of all manufacturing output was produced by the hundred largest enterprises' (Prais 1976: 7). It remains to be seen whether his prediction that by about 1990 the 100 largest enterprises in manufacturing 'would be responsible for about two-thirds of manufacturing net output' (Prais 1976: 7) comes true. Of course, not all of Prais' 100 enterprises are (or were) multinationals, but clearly they are the dominant force in this concentration process. Nevertheless, concentration in the slow-growth or deindustrializing UK economy in the past decade has undoubtedly meant less growth for most UK multinationals than it would have had they been located in Japan, West Germany or even the USA.

11.4 The shift of UK-owned multinationals overseas

It was suggested above that the movement of business abroad by UK-owned multinationals might have contributed to manufacturing decline in the UK. The lack of official data collection on, and very little research into, this question seriously constrain meaningful assessment of the scale, nature and impacts of such movement. Yet UK companies are believed to own some 7,000 subsidiaries abroad (Campbell 1981). Until the deindustrialization crisis really began to bite in the UK in the late 1970s, pioneering work had been conducted in this field by Owens (1980) who investigated the spatial implicatons for the UK of foreign direct investments in Ireland (Eire) by British industrial firms. More recent evidence, brought to light on employment change in leading UK industrial firms at home and overseas during the 'deindustrialization years' 1979 to 1982, is set out in Table 11.4. The forty-five firms surveyed employed 1.25 m. workers in Britain at the end of 1982, more than one-fifth of the manufacturing labour force; their combined overseas employment exceeded 800,000.

During the four years 1979 to 1982, these multinationals made some 307,200 net jobs abroad. A great deal more research is necessary into this and other measures of change, given the limitations of employment as an indicator. Nevertheless, the evidence is clear: these forty-five firms, which are the vanguard of British manufacturing, created net redundancies in the UK equivalent to approximately 15 per cent of the net number of workers made unemployed in these four years. (Of course, this does not mean that the firms actually created 15 per cent of the unemployed or of the long-term unemployed.)

Table 11.4 shows, moreover, that of the forty-five leading UK multinationals for which employment change data is available:

21	decreased jobs in the UK and increased them overseas;
12	contracted employment in the UK more than they did so abroad;
3	experienced equal relative or absolute contraction at home and overseas;
3	decreased jobs less in the UK than overseas;
5	expanded in both the UK and overseas (but of these two grew faster overseas than in the UK);

Clearly, then, thirty-three multinational corporations were mainly responsible for the redistribution of their activities to non-UK locations, while just six firms could be said to have contributed positively to employment creation in the UK in these years.

Only two broad groups of industries exhibit unison in the behaviour of their leading constituent multinationals: the chemicals and rubber sectors in which all firms contracted their

Table 11.4 Changes in employment in the UK and overseas 1979–82 by sectors and leading UK-owned multinationals

Multinational firm and sector	Changes in employment			
	Located in the UK		Located overseas	
	('000)	Approx. change (%)	('000)	Approx. change (%)
Tobacco: Total	−2.7	−12	+19.3	+12
BAT Industries	+10.6	+23	+5.9	+4
Imperial Group	−13.3	−15	+13.4	+94
Food & drink: Total	−20.1	−15	−5.8	−14
Rank Hovis McDougall	−10.4	−20	+0.8	+18
Cadbury Schweppes	−5.6	−19	−4.6	−26
United Biscuits	−1.0	−3	0.0	0.0
Rowntree Macintosh	−3.0	−13	−0.4	−4.5
Engineering and metals: Total	−96.4	−29	+7.3	+6
Guest Keen & Nettlefolds	−31.6	−46	+0.6	+2
Hawker Siddeley	−5.6	−14	+5.2	+44
Tube Investments	−23.1	−38	−1.7	−17.5
Babcock International	−5.7	−26	+3.4	+23
Davy Corporation	−2.8	−25	+3.1	+47
Northern Engineering Industries	−4.8	−15	+3.6	−81
Vickers	+3.6	+21	−2.6	−28
IMI	−7.0	−26	−4.1	−62
Delta Group	−8.6	−37	0.0	0
Associated Engineering	−9.1	−39	−0.4	−7
Simon Engineering	−1.7	−35	+0.2	+8
Electrical/electronics: Total	−41.9	−10	+32.6	+34
General Electrical Company	−9.7	−6	+16.5	+61
Thorn EMI	+5.0	+7	+13.0	+203
BICC	−4.2	−16.5	+4.5	+28
Lucas Industries	−20.2	−29	+2.5	+16
Plessey	−5.3	−14	−5.3	−35
ICL	−8.4	−34	−2.1	−22.5
Racal Electronics	+6.6	+108	+2.9	+85
Smiths Industries	−5.7	−30	+0.6	+37.5
Chemicals & rubber: Total	−54.5	−35	−16.4	−13
Imperial Chemical Industries	−24.7	−27	−2.5	−4
BOC Group	−8.1	−39.5	−7.7	−22
Fisons	−2.7	−25	−0.2	−10.5
Dunlop	−19.0	−39.5	−6.0	−11.5
Pharmaceuticals: Total	−3.3	−8.5	+3.4	+8
Beechams Group	−0.3	−2	+1.8	+11
Glaxo Holdings	−2.7	−17	0.0	0
Wellcome Foundation	−0.3	−4	+1.6	+15.5

Table 11.4 (contd)

Multinational firm and sector	Changes in employment			
	Located in the UK		Located areas	
	('000)	Approx. change (%)	('000)	Approx. change (%)
Paper, printing & publishing: Total	−16.6	−13	+1.2	+3
Reed International	−6.2	−12	−5.3	−32
Bowater	−3.7	−17	+2.1	+14
Pearson	−0.9	−3	+2.1	+87.5
DRG	−5.8	−29	+2.3	+46
Textiles & clothing: Total	−55.9	−38	−13.6	−19
Courtaulds	−40.4	−39	−2.6	−13
Coats Patons	−6.3	−26	−10.7	−25
Tootal	−9.2	−45.5	+0.3	+3.5
Building materials: Total	−7.1	−24	+13.8	+72
Pilkington Brothers	−5.1	−22	+10.9	+92
Redlands	−2.0	−28	+2.9	+39
Miscellaneous: Total	−8.7	−14	−3.0	−5
Metal Box	−8.3	−24	−7.1	−24
Steetley	+2.1	+55	0.0	0
Chloride	−2.8	−26	+0.4	+4
Reckitt & Colman	+0.3	+2	+3.7	+17.5
Total	−307.2	−19.5	+39.5	+5

Source: Labour Research April 1983

UK employment more dramatically than they did their overseas employment; and the building materials and glass sectors in which more jobs were added abroad than were lost in the UK. Lower costs and stronger markets account for the former while the very depressed UK construction market acted as a strong push factor in the latter industries. In all other sectors, multinationals behaved in varying ways, although most exhibit a dominant trend. Decline at home and expansion overseas typified the net pattern in tobacco, engineering and metal manufacturing, electrical/electronics, pharmaceuticals, paper, printing and publishing. Rates of decline were roughly equal in the UK and overseas amongst the aggregate of food and drink multinationals, while those in textiles recorded faster rates of job loss in Britain than overseas.

Several firms, however, produced changes which were contrary to these broad patterns between 1979 and 1982. The following examples illustrate some of the reasons for this. Rarely is there a close identity between a sector and a firm. In contrast to other tobacco multinationals, BAT Industries increased its UK payroll significantly more than its overseas payrolls mainly because of the corporation's strategy to diversify through acquisitions into other manufacturing and into services. The very broad 'engineering and metals' sector includes a range of specialized firms such as Vickers, Tube Investments and IMI which, for a variety of reasons, retrenched sharply their overseas operations; only Vickers expanded jobs in the UK, mainly through acquisition of Rolls Royce cars and partly in response to new contracts arising from the Thatcher government's increased defence spending. Multinationals in the electrical – electronics industries also displayed considerable diversity of behaviour as a result of their differentiated scales, product lines, cash-flow situations, competitiveness, management quality and capability, age or maturity. While larger long-established firms such as GEC and Thorn-EMI made significant expansions overseas, Plessey and IGL retrenched jobs everywhere in the face of stronger competition, and Racal Electronics consolidated its UK production base mainly through acquisitions. In papermaking, Reed International had to contract its overseas operations more than other UK firms in the sector. Glaxo stands out from other pharmaceutical firms in having no overseas expansion, but then Glaxo – unlike Beecham and Wellcome Foundation – has considerable interests in food manufacturing and is thus broadly in line with trends in *that* sector. Finally, Tootal the clothing firm shows distinctive trends from the textile multinationals in that it expanded jobs overseas, especially in low labour-cost countries, whereas the fibre manufacturers shed labour everywhere in the face of automated competition as they themselves replaced labour by machines.

Most research into multinational enterprises has examined the patterns of growth, development and change *abroad* and of impacts in *host* countries. The virtual neglect of the *home-country* effects of fdi overseas explains why case studies are presented here to support and to complement the highly aggregated available data. The most salient feature is that the IMF – OECD Common Reporting System on Balance of Payments data – with the provisos that the UK data include neither reinvestment profits nor the petroleum industry – show the UK sustaining the most consistent (even slightly rising) proportion of global fdi overseas amongst developed OECD countries (8.7 to 9.2 per cent) throughout the entire 1961 to 1981 period. This suggests three important findings:

1. Outward fdi from the UK most probably increased its share of all UK investment in manufacturing in these two decades.
2. As the UK proportion of global fdi began to rise after 1974, it may have accelerated – and been further accentuated by – the 'de-industrialization' of the economy, although abolition of all exchange controls in 1979 has made overseas investment much easier.
3. Outward fdi has become increasingly associated with the export of manufacturing jobs from the UK since 1974 whereas earlier it was still either contributing to UK job creation or to cushioning the scale and rate of job loss by maintaining manufacturing exports.

The key point here is the extent to which outward fdi and technology transfer retain the need for the UK facilities to supply the capital equipment, intermediate goods, components, subassemblies or finished products in knocked-down form, i.e. manufactured exports, from the UK; and how far does outward fdi substitute production in, and exports from, locations abroad for UK manufactured exports directly. That the UK has moved into an era since 1980 of importing more manufactures by value than it exports – probably for the first time since the advent of the industrial revolution – is aggregate evidence of the weakening of the UK manufacturing base. Although, of course, this reflects the weak competitiveness of British industry as a whole, it also embodies certain levels of inability or unwillingness on the part of UK-owned (and foreign) multinationals to meet UK demand from production sites located in the UK.

From Table 11.3, which presents some indicators for the top sixty UK industrial firms in 1982, it is clear that multinationals in a wide variety of sectors already manufacture very significant proportions of their total sales at locations abroad. The data support Dunning and Pearce (1981) who claimed that UK MNEs exhibit overseas production ratios which are amongst the highest in their world sample of 523 corporations in 1977. Of the forty-six for which such data are to hand, twenty-eight produce at least one-third of their sold products outside the UK, twelve of them manufacturing more than 50 per cent abroad. It is also fairly certain that one could add all the top three firms (for which data are not available, i.e. Royal Dutch Shell, BP, Unilever) and the other petroleum producers (Burmah Oil, Ultramar) to these. Significantly, it is the multinationals that process raw materials from areas abroad (e.g. tobacco, rubber) or bulky materials (e.g. building materials, timber) that have located (or acquired) the greatest proportions of production overseas, i.e. in the tobacco (BAT Industries, Rothmans International), petroleum, chemicals (ICI, BOC International), rubber (Dunlop), textiles (Coats Paton, Courtaulds), paper (Bowater), glass (Pilkington Brothers), pharmaceuticals (Beecham, Wellcome Foundation, Glaxo) and selected food industries (Cadbury Schweppes, Dalgety, Rowntree-Mackintosh, Reckitt & Coleman), although market access is important too. The shift of engineering firms (Babcock International, GKN, Hawker-Siddeley) and electrical/electronics manufacturers (GEC, Thorn-EMI, BICC) into foreign locations has been a comparatively recent phenomenon, but their overseas facilities already account for more than one-third of the output of these leading firms. Least overseas production characterizes selected drinks manufacturers (the brewers Bass and Whitbread and Distillers), Rolls Royce aeroengines, and British Leyland following sales of most of its overseas subsidiaries in the 'crisis years' of the late 1970s.

As is evident from Table 11.3, and indeed expected, there is broadly an inverse relationship between the two right-hand columns, i.e. between production abroad as a percentage of total sales and UK manufactured exports as a percentage of UK production. Most UK multinationals which produce more than one-third of their sales in overseas plants tend to export less than 10–15 per cent of their UK output (e.g. BAT Industries, Thorn-EMI, BOC International, Pilkington Brothers), although there are significant exceptions such as ICI, GEC, John Brown, Courtaulds, Glaxo and Wellcome Foundation which export much higher percentages. On the other hand, firms with limited overseas production facilities tend to export much larger percentages of their UK output, e.g. British Leyland, Rolls Royce, Lucas Industries, Tube Investments, Northern Engineering Industries, Vickers and IMI. Significantly the former group contains producers of higher standardized or bulky goods while those with larger exports exhibit technological advantages especially in engineering, electrical and pharmaceutical products. The overall picture, however, is blurred by the presence of large numbers of foods and drinks multinationals whose exports tend to be low irrespective of the percentage of overseas or UK production.

Further light can be shed on the characteristics of the UK and overseas facilities of these leading UK multinationals by an examination of the ratio of the percentage of total employment accounted for by UK employment to the percentage of total sales accounted for by UK production. Although a rather crude measure and calculable only for thirty-two of the firms listed in Table 11.3 it does provide further insights. Twelve multinationals exhibited almost equal percentages in 1982, i.e. a ratio of approximately 1, implying attempts to bring activity levels roughly into line with market shares at least at the very broad UK/overseas level; such features apparently typified BAT Industries at one extreme with 74 per cent of its production and 73 per cent of its workforce abroad to Rank Hovis McDougall at the other with 88 per cent of its production and 89 per cent of its labour force located in the UK. The largest number of firms (17), however, exhibited at higher concentration of employment than production in the UK. In most cases the percentage of employment rarely exceeded production by more than 4–10 percentage points, and may simply indicate localization in the UK of headquarters management, administrative, financial, R and D and advertising staff, and skill-intensive production functions, as compared with a predominance of more standardized manufacturing overseas. In a small number of cases, (e.g. GEC, Thorn-EMI, Courtaulds, Hawker-Siddeley, Rowntree-Mackintosh) the percentage of employment in the UK exceeded the percentage of UK production in total sales by 12–20 percentage points and may additionally reflect other characteristics such as overmanning in older UK production facilities and/or 'captive imports' from more modern factories located in lower-wage areas abroad. Clearly these conjectural possibilities need further research. In three cases, however, the reverse was true, with a higher percentage of sales coming from UK production than the percentage of UK employment: GKN, Glaxo and the Imperial Group. Again there are a number of explanations which require further investigation, including highly efficient UK production which competes successfully abroad and operation of more labour-intensive production units and sales and distribution outlets overseas.

GKN: leader of the Engineering multinationals

In broad terms, the UK engineering industry is a 'latecomer' in becoming significantly transnational or multinational. This is especially apparent when engineering firms are compared with those operating in sectors using mostly foreign-located natural resources (e.g. such industries as petroleum refining, petrochemicals, rubber, paper, and some foods) or selling differentiated products in consumer markets under conditions of oligopolistic competition (e.g. vehicles, electrical appliances, household chemicals). It may be less so, however, in comparison with firms in other sectors which mainly in the last decade have been seeking low-cost labour (e.g. electrical appliance assembly or components manufacture, shoes, textiles and clothing) in locations overseas. The search for cheaper labour, though, is not the major reason for the rapid multinationalization of engineering: the reasons must be sought elsewhere.

Table 11.5 Indices of the shift overseas of UK engineering firms 1973–82

Firm	Turnover 1982 (£bn rounded)	Growth in turnover 1973–82, (%)	UK exports as a percentage of turnover		Sales from overseas subsidiaries as a percentage of turnover	
			1973	1982	1973	1982
Guest Keen & Nettlefolds	1.85	131	9	9	30	35
Hawker-Siddeley	1.4	154	22	25	31	41
Lucas	1.25	206	16	20	32	50
Babcock International	1.0	395	17	15	24	56
Tube Investments	0.9	119	20	15	12	27
Northern Engineering Industries*	0.85	92*	26	20	16	29
Davy Corporation	0.75	311	17	28	38	56
Vickers	0.7	208	19	27	35	33
John Brown	0.66	419	34	25	17	42

Source: The Economist Newspaper Ltd, 8 October 1983, p. 82.
* Company formed 1977, therefore the figures relate to 1978–83.

Tardiness in the spread overseas of the UK engineering industry has many explantions. Basically these are rooted in: the diversity of engineering products; an historical predominance of smaller firms; the importance of capital-goods markets with relatively slow growth which often required custom-built goods and which, wherever located, could be supplied competitively from factories in the UK; ample opportunity for concentration of firms to take place via acquisitions within the UK. Today, as Table 11.5 shows, the leading UK firms are well established as multinationals. Of the nine listed, all except one (Vickers) have increased very substantially the percentage turnover of sales produced by overseas-located subsidiaries and four firms – Babcock International, Tube Investments, Northern Engineering Industries and John Brown – in fact doubled or more than doubled that percentage, resulting in a significant reduction in the share in their total turnover of exports from their UK production facilities.

These shifts imply substantial reduction or restructuring of these engineering firms' UK operations and employment simultaneously with a fairly rapid expansion, relative shift or relocation of functions overseas. They are responses to two major sets of forces: unfavourable business climate in the UK; and profit-yielding opportunities in foreign markets and at foreign production sites.

The unfavourable business climate in Britain has many facets and is well documented elsewhere in this volume. High inflation, oil-related overvaluation of sterling, low investment in modernization, made British industry uncompetitive at home and abroad in the 1970s and led to a speedy shrinkage of manufacturing output and exports. This had particularly negative multiplier effects on demand from engineering for factory equipment and machine tools. Severe problems in Britain's automotive firms (British Leyland, and specialized truck builders such as Seddon-Atkinson, Foden or ERF) and overseas development strategies of major US automobile corporations (involving greater shifts into the continental EEC) dramatically cut back the home market for automotive parts. Yet the poor quality and unreliability of UK-produced parts has also led manufacturers like BL to buy progressively more components from European, Japanese and US firms. Engineering output has thus fallen since 1978 by 20 per cent while in the same period capacity has been reduced by 30 per cent (reflecting large surplus capacity even before 1978) and direct employment has been cut by 25 per cent. The larger firms have sought to compensate for poor UK market prospects and high production costs by searching for profitable outlets abroad. This process has often provided them with the resources and market sales to facilitate simultaneously drastic restructuring of their UK operations. Frequently, overseas outlets, operations and linkages have changed too. Traditional Commonwealth (or once-Commonwealth) markets in Australia and South Africa, and the newly mushrooming Middle Eastern markets, for UK-produced engineering exports have begun to stagnate. Leading companies have responded recently, therefore, by entering or penetrating the more dynamic markets of the densely industrialized countries – especially of

North America and, to a lesser extent, Western Europe. There, in addition, they have often been able to acquire established engineering firms which may be on sale relatively cheaply because of surplus capacity or cash-flow problems but which also offer know-how, sales outlets and growth potential.

These trends can be illustrated by reference to the changes made since 1977 at home and abroad by the UK's largest engineering firm, Guest Keen and Nettlefolds (GKN). Figure 11.1 outlines the aggregate scale and geographic distribution of employment cutbacks in the UK. In just five years, GKN has almost halved the number of employees working in its UK establishments, although this reduction is somewhat exaggerated by the 3,860 workers in GKN's General Steels Division which was transferred in 1981 to Allied Steel and Wire (Holdings), an associated company owned jointly by GKN and BSC. The West Midlands, GKN's 'core region', bore the brunt of the firm's redundancies (Fig. 11.1), the firm thus contributing significantly to the rise of the region's unemployment rate from well below the UK average in 1978 to well above it in 1983. Figure 11.1 clearly shows, however, that the recession and corporate restructuring in response to it have led to a widely dispersed pattern of redundancies, plant contraction, plant closure and plant sales in most regions of England, and in Wales and Scotland – a feature illustrating the contemporary complexities of interregional multiplier effects both in, and emanating from, those larger corporations that grew and diversified in the 1960s by acquisitions and, encouraged by government policies, by setting up branches in peripheral areas.

Decisions by GKN management in 1977 to restructure the firm broadly required three changes which directly or indirectly explain these trends. First, the shedding of more traditional production lines, particularly special steels, forgings (e.g. vehicle body parts, bathroom fittings) and fasteners (nuts, bolts, screws) account for most of the retrenchment in the West Midlands and at Cardiff, Wrexham, Ayr and Rotherham. Second, the need to rationalize and to modernize output of automotive parts led to closure of plants in Burnley, Newton Abbot, Durham and Newtown (Powys) and to the retooling and automation (involving redundancies) of factories at Birmingham and Leeds. These major changes express GKN's decision to specialize on certain automotive components – related mainly to axles and transmissions for front-wheel drive vehicles – in which the firm has a world lead and faces a rising demand from major automotive firms. Indeed, the proportion of GKN sales from vehicle parts rose from 35 per cent in 1979 to 46 per cent in 1983. That is linked directly to a third change, namely, the adoption of a new strategy by GKN to shift from a dependence on ailing or stagnant UK vehicle producer markets to serve world markets in which French, Italian, German and US automotive corporations at least provide expanding sales opportunities. **That** shift requires factories located overseas to play a major part in GKN operations. Such a policy change explains why the firm today employs almost as many workers abroad as it does in the UK. Indeed, the proportion of GKN capital expenditure outside the UK rose from 38 per cent in 1978 to 49 per cent in 1981, with the increased capital flowing into three new factories; two in North Carolina, USA, to supply markets in the Americas and one in Lorraine, France to supply continental EEC automotive producers. Factory operation in these two arenas is being supported by newly-purchased wholesaling and distribution networks, i.e. Parts Industries Corporation (Memphis, Tenn.), Gallipolis Parts Warehouse (Ohio) and Maremont Corporation's parts division in the USA; Armstrong Equipment, Unigep and G. Borderieux in France. In 1984, GKN consolidated its position by buying a one-third stake in Carraro of Padua, a leading Italian axle supplier to European and US manufacturers.

Restructuring yielded GKN in 1983 its best profits since 1979, 'proving' the correctness of decisions made in 1977. Yet the fact that the rate of profit derived by GKN in the UK in 1983 was only 0.25 per cent on £1 bn turnover compared with 12 per cent on £480 m. in Western Europe, 11 per cent on £300 m. in the USA and 4 per cent on £85 m. in the rest of the world (*Financial Times*, 15 March 1984) suggests that the firm will continue to shift its operations abroad to penetrate the overseas markets at the expense of exports from the UK as a means of sustaining its improved profitability.

Much that has been outlined above for GKN could also describe the broad causes and effects – though not the details – of changes in the thirty-two other leading UK multinationals that shifted their operations overseas in varying degrees. Such shifts have had major negative multiplier effects on regional economies in the UK through the shrinkage, truncation, substitution or disappearance of former national and regional linkages (Hamilton and Linge 1983).

The shift of UK – owned multinationals overseas

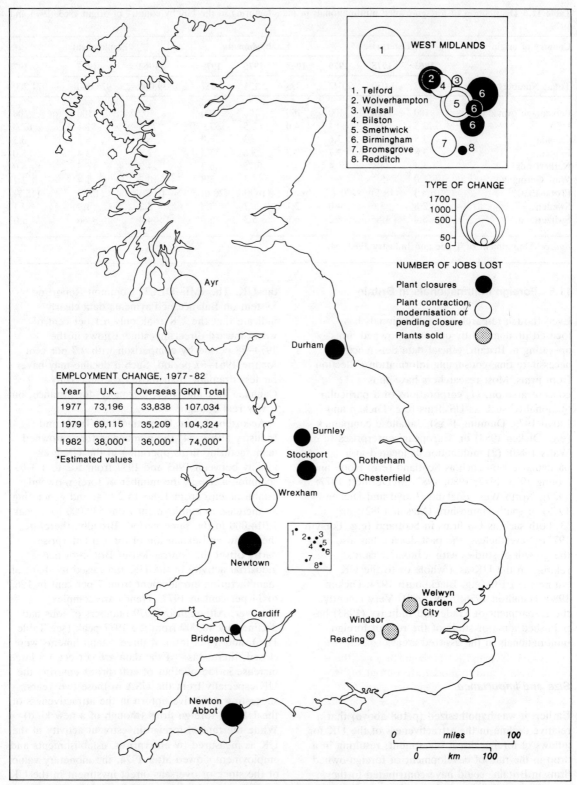

Fig. 11.1 The locational pattern of employment change in GKN, 1978–82

Table 11.6 Distribution of foreign-owned multinationals in manufacturing in the UK by country of origin 1963, 1975 and 1979

Country of origin	Enterprises			Establishments			Employment		
	1963	1975	1979	1963	1975	1979	1963	1975	1979
Total: Numbers	502	1,030	2,042	1,098	2,121	2,657	539,000	925,000	974,200
Percentage owned by	100	100	100	100	100	100	100	100	100
USA	73.5	64.5	60.4	74.0	62.5	57.9	75.3	71.1	67.9
Canada	3.8	3.7	3.6	3.7	6.3	6.0	6.4	6.5	6.5
France	3.8	4.75	3.7	2.7	3.6	3.3	2.9	3.4	5.0
Netherlands	3.4	4.1	5.4	4.8	5.75	6.3	4.2	6.8	6.0
West Germany	1.0	5.1	6.5	0.5	4.4	5.8	0.2	1.4	2.2
(Total EEC)	(n.a.)	(18.1)	(20.4)	(n.a.)	(16.8)	(20.6)	(n.a.)	(12.7)	(15.7)
Sweden	2.8	3.6	4.0	2.1	2.7	3.8	2.2	1.9	2.1
Switzerland	5.4	4.85	5.2	6.6	5.3	5.0	4.4	4.66	5.0

Source: Department of Trade and Industry 1983: 54.

11.5 Foreign multinationals in Britain

Even though research on multinationals has focused predominantly on foreign-owned firms operating in Britain, official data deficiencies necessitate time-consuming information collection from firms. Most researchers have thus concentrated on: (1) corporations of a particular 'nationality' such as US firms (e.g. Dicken and Lloyd 1976; Dunning 1958), Japanese companies (e.g. Dicken 1983) or 'European' enterprises (e.g. Watts 1980); (2) multinationals located in a particular region such as Scotland (e.g. Hood and Young 1976; 1977; 1980; 1982; McDermott 1977; 1979), North West England (Lloyd and Dicken 1979) or south Hampshire (Mason 1982); or (3) both such as US firms in Scotland (e.g. Forsyth 1972). Nevertheless, the past decade has also seen the growth of studies with a broader canvas relating to the UK as a whole or to the UK within the EEC (e.g. Blackbourn 1974; Dicken 1980; Hamilton 1976; Law 1980). Very recently, the Department of Trade and Industry (1983) has published a major study of the role of foreign multinationals in the assisted areas.

Size and importance

Earlier, it was hypothesized (p.168 above) that a relative decline in the attractiveness of the UK for inflows of foreign direct investment, resulting in a drop in the rate of development of foreign-owned firms in Britain, could have contributed to the stagnation or shrinkage of manufacturing located in the UK. The IMF–OECD Common Reporting System on Balance of Payments data clearly indicate that the UK took only 6.1 per cent of world inward direct investment flows in the 1974–78 period by comparison with 9.7 per cent for the 1961–67 period. Such a decline may have far less dramatic than those experienced by Canada, Italy and West Germany, for instance, but it was real enough.

According to the Department of Trade and Industry (1983: 54) the number of foreign-owned manufacturing firms operating in the UK grew rapidly between 1963 and 1977 from 502 to 1,370, spanning a rise in the number of foreign-owned establishments from 1,098 to 2,654, and generating an increase in employment from 539,000 to a peak 1,014,000 in the same period. Broadly, therefore, the inflow and expansion of foreign enterprise partly offset the contraction of British-owned industrial activity in the UK and raised its share of manufacturing employment from 7 per cent in 1963 to 14 per cent in 1977. Trends are complex, however. Although by 1979 numbers of jobs had declined by 40,000 from the 1977 peak (see Table 11.6), and a net total of three establishments were closed, there was – if the data are correct – a large increase in the number of enterprises entering the UK especially from the USA in these two years, suggesting a relative upturn in the attractiveness of the UK for foreign firms (though of a new kind). While the rate of foreign investment activity in the UK as measured by enterprises, establishments and employment slowed after 1974, the monetary value of the stock of overseas direct invstment in the UK actually trebled between 1974 and 1983. This was mainly due to inflation and to a four-fold rise in North Sea oil-related investments by foreign companies: in 1982, foreign oil company investment accounted for 38 per cent of the stock of fdi in the UK (£11.8 b out of £30.8 b) compared with 30 per cent in 1970 (£1.45 b). In real terms, however, it is estimated that the stock of fdi in Britain actually declined between 1979 and 1982, mainly because of the cumulative effects of disinvestment by several major US corporations located in England and Scotland (Du Pont, Goodyear, Honeywell, Hoover, Monsanto, NCR and Singer) which was not compensated for by fdi by firms from other countries (Department of Trade and Industry 1983: 17–18, 53).

Most studies of fdi in host countries stress that foreign firms operate much larger-scale facilities on average than do local manufacturers because bigger firms have a higher propensity to expand overseas. The ratio in employment per establishment in the UK in 1979 was 6:1, foreign-owned facilities having an average of 367 workers in contrast to an average of 64 workers for all plants. The utility of such comparisons is highly questionable. More meaningful would be an evaluation of foreign-owned facilities with those owned in the UK in the same industrial sectors by UK multinationals, but insufficient data is to hand to do this. Measurement of size by employment is suspect anyway in the absence of comprehensive and comparable data on output, capacity utilization, productivity and value added.

Origins of inward investment

With these provisos, one must be content to use what data are available on enterprises, establishments and employment (Table 11.6). Clearly, US-owned corporations have continued to dominate fdi in the UK despite a relative decline in these indicators from about 75 per cent across the board in 1963 to 58–68 per cent in 1979. In the latter year, American firms engaged 9.5 per cent (661,500) of all UK manufacturing workers as compared with 5 per cent (406,200) in 1963. Dunning (1958: 56) found that 205 US firms employed 246,200 people or 2.8 per cent of the British manufacturing labour force in 1953 and estimated that this had reached 340,000 by 1956. That little growth occurred between 1956 and 1963 is explained by the formation in 1958 of the EEC 'Six' which diverted substantial US investment from the UK to continental Europe. Canadian manufacturers, however, have retained a remarkably stable aggregate share of fdi in UK manufacturing and, with 63,400 workers in 1979, were in second place.

After 1968 there was a significant rise of inward fdi from West European headquartered companies: their combined labour force in Britain grew from 80,000 in 1963 to more than 221,000 in 1979. Several powerful economic arguments explain this upsurge: consolidation of many EEC-based firms into strong competitors by the late 1960s; the market attraction of the UK then still outside the EEC (and uncertain about entry) but offering the largest production platform to EEC firms within the EFTA 'Seven'; upward valuation of the Dutch guilder, Deutschmark and Swiss franc against the £ sterling, encouraging fdi in the UK by companies from these countries; and the increasing lag of UK wage rates behind those in Benelux, France and West Germany. Nevertheless, a neglected but significant set of 'push' factors from the Continent were the far more widespread 'near revolutionary' socio-economic environments involving workers there than in the UK in the late 1960s and possibly also fears connected with the Soviet invasion of Czechoslovakia.

Japanese fdi also grew in Britain, especially after 1972, but this must be kept in perspective. Though estimates of employment in Japanese-owned factories have varied from 'approaching 15,000' (*The Sunday Times*, 6 December 1981: 56) to 'less than 6,000' (Dicken 1983: 278), it is in fact nearer to the latter and more comparable to the numbers of jobs created (1979 figures) by Danish (5,100 workers) and Irish (10,000) manufacturers in the UK. It is much less than employment in Swedish (20,000), West German (22,300), Swiss (48,500), French (49,200) or Dutch-owned (58,700) firms. The publicity given to the attraction of Japanese investment currently is great, of course, because many of the larger, mature US firms seem to have lost investment momentum, or are being attracted to newly-industrializing countries, and West European capital has been increasingly enticed away from the UK to the USA and, in part, to Ireland. On the other hand, Japan has become a major and growing potential source of fdi not least because of EEC threats to limit imports from Japan unless efforts are made to manufacture *within* the EEC, and Japanese technology, productivity and management promise sorely needed for industrial renewal, modernization and competitiveness *in* Britain, while simultaneously threatening the business of British-owned manufacturing.

The quality of foreign direct investment

The most comprehensive data published on fdi relate to US firms overseas in 1977 (US Department of Commerce 1981). They indicate that American investment in UK manufacturing 'bottomed out' as a percentage of total US industrial investment in the EEC in 1976 (when it was around 30 per cent compared to 58 per cent in 1961). Recently, some relative increase has occurred partly because 'in 1979 for the first time in the 70s the rate of return [on investment] in Britain exceeded that of the [European community] as a whole' (Department of Trade and Industry 1983: 24). Some interesting contrasts emerge between American-owned manufacturing in the UK and that located elsewhere in the EEC in 1977:

1. The value of employee assets, sales and remuneration where higher in all EEC countries (except Ireland) than in the UK; indeed, more than 50 per cent higher in the case of remuneration per worker; but
2. The ratio of the value of sales to remuneration per worker was US$7.86 per dollar in the UK, 16 per cent above the EEC average of US$6.77 per dollar, a fact which possibly helps explain that:
3. UK-located affiliates produced 52 per cent of all manufactures imported into the USA from the EEC by US firms; and yet:
4. Although the UK localized 40 per cent of jobs in US subsidiaries in the EEC, Britain contributed only 29 per cent of their sales by value and earned less than 25 per cent of their wage bills.

Points (2) and (3) suggest the competitive strength of US-owned manufacturing located in Britain. Points (1) and (4) indicate a much less positive evaluation: that the UK concentrates upon plants using more labour-intensive methods and semiskilled labour to produce standardized, lower value-added goods and components in which the USA has lost its comparative advantage but which – because of their mature stage in their product life-cycles – have become increasingly vulnerable to international competition, especially from producers in NICs and southern European countries. Subsequent closures of US plants in the UK tend to confirm this. So, too, does evidence from sample multinationals which showed that a *higher* proportion of European-owned firms (than US firms) had developed facilities of lower capital intensity in the UK than in their home countries (Department of Trade and Industry 1983: 116–19). By contrast, American-owned facilities located in the continental EEC appear, from the macro-economic data, to manufacture higher-value technologically more advanced products for the EEC market and employ higher-paid, skilled labour. Detailed cross-national comparative research still needs to be carried out but the most likely explanations are that:

1. On average, American-owned plant in the UK is much older than on the Continent: most pre-Second World War US investment in Europe, and more than 60 per cent of it between 1945 and 1960, came to Britain; thereafter this situation was reversed.
2. For that reason, and because British incomes declined considerably relative to the EEC 'Six' and Scandinavia in the 1960s, the sectoral mix of US manufacturing in the UK has tended to remain lower-to-medium (rather than medium-to-high) in knowledge intensity: this, of course, reinforces the lower research intensity of UK multinationals referred to earlier, even though the latter has permitted the substantial dominance in the UK by US firms in medium-to-higher research-intensive industries (such as vehicles, tyres, pharmaceuticals, electrical/electronics, machinery).
3. The policies of the EEC 'Six', especially the Common Agricultural Policy, combined with rapid industrial growth and development to raise incomes (and wage rates) relative to the UK in the 1960s and 1970s, simultaneously: offering a richer market to investors; requiring new US entrants to, and existing US firms located on, the Continent to introduce labour-saving machinery (to yield higher productivities) and to use more recent technologies and product designs (to yield higher value-added, higher-quality goods); and
4. Work practice and labour relations' differences broadly speeded the introduction of new technology and production organization on the Continent and hindered it in the UK.

11.6 Regional impacts of foreign multinationals

The foregoing considerations lead on naturally to an evaluation of the impacts of foreign multinational corporations in creating jobs, providing various occupations, generating incomes, stimulating growth and development multipliers in

UK space; and to a brief assessment of the effects of government policies in shaping multinational corporate behaviour within that space. At a general level there are three aspects to this: first, the distribution of foreign-owned manufacturing amongst UK regions; second, the importance and salient features of multinational enterprise for the economies of the individual regions; and third, how both have changed through time. Unfortunately, the data constraints on the elaboration of these three facets are severe and figures for years more recent than 1979 impossible to come by.

Regional dynamics

Table 11.7 shows that the industrial structures of six regions (with location quotients 1.0) concentrated a greater proportion of foreign-owned manufacturing (in job terms) in 1979 than they do UK manufacturing in general. These regions fall into two groups: South East England (location quotient: 1.4) and East Anglia (1.5), non-assisted areas constituting the 'centre' of the UK economy and its industrializing margins; and North West England (1.01), Wales (1.3), Scotland (1.2) and Northern Ireland (1.7), all assisted areas making up most of the western and northern periphery of UK space. Regions intermediate between the centre and the periphery, on the other hand – the East and West Midlands, Yorkshire and Humberside with location quotients respectively of 0.5, 0.6 and 0.65 – have clearly failed to attract the same level of fdi either by new foreign entrants or by acquisition of their local firms by overseas-headquarted corporations. Nor have the North (0.82) and South West (0.7). The limited presence of foreign-owned manufacturing in these last five regions can be explained by their

1. Relative isolation from modern port facilities for bulk freight movements in and out of Britain and from airports with high international accessibility for business travel by management, executives and other skilled personnel.
2. Limited industrial traditions (e.g. the South West) or dominance by older, nationalized industries and semiskilled labour force (e.g. the North) or historically competitive, highly specialized and tightly-knit regional systems of UK-owned firms employing mainly skilled, high-wage and often scarce labour (e.g. the East and West Midlands) which may have discouraged new entrants to the UK from locating new facilities in these regions or from acquiring UK firms there.
3. The general lack (except in the North, which has the highest location quotient of foreign manufacturing of the five regions) of regional aid incentives

Table 11.7 UK regional pattern of foreign-owned manufacturing 1979*

Region	Units		Employment ('000)		Employment size per unit	
	Foreign-owned	All Industry	Foreign-owned	All Industry	Foreign-owned	All Industry
UK total	3,593	118,726	919.7	6,746.5	255	56
Percentage in:	100	100	100	100	—	—
South East	33.5	33.8	34.4	24.7	262	41
South West	5.9	5.7	4.3	6.1	185	60
East Anglia	3.7	2.8	4.3	2.8	299	58
East Midlands	6.0	7.6	4.3	8.4	185	62
West Midlands	8.4	11.1	8.1	13.5	246	68
Yorkshire & Humberside	6.8	9.4	6.5	10.0	224	60
North West	12.8	12.7	14.0	13.8	280	61
North	5.0	3.6	4.9	6.0	248	95
Wales	5.6	3.9	5.9	4.5	268	65
Scotland	10.3	8.1	10.2	8.3	253	57
Northern Ireland	2.1	1.3	3.2	1.9	395	86

Source: Business Monitor PA1003, 1979

* These data differ from those given in Table 11.8 because they are based on analyses of manufacturing *units* which differ statistically from *establishments*.

Yet the regional distribution of foreign-owned industry is far from static. Law (1980) has shown how the important initial fdi in Britain was localized before 1944 in the South East (mainly suburban London: 52 per cent before 1918; 70 per cent 1918–44) and the North West (respectively 24 and 12 per cent); clearly, access via the country's premier ports to the major urban market regions explained this localization. Since 1945 the locational pattern has been quite different; interregionally it is far more 'even'. The attraction of the North West tailed off to only 7–10 per cent in the 1960s and 1970s. Except for the 1966–75 period, when it drew in 25–36 per cent of new plants, the South East has generally pulled in less than one-fifth of fdi in UK manufacturing. Until the early 1970s, regional development grants policies were diverting considerable new fdi into the Development Areas, primarily Scotland which – through the publicity of the Scottish Council – attracted the largest number of new foreign-owned plants between 1945 and 1966 and then again 1972–75 (Forsyth 1972; Law 1980). Yet the most prominent trend in the 1970s was the fastest growth of foreign-owned manufacturing in the 'intermediate' areas (Table 11.8) at the expense of relative decline in the South East and stagnation in the Development Areas. 'Deglomeration' of port facilities from London along the east coast 'facing' the continental EEC, completion of motorway links between the Humber and central England, upgrading of Midlands airport facilities and improvements and growth in medium-sized cities (in the general redistribution process of urbanization) have all contributed recently to the attraction of greenfield development and acquisitions by foreign firms in East Anglia, the East Midlands, Yorkshire and Humberside and, to a lesser extent, the West Midlands. Similar upgrading of national and international access, combined with industrial traditions and scientific rsearch, have enabled the Bristol – Swindon zone to pull in recently several foreign firms in high research-intensive industries (e.g. the US electronics firm, Hewlett-Packard, setting up in Bristol its first research laboratory outside Silicon Valley), so extending the M4 'high tech' corridor from London into the South West region. On the other hand, the late 1970s and early 1980s saw drastic international rationalization by US firms especially in Scotland, Northern Ireland and, to a degree, the North West, although the 'Japanese factor' sustained the importance of foreign-owned industrial growth in Wales; nevertheless, employment growth in the Development Areas in the 1970s dropped to the national average. Evidence (e.g. Smith 1980; Department of Trade and Industry 1983) suggests that the relative decline of foreign-owned manufacturing in the South East in the 1970s was less than might be expected because foreign firms, especially from Western Europe, have been most active in this region in acquiring UK firms.

Data for 1977/79 (Table 11.9) indicate that there are no significant regional variations in nationality ownership patterns: apart from a stronger presence of Dutch (e.g. Philips) and West German firms in North West England and Northern Ireland, the proportion of jobs in UK-owned facilities in all other regions is around the UK average in the UK as a whole. Nevertheless, in the absence of more up-to-date figures it can be postulated that the international trend for firms headquartered in more countries and operating in more industrial sectors to become multinational is permeating the UK industrial system, too. The populations of foreign-owned firms in the UK regions are becoming more diversified nationally and sectorally. Foreign acquisitions are not any longer solely of UK firms for there is an increasing trend for new foreign entrants to the British market to take over existing foreign-owned firms. In part this involves the retreat of some US multinationals and their acquisition by West European firms. Examples from the vehicle industry are the well-known takeover of Chrysler's West Midlands and Scottish factories by Peugeot-Citroen PSA in 1975 and, very recently in the North West, the acquisition from International Harvester of Seddon-Atkinson by ENASA, the Spanish state-owned truck

Table 11.8 Changing regional shares of foreign-owned manufacturing employment in the UK 1971–79

Region	1971 (%)	1975 (%)	1979 (%)
South East	41.1	37.6	35.3
Development Areas*	36.4	36.6	37.2
Rest of UK†	22.5	25.8	27.5
Total	100	100	100

Source: Department of Trade and Industry 1983: 70.
* This includes the North West and North regions in England, Wales, Scotland and Northern Ireland.
† This group comprises East Anglia, South West, East Midlands, West Midlands, and Yorkshire and Humberside.

Table 11.9 Foreign ownership in manufacturing in selected UK regions by country of origin, 1977

Country of origin	Percentage of total employment in foreign-owned units						
	UK	South East	North West	North	Wales	Scotland	Northern Ireland
Canada	6.8	4.4	4.7	8.7	9.4	8.2	4.9
USA	70.0	71.8	65.1	69.6	71.7	70.1	61.3
North America: Total	76.8	76.2	69.8	78.3	81.1	78.3	66.2
Denmark	0.3	0.3	0.8	n.a.	n.a.	n.a.	—
France	3.1	3.4	1.5	n.a.	3.8	8.2	n.a.
Ireland	1.0	0.3	2.3	n.a.	n.a.	n.a.	—
Netherlands	6.3	6.6	12.4	8.7	—	5.2	9.7
West Germany	1.9	0.9	1.6	4.3	3.8	1.0	6.5
Other EEC	0.9	0.4	2.3	2.2	1.8	1.1	12.8
EEC: Total	13.5	11.9	20.9	15.2	9.4	15.5	29.0
Sweden	2.0	3.1	0.8	2.2	1.9	1.0	n.a.
Switzerland	5.4	6.6	3.1	2.2	1.9	3.1	n.a.
Australia	1.2	0.9	3.1	—	3.8	—	n.a.
All other countries	1.1	1.3	2.3	2.1	1.9	2.1	n.a.
Rest of the world	9.7	11.9	9.3	6.5	9.5	6.2	n.a.
World: Total	100	100	100	100	100	100	100

Source: Department of Trade and Industry 1983:59.

manufacturer. Similarly, the abortive bid for Hymac, a Welsh excavator producer, by Daewoo of South Korea indicates the willingness of multinationals from NICs and LDCs to enter UK manufacturing.

Regional variations in multinational quality

That in 1979 two-thirds of employment in foreign-owned manufacturing in Britain were in chemicals and pharmaceuticals, in vehicles, and in mechanical, instrument and electrical engineering indicates the importance of foreign multinationals in the medium and high research-intensive industries. (This compares with only two-fifths of total UK industrial employment in these sectors). Although foreign firms operate facilities in these industries in most regions, some regional specialization is evident in the prominence of foreign-owned food industries in Northern Ireland, chemicals in all regions (through especially Northern England), mechanical engineering in the peripheral regions (mainly Scotland and, less so, in Wales and the North), vehicles in the South East and North West, and electrical engineering (including electronics) in the South East and Scotland.

Such sectoral variations hint of an important role that foreign multinationals play in the socio-economic differentiation of UK space. Yet any assessment of the costs and benefits of foreign multinational enterprise to the national and regional economies, as part of the 'external control debate', must remain subjective in the absence of fully comprehensive research and data (for discussions of some of the international, national and regional dimensions of this problem, see Hamilton and Linge 1981; 1983). Only the lines along which such assessment might be made can be sketched in here using recent published evidence.

Given the variety of theoretical and empirical interpretations of multinational enterprise, the extreme views are the 'business school' approach that argues the benefits such enterprise inevitably brings to regional economies because it is globally the most efficient form of enterprise, while the 'Marxist' school emphasizes the inevitability of inequality that must arise between capital and labour, core and periphery from the activities of multinational enterprise as, in Lenin's words (see

Hamilton 1976), 'imperialism is the highest stage of capitalism'.

Between these extremes it is arguable that the locational behaviour of multinationals, if given full rein, can and does contribute to interregional imbalances: the pre-1939 localization of US investment in newer industries in the London region and their almost complete absence elsewhere (except the North West) added significantly to the structural – spatial dichotomy of Britain in the 1930s (Hamilton 1976). Where state policies are applied to correct such inequalities, multinationals are more likely to be responsive than local firms to regional incentives (Blackbourn 1974; Hamilton 1974) because of their more thorough search procedures and the fact that their vertical intrafirm integration greatly weakens their need for regional ties (with labour, buyers or suppliers): the success of the Development Areas, especially Scotland, in attracting fdi in manufacturing after 1945 confirms this.

The survey of 140 foreign multinationals in the early 1980s by the Department of Trade and Industry (1983) found that every '18 out of 20 firms incorporated Regional Development Grants in their investment appraisal' (p. 93). The report (pp. 94–5) suggests that whereas labour availability in the assisted areas was by far the strongest incentive for locating there in the 1960s, regional aid became the most influential factor in the 1970s, with labour availability dwindling to insignificance. The reasons, which the report does not directly state, were that in the 1970s and 1980s corporations with a relatively dwindling supply of internationally mobile manufacturing capacities (certainly in job terms) have been facing, on the one hand, ever-widening regional choices for that capacity as rising unemployment reduces or eliminates the locational significance of labour availability; and, on the other, greater interregional differentiation in capital costs as inflation-raised investment costs and escalating interest rates made borrowing exorbitant. Hence the enhanced locational influence of regional grants in subsidizing capital costs.

Although regional incentives have contributed towards a more 'even' geographic distribution of foreign-owned industrial activity in Britain since 1945, at least as measured quantitatively in employment at regional levels, recent research has begun to uncover some aspects of the qualitative spatial inequalities that it may hide. These are rooted in three closely inter-related facets of multinational enterprise: the mode of entry into, and further development in, the UK by foreign firms; the technology they transfer; and the character of their UK operations in the wider contexts both of their entire corporate structures and of the UK industrial system in the international division of labour.

Modes of entry

The extreme cases are, on the negative side, acquisition of UK firms by foreign companies to eliminate competition, resulting in early closure or rundown of UK-located production facilities, with redundancies, as the market is supplied increasingly from abroad via an acquired UK sales network. This has happened in the entry of some European firms, particularly West German chemicals producers. On the positive side, new plant using the latest technology, manufacturing the newest products, is set up on greenfield sites to supply the UK and wider EEC or world markets, generating significant direct and indirect employment and paying higher wages: development of very advanced micro-chip production in Scotland by the US firm National Semiconductor could be a case in point. This is not to imply, however, that new plant development is necessarily preferable to acquisition in regional or national development: development of Japanese TV and electrical appliance assembly in Britain, and the impending greenfield plant construction by Nissan in northeast England are examples of growth and development in certain regions which has eliminated, or threatens to eliminate, competition and jobs, or cause drastic restructuring – including flight of production overseas – of British and other foreign firms located in the same or in other regions of the UK.

It appears that the balance may have shifted relatively from peak plant development in the 1950s and 1960s during the growth era when the greatest scale of US corporate entry and expansion occured, to higher levels of foreign acquisitions of UK (and other foreign) firms in the 1970s and 1980s, particularly as West European companies seek in the UK: (a) to find unexplored sales market outlets (i.e firms not previously selling in the UK) or to eliminate British competition from a saturated EEC market; (b) to minimize risk by taking over 'going concerns' with marketing networks given their relative inexperience in multinational production and, as with many West German firms, also quite small size and financial resources; or (c) to achieve rapid expansion in new product fields.

It has also been suggested that acquisition entry to the UK is also most likely among multinationals with multiplant networks in continental Europe (Department of Trade and Industry 1983). This relative shift in the mode of entry probably helps to explain the changing regional incidence of foreign ownership in manufacturing, i.e. its more rapid growth in the areas 'between' the South East itself and the Development Areas. These more densely industrialized English regions – the East and West Midlands, Yorkshire and Humberside, the North West, as well as the South East itself – contain many medium-sized and larger firms with acquisition potential on any of the above criteria. Some support for this has been given by Smith (1980). By contrast, government financial incentives, at least in the 1970s, sustained considerable greenfield development in the peripheral regions where, anyway, there were fewer indigenous firms which could be acquired.

Deeper research is needed into the development implications of different modes of entry and subsequent behaviour by multinationals of different country of origin, age, sector and structure in a variety of regional environments and in various time-periods. Acquisition, however, is often used as a route to product diversification and new market entry (for 'old' or 'acquired' products) and, if one accepts Wilson's (1980) findings, this becomes a built-in corporate strategy. Strength in marketing combined with product diversity might be thought to make such multinationals 'secure', or 'stable' influences in, the regional economy. Yet dependency mainly on their trading skills may also impart to them high propensities to divest less profitable subsidiaries (and product lines), bringing local instablity and to develop the more routine occupations associated with production of (in effect) standardized, branded goods. On the other hand, corporations with a technological lead are more likely to build new plants for new products and to apply new process technologies; although their capital intensity is usually higher and their job-generating capacity lower (than the 'trading' multinationals), they are more likely to bring to a region diverse, well-paid jobs with better long-term prospects. Wilson (1980) observed that Japanese-owned firms show far less inclination to acquire than did firms headquartered in other countries; this is related both to technological lead and to culture-specific forms of business organization.

Technology transfer

Implicit here are differences in technology transfer. As hinted earlier, foreign-owned multinationals tend to bring to the UK product and process technology which is 'intermediate' between that of lagging British firms and that of lead multinationals or of the lead facilities of the same multinationals in the continental EEC. Short of very detailed surveys, clues to regional variations within the UK in the technology transferred from abroad must be found in a number of ways.

First, tendencies towards sectoral specialization amongst foreign-owned multinationals in certain regions form one set: the pre-eminence of the South East and Scotland in the newer micro-electronics and higher research-intensive industries are examples. Indeed, according to the Secretary of State for Scotland – with the establishment of plants by Burr Brown, Hughes Microelectronics, General Instruments, Motorola, IBM, National Semiconductor and Hewlett-Packard from the USA, and NEC and Shin Etsu Handotai from Japan – Scotland, with 42,500 workers in micro-electronics alone, 'has the biggest concentration of 'high tech' industries in Europe'. One wonders, however, whether this is not a path towards repeating regional history in product overspecialization, but this time with a very high dependency on foreign firms setting up their 'springboard' European factories and in industries in which product life-cycles seem to be being shortened from decades to years or even to months.

Second, the presence or absence of R and D facilities is an important indicator of the technological level of facilities, higher skill and pay and longer-term commitment of a firm to operations in a particular place. While the South East and M4 corridor has pre-eminence in R and D, it would be interesting to know how far the recent influx of 'high tech' foreign firms into Scotland really has changed the finding of Haug *et al.* (1983) that the region had attracted very little R and D, most of which was concerned with routine problem-solving, not with product or process innovation.

Third, regional development grants have tended to attract capital-intensive industries and 'larger investments than would otherwise be the case' (Department of Trade and Industry 1983: 174) to the Development Areas. While this may explain the very recent upsurge of US and Japanese investments in some of these regions, it would be erroneous to correlate high capital intensity with high research intensity: the former is often associated with labour-replacing technologies in the manufacture of low or medium research-intensive products. Foreign-owned chemicals industries, for instance, have high capital intensity but in manufacturing homogeneous products, they also have very low R and D and management status. That the twenty subsidiaries surveyed in Wales had the largest sales turnover per employee amongst the 140 multinationals sampled by the Department of Trade and Industry (1983) suggests the high efficiency of new automated production lines, amongst them Japanese-owned ones.

Character of foreign-owned operations

These aspects also form components of the character of the UK operations of foreign multinationals, but there are others. Host nations and regions can become 'branch plant economies': subsidiaries assembling products for the UK national market (e.g. TV, VCRs) locally add little value, pay relatively low wages, and have few multiplier effects through local backward or forward linkages, most value-added being imported as components or knocked-down kits from parent facilities in the home country (e.g. Japan) or from overseas subsidiaries (e.g. Singapore). Branch plant managements generally have little decision-making power or R and D opportunities, being subordinated to external control from headquarters overseas. Combined with few or no local ties with suppliers or with skilled labour pools, it is argued that this makes branch plants particularly vulnerable to closure or sale in times of economic recession and changing corporate investment strategies. Generally, UK firms have more management and R and D functions at UK units than other foreign multinationals. Nevertheless, McAleese and Counahan (1979) have rightly distinguished between those multinationals which are 'stickers' in host environments, committed to building up their business presence over a long term in a local community, and those which are 'snatchers', concerned mainly with quick profits and thus just as liable to shut or sell subsidiaries as develop them (especially when government grants are available for limited periods, as has been the practice in Development Areas; see also Watts 1980: 6). One can hypothesize an enmeshing of this distinction with that between 'acquirers' and 'greenfield developers' cited above.

For all these reasons the UK government, following the example of initiatives in the US and some NICs, has recently begun to attempt to insist on certain minimum levels of 'local (i.e. UK-made) content' in the products that foreign multinationals sell in Britain: the goal is to increase the percentage of value-added actually manufactured in the UK so as to stimulate recovery of jobs and incomes through intensified regional and interregional backward linkages within the country. Government pressure may have begun to have an effect, at least in *ad hoc* ways. For instance, the foreign-owned vehicle manufacturers Ford, General Motors and Peugeot-Citroên PSA (known as Talbot) in the 1970s shifted investment markedly to new facilities in the continental EEC (and Iberia), involving production capacity, higher-value vehicles, R and D and management functions. It resulted in the UK becoming a major importer of Ford, GM and Peugeot-Citroên cars; in 1984 all three have commenced significant investments in new UK facilities (e.g. Ford at Halewood, GM at Luton, Peugeot-Citroên in the West Midlands); in the Peugeot-Citroên case the decision came remarkably quickly after the firm experienced serious labour troubles in its main plants in Paris whereas industrial relations have remained excellent in the West Midlands. Some Japanese firms are already increasing UK content at their own or licensed plants, e.g. C. H. Beazer in Bridgwater (Somerset) has raised the UK manufactured content of Hitachi excavators from 20 to 60 per cent, while Yamazaki intends to do likewise at a machine tool factory in Worcester.

The character of fdi in UK facilities can be assessed also by the quantitative and qualitative nature of their linkages. Most multinationals appear to operate capacities and apply technologies to manufacture products which reduce the costs in the UK to the corporation 'rather than to match competition or to ensure survival of the affiliate'. (Department of Trade and Industry 1983: 126): foreign firms do just enough to keep ahead of weak UK competition. Generally, in line with international trends, existing foreign-owned facilities have been retooled (e.g. Ford Dagenham) and newly constructed units designed (e.g. Ford Bridgend) to be more specialized in their functions and to be more strongly integrated through intrafirm trade in products, components, information and financial transactions with other units belonging to the corporation located in the host, other overseas and home countries. Former subcontracting links with indigenous firms in the same or in other regions of the UK have thus been 'truncated' and substituted relatively (or totally) by international flows. The logic of the rise of intrafirm trade across frontiers resides in the theory of internalization (Rugman 1981), i.e. the possibilities offered for high levels of internal control of prices, profits and tax payments by corporations and the concomitant lessening of their dependency in such financial matters on the market or on governments. The Department of Trade and Industry found that 70 per cent of its sample of 140 multinationals obtained some inputs from other corporate units through intrafirm trade (significantly US firms bought one-third from US-located sources, European firms a similar proportion from continental European sources), but only 40 per cent sold their UK products to other facilities via intrafirm trade. This suggests that the multinationalization of operations in Britain primarily involves assembly and manufacture of final products for the UK market and could be a key component in import penetration and the conversion of the UK from a net exporter to a net importer of manufactures.

Yet the survey found only a minority of (mainly US) firms used Continent-wide and global resourcing policies involving substantial intrafirm trade, mainly associated with long-established integrated operations, a developed European production-system network, and the need to rationalize to meet global competition involving technological change (as, for instance, in vehicles). Although this worked against UK regions in the 1970s as US firms (using Britain as their first or major foothold in the European market) moved functions into the continental EEC, British regions can now perhaps benefit from the improved national business climate insofar as multinationals maintain a significant UK network of units forming a subsystem of the wider EEC system (e.g. Ford, GM). Some credence is given to this by the observation that 'some of the most dominant [intrafirm] trade relationships seemed to be between UK plants rather than involving cross-frontier transfers' (Department of Trade and Industry 1983: 128). The question is one of how much and what kind of intrafirm trade is intranational and what is international in which direction. By contrast, European and Japanese firms import inputs from parent plants to their single UK branches. Little evidence exists that the UK has been evaluated as a suitable export platform for foreign multinationals; only time and research will show if the most recent entrance by higher-technology US, European and Japanese firms (e.g. micro-electronics in Scotland, Pirelli fibre optics in the South East) will significantly alter this.

11.7 Raising a setting sun?

The focus of this chapter has been restricted to the largest UK multinationals and to broad trends in foreign direct investment to illustrate selected aspects of the roles of multinationals in the British industrial system. The top sixty UK firms employ about two million workers at home, roughly double the labour force working in more than 1,300 foreign-owned manufacturing firms located in UK space. World-wide, the top sixty firms have a payroll approaching five million, equivalent to almost four-fifths of the total British-located manufacturing workforce. Although treated separately above, the two sets of multinationals are often interdependent in several ways.

First, many British-owned multinationals buy capital equipment or inputs directly from foreign-owned units located in the UK (e.g. British Leyland purchases truck parts from the US-owned Eaton Corporation) and vice versa (e.g. Ford fits Dunlop tyres on some vehicles).

Second, though the size and marketing advantage at home of low research-intensive multinationals may somewhat deter foreign-owned food or tobacco manufacturers, competitive weakness or technological backwardness amongst British firms in medium and high research-intensive sectors has allowed comparatively easy entry to, and penetration of, the UK market by foreign firms in these sectors. Global competition amongst multinationals in the low-to-medium research-intensive functions, however, has forced many UK multinationals to relocate or expand production abroad as place-specific comparative advantages shifted from the UK overseas.

Third, interdependencies *within* the population of foreign firms can also influence the behaviour of 'home' multinationals (and vice versa). A simple example involving US firms in the UK illustrates this. 'Follow-the-leader' entry into the UK by foreign firms in one industry from one country (e.g. Ford, GM, Chrysler) generates multiplier effects through the entry and growth of 'clusters' of linked manufacturers from the same country (e.g. Goodyear, Uniroyal in tyres; AC-Delco, Borg-Warner, Dana, Baton, Fruehauf, Westinghouse etc. in automotive parts) and of linked service firms (e.g. Hertz, Budget, Avis; insurers; hire-purchase financiers; petroleum and lubricants suppliers; and their bankers and lawyers). Such competition and market penetration induced UK manufacturers and their larger, linked suppliers (e.g. British Leyland, GKN, Dunlop) to invest and to produce in markets abroad where they could gain certain monopolistic advantages. It is a mistake to think that this phenomenon is confined to periods of economic growth (such as the spate of US entrants to the UK in the 1950s and 1960s); it has also occurred amongst firms in new growth sectors during recessionary periods (e.g. US consumer goods firms and the forerunner of IBM in the 1930s; the new wave of US and Japanese electrical/electronics firms in the late 1970s and 1980s).

Fourth, foreign-owned firms employ a rising proportion of British manufacturing labour, having increased their payrolls between 1966 and 1977 (when British-owned firms cut their labour force quite markedly) and decreased it less (than UK firms) after 1977. Possibly, therefore, the relative growth in foreign acquisitions of British firms in the last decade could have slowed the rate of job loss in manufacturing as a whole, injecting foreign money into ailing companies which otherwise might (perhaps without government support would) have joined the lengthening company queues at British bankruptcy courts. Fdi could thus have partially compensated *in the UK* for redundancies in UK multinationals, certainly insofar as acquisitions occurred in the industrialized regions where those multinationals cut back labour. But because rising unemployment meant labour availability throughout Western Europe, such redundancies themselves did not prove to be a locational determinant of fdi in UK regions. Acquired British companies undoubtedly offered multiple benefits (skills, market networks etc.) to the foreign acquirers, yet the relative growth of this kind of acquisition activity may well have resulted from the very substantial cheapening of UK assets when calculated in the acquirers' home (i.e. West European, lately US) currencies following devaluation of the £ sterling. Far more research is needed, however, to unravel the comparative direct and indirect, short- and long-term effects for the UK and its regions of foreign acquisitions compared to fdi in new facility construction.

The foregoing suggests some significant relationships between the relative global decline of UK multinationals, their substantial shift of productive capacity overseas, and the enhanced role of foreign firms in the British industrial system. Despite improved cost and price advantages, the UK manufacturing trade balance has continued to deteriorate (see Ch. 13). Economics research shows that, amongst twenty-six OECD industrialized and industrializing countries (1953/57–1970/72), the UK had the highest income elasticity of demand for manufactured imports and faced a low price elasticity of world demand for its manufactured exports. The meaning is clear: industrialists located in the UK responded 'rather ineffectively to changes in domestic and foreign demand' (Singh 1977: 131) because of 'the lower quality, design and general performance of [their] products relative to other countries' (p. 132) as a result especially of higher rates of investment abroad than at home by UK multinationals (with a similar UK/overseas split amongst foreign investors). Already by 1973 this meant that: 'the value of production abroad by UK multinationals . . . was more than twice the value of their exports from the UK . . . (even though) because of high concentration in the UK export trade, these multinationals account for the bulk of the country's manufacturing exports' (Singh 1977: 132)

By contrast, Holland (1976) found the corresponding proportions for West German and Japanese multinationals were less than two-fifths. The large number, size and high percentage of 'offshore' manufacturing of UK multinationals weaken further the ability of the UK to export manufactures. Hymer (1960) observed that the establishment abroad of production subsidiaries not only substituted exports from, but often generated manufactured imports for, the home country. Kaldor (1977) offers two further contributions to this debate. First, Keynesian 'full employment' or 'consumption-led' growth policies of post-1945 Britain sucked in manufactured imports, leading to balance-of-payments crises and 'stop–go' government policies in response which further inhibited productive investment in the UK. Second, rising real incomes have invariably been associated to date with a rate of growth of *exports* of manufactures, which . . . was considerably higher than the rates of growth of the *total* output of manufactured goods' (Kaldor, 1977: 202). The implications for UK government policy provide a framework for some concluding remarks.

Multinationals are a fact of capitalist life, purveyors and protectors of self-perpetuating, firm-specific, monopolistic (rather than 'superior') advantages in financial resources (and access thereto), management, marketing and technology. These, transcending the territory of government, are applied to extract the most from real and perceived place-specific attributes in human and natural resources. Governments must bargain and compromise as firms can usually threaten to locate facilities in another government's territory.

The beginning of the chapter indicated that the UK lacked government policies towards multinational enterprise. The reasons are not hard to seek. As the UK is both a major source and a major recipient country of fdi, British governments feel uneasy about discriminatory policies for fear of retaliation against UK-owned firms abroad. Most UK governments since 1951 have been conservative (with either a big 'C or a small 'c') and have generally not even conceived, let alone attempted, comprehensive national planning or coordinated economic industrial and regional policies. Rather, state intervention has been *ad hoc* and broadly preoccupied with protecting older industries, with maintaining employment (rather than productivity-related technical and infrastructural change), and with so-called regional policies. These policies, of course, have all contributed to preserving an environment more conducive to lower rather than to higher research-intensive firms and functions (compared to West Germany, USA, Japan or Sweden). Indeed, apart from the work of Hodge (1975), virtually no analysis exists even of the *interactions* between UK governments and multinationals. In emphasizing the significance of regional development grants as *some* stimulus to the inflow of fdi, the Department of Trade and Industry (1983) broadly suggests three lines of policy to strengthen the international appeal of the UK to investors:

1. New national incentives to attract multinationals, otherwise the UK will not maintain even its share of diminishing worldwide fdi, for which competition is already fiercer from countries within both North and South.
2. The incentives should be multipurpose and more broadly based to encourage, in complementary ways: greenfield site development by new entrants; expansion of foreign-owned firms or facilities already located in Britain; and acquisition of UK firms.
3. There should be more flexibility in meeting specific managerial requirements, such as the promotion of selected areas for Japanese firms which wish to 'cluster' spatially to support common infrastructure and services for Japanese workers and their families.

Yet fdi in manufacturing has been mainly to supply the UK market with finished goods, in effect displacing indigenous firms abroad or altogether. While, therefore, it may be a necessary condition to improve the British manufacturing trade balance by attracting new fdi or by retaining foreign firms already located in Britain, it is certainly not a sufficient condition. Efforts must be made to stimulate foreign firms to orientate new or to reorientate existing production capacities towards export markets. Enterprise-zone policy is one move in this direction but it is also insufficient: these zones could become merely entrepôts where little value is added, contributing to further conversion of the UK to a branch-plant economy. If, however, one sustains the thesis that UK enterprise zones can only become globally competitive if labour cost productivities match those in enterprise zones in ASEAN countries, then the UK zones may only be populated by foreign firms adding high value in sophisticated products. The point is that exports of manufactures (including the products of the information sector) must be of high value, embodying very substantial UK value-added. Government policies should thus select the types of industries, activities and segments of production-system chains that the UK would welcome foreign firms to establish. As Kaldor (1977: 204) argues: 'The secret of "successful" industrialization . . . appears to be an "outward" strategy to develop the ability to compete in export markets in selected fields . . . and to keep the growth of export capacity in line with the growth of industrial activities.' Perhaps today that quotation should be modified, with 're-industrialization' replacing 'industrialization' in the first line while adding 'and information' after 'industrial' in the last one. The difficulties are to select the industries, activities or chains and then to persuade the firms to set up that will be export market 'winners'; and not to fall into the trap of selecting the same industries as other countries, which can only promise devastating future competition (Hamilton 1984).

Again, one must be aware of possible future interdependencies between foreign and other firms. There is global evidence of a shift away from wholly-owned subsidiaries or even acquisitions by multinationals towards more joint ventures. The consortia of European, Japanese and US firms emerging in the newest 'high tech' sectors suggest that government should be paying more attention to prospects for link-ups between UK and foreign firms. Far more emphasis should also be placed on inducing *British* multinationals to retain more of their capacities within UK space and to expand manufactured/information-processed exports. This is not to advocate protectionism; quite the reverse. Policies should stimulate restructuring of British firms into new product lines in their UK units in which they will have certain global advantages. This requires a significant shift into higher research intensities while relinquishing much more (than now) of the low and medium research-intensive functions to less developed countries. Selecting the fields into which several UK multinationals can apply their advantages requires enlightened government leadership and partnership with the business community to steer desired development in new directions. There should be encouragement to smaller firms as these, too, are becoming multinationals in highly specialized fields. At present, the Conservative government's policies of monetarism and of undermining all forms of democracy is taking Britain back to the Dickensian era, but this time without the monopolistic advantage that British capital then enjoyed: currently society is being subordinated to an economic materialism largely imposed by the technical, product and cultural changes made by foreign multinationals in the world at large. The latest US and Japanese entrants making these changes appear, however, to be using the UK more as an export production platform than their predecessors. Only time will tell, though, if this is simply an initial advantage which will fade as they consolidate a European-wide production network in the future. Hence the necessity of support for UK-owned firms. British governments must invest vigorously in human and social priorities central to the quality of life, in the sophistication of the environment, and in the creation of a highly educated and skilled labour force to expand future R and D *in* the UK and to create wider markets for a new generation of advanced products and information in which UK multinationals (current, embryonic and future) can excel and export world-wide. In this way the country will earn sufficient to restore everyone's human dignity in Britain and to help achieve a living global environment worthy of the year 2000.

References

Blackbourn, A. 1974 The spatial behaviour of American firms in Western Europe. In Hamilton, F. E. I. (ed.) *Spatial Perspectives on Industrial Organization and Decision-making*. Wiley, 245–64.

Campbell, M. 1981 *Capitalism in the UK*. Croom Helm.

Department of Trade and Industry 1983 *Multinational investment strategies in the British Isles*. HMSO

Dicken, P. 1980 Foreign direct investment in European manufacturing industry: the changing position of the United Kingdom as a host country. *Geoforum* **11**: 289–313.

Dicken, P., 1983 *Overseas investment by UK manufacturing firms: some trends and issues*. Discussion Paper 12. North West Industry Research Unit University of Manchester.

Dicken, P., Lloyd, P. 1976 Geographical perspectives on United States investment in the United Kingdom. *Environment and Planning A* **8**: 685–705.

Dicken, P., Lloyd, P. 1980 Patterns and processes of change in the spatial distribution of foreign-controlled manufacturing employment in the United Kingdom, 1963–1975. *Environment and Planning A* **12**:1405–26.

Dunning, J. H. 1958 *American Investment in British Manufacturing*. Allen & Unwin.

Dunning, J. H. (ed.) 1971 *The Multinational Enterprise*. Allen & Unwin.

Dunning, J. H. 1982 The history of the multinationals during the course of a century. Unpublished paper, Internatinal Conference on Multinationals in Transition, CREPA & IRM, Paris, November.

Dunning, J. H., Pearce, R. D. 1981 *The World's Largest Industrial Enterprise*. Gower: Farnborough.

Forsyth, D. J. C. 1972 *United States Investment in Scotland*. Praeger, New York.

Hamilton, F. .E. I. (ed.) 1974 *Spatial Perspectives on Industrial Organization and Decision-Making*. Wiley, 2nd edn, 1981.

Hamilton, F. E. I. 1976 Multinational enterprise and the E.E.C. *Tijdschrift voor Economische en Sociale Geografie* **67**: 258–278.

Hamilton, F. E. I. 1984 Industrial restructuring: a world problem. *Geoforum* **15**: 349–364.

Hamilton, F. E. I., Linge, G. J. R. (eds) 1981 *Spatial Analysis, Industry & the Industrial Environment Volume 2: International Industrial Systems*. Wiley.

Hamilton, F. E. I., Linge, G. J. R. (eds) 1983 *Spatial Analysis, Industry and the Industrial Environment Volume 3: Regional Economies and Industrial Systems*. Wiley.

Haug, P., Hood, N., Young, S. 1983 R & D intensity in the affiliates of US owned electronics manufacturing in Scotland. *Regional Studies* **17**: 383–92.

Hodge, M. 1975 *Multinational Corporations and National Government; A Case Study of the United Kingdom's Experience 1964–1970*. Saxon House, New York.

Holland, S. 1976 *Capital versus the Regions*. Macmillan.

Hood, N., Young, S. 1976 US investment in Scotland – aspects of the branch factory syndrome. *Scottish Journal of Political Economy* **33**: 279–94.

Hood, N., Young, S. 1977 The long-term impact of multinational enterprise on industrial geography: the Scottish case. *Scottish Geographical Magazine* **93**: 159–67.

Hood, N., Young, S. 1980 *European Development Strategies of US Owned Manufacturing Companies Located in Scotland*. Allen & Unwin.

Hood, N., Young, S. 1982 *Multinationals in Retreat: The Scottish Experience*. Edinburgh University Press.

Hymer, S. 1960 *The International Operations of National Firms: A Study of Direct Investment*. Unpublished PhD dissertation, MIT.

Kaldor, N. 1977 Capitalism and industrial development: some lessons from Britain's experience. *Cambridge Journal of Economics* **1**: 193–204.

Law, C. M. 1980 The foreign company's location investment decision and its role in British regional development. *Tijdschrift voor Economische en Sociale Geografie* **71**: 15–20.

Linge, G. J. R., Hamilton, F. E. I. 1981 International industrial systems. In Hamilton, F. E. I. and Linge, G. J. R. (eds) *Spatial analysis, Industry and the Industrial Environment*, Vol. 2. Wiley.

Lloyd, P., Dicken, P. 1979 The contribution of foreign-owned firms to regional employment change: a components approach with reference to North-West England. North West Industry Research Unit Working Paper, University of Manchester, 8.

Mason, C. 1982 Foreign-owned manufacturing firms in the United Kingdom: some evidence from South Hampshire. *Area* **14**(1): 7–17.

McAleese, D., Counahan, M. 1979 'Stickers' or 'snatchers'? Employment in multinational corporations during the recession. *Oxford Bulletin of Economics and Statistics* **41**: 345–58.

McDermott, P. 1977 Overseas investment and the industrial geography of the United Kingdom. *Area* **9**: 200–7.

McDermott, P. 1979 Multinational manufacturing firms and regional development: external controls in the Scottish electronics industry. *Scottish Journal of Political Economy* **26**: 287–306.

Owens, P. R. 1980 Direct foreign investment – some spatial implications for the source economy. *Tijdschrift voor Economische en Sociale Geografie* **71**(1): 50–62.

Prais, S. J. 1976 *The Evolution of Giant Firms in Britain*. Cambridge University Press.

Rugman, A. M. 1981 *Inside the Multinationals*. Croom Helm.

Singh, Ajit 1977 UK industry and the world economy: a case of de-industrialisation? *Cambridge Journal of Economics* **1**: 113–36.

Smith, I. J. 1980 Some aspects of inward direct investment in the UK with particular reference to the Northern region. Centre for Urban & Regional Development Studies Discussion Paper (University of Newcastle upon Tyne), 31.

Smyth, D. J., Boyes, W. J., Peseau, D. E. 1975 The measurement of firm size: theory and evidence for the United States and the United Kingdom. *Review of Economics and Statistics* **57**(1): 111–14.

Watts, H. D. 1980 The location of European direct investment in the United Kingdom. *Tijdschrift voor Economische en Sociale Geografie* **71**(1): 3–14.

Wilson, B. 1980 The propensity of multinational companies to expand through acquisitions. *Journal of International Business Studies* Spring/Summer: 59–65.

CHAPTER 12

Public sector industries

Graham Humphrys

12.1 Introduction

In any reasoned discussion of recent change in the industrial geography of the UK, it is necessary to deal separately with public sector enterprises. This is because in taking location decisions which have spatial results, such enterprises are directly influenced by the policies of the state which owns and controls them. The geographical distribution of their activities differs, therefore, from that of enterprises in the private sector. This kind of reasoning is part of the recent development of thinking in industrial geography. Earlier, industrial location theory over-emphasized the influence of the external environment on the creation of spatial patterns. This has now been balanced by recognition of the significance for location decisions of the internal features of enterprises such as ownership and organization. What recent work suggests is that in the same external economic environment, enterprises with different organization and control will exhibit different geographies. In explaining location patterns, the internal characteristics of an enterprise, including its organization, control and ownership, historical evolution and aims, will thus be at least as important as features of the external economic environment such as the costs of land, labour and capital.

Hamilton (1967) was among the earliest writers to suggest a more useful division of enterprises into three groups for modern location analysis. The first is the small firm, often with only one plant and managed by a single owner–entrepreneur. The second type is the large enterprise, corporate owned, multisite and often multiregional and mutinational, with employee managers operating the factories and with spatially-segregated headquarters location. The third type of organization is the public sector enterprise, again often large, multisite and multiregional, though only sometimes multinational. The significant difference between the first two and the last is that the public sector industries are required to conform with the policies of the government in power. As the owner, the government has responsibility for their decisions and behaviour (*The Nationalised Industries* 1978).

An underlying assumption of the recognition of public sector industries as a separate group for location studies is that the interests of the public and private sector industries are not exactly the same. Put simply, what is good for Imperial Chemical Industries (ICI) is not necessarily good for the UK. This important point can be illustrated with the following simple hypothetical example.

A car manufacturer decides that a new car-assembly plant is needed in the UK to provide additional car production capacity. Two locations are considered, the company operating costs being the same at each. Location One requires an investment by the company of £100 m. for site acquisition, construction and capital equipment, while at Location Two the investment needed is £125 m. However, Location One is in a congested urban area where existing public services and utilities are already used to capacity and where suitable labour is in short supply. At this location, public investment would be necessary to provide such things as gas, electricity and water supplies and road and rail access. The jobs at the factory would attract workers from elsewhere, leading to an influx of additional population which would require investment to cope with their needs. The total public sector investment required by the new factory at Location One will amount to £120 m. Location Two is in an area where there is sufficient spare capacity in utility provision already available and where there is a surplus of labour. Some additional public sector investment will be required to supply some services to the site and for some upgrading of local social facilities, but this amounts to only £20 m. Total costs, that is investment by the company plus the investment by the public sector, amounts to £245 m. at Location One but only £145 m. at Location Two. For a car manufacturer in the private sector, Location One would be the obvious choice. If the car manufacturer is in the public sector, then the obvious choice would be Location Two. In this example, the state would be justified in giving a £30 m. subsidy to make Location Two more attractive, since the total cost to the national economy would still be much less than at Location One.

In the example used, the critical feature is how much of the total cost of an industrial operation has to be borne internally by the enterprise and how much can be passed on to the public sector. For a private sector enterprise, one way to maximize profits is by minimizing the proportion of total costs it has to bear. This can be done by the obtaining of direct state subsidies in the form of such things as grants and loans, and indirectly by taking advantage of state provision of education and health services and the passing on of disutility costs such as pollution to individuals (Smith 1979). For a public sector enterprise there is a moral obligation to take decisions on the basis of total costs and not to seek to maximize its profits by passing on the costs generated to other parts of the public sector or to individuals. This raises one of the major problems of dealing with public sector enterprises in the UK. Some governments seem to have adopted the view that these should operate as if they were in the private sector, the only difference being that they are owned by public funds. They are required to maximize their individual profits, if necessary by externalizing as much of their costs as they can, even if this is to the net detriment of the national economy. On this basis, for example, the Central Electricity Generating Board (CEGB) would buy the cheapest coal available to supply coal-fired power stations. This would mean a much larger proportion of the coal burned would be imported, since this would be cheaper to use than domestically-mined coal. The effect of such a policy would be a decline in the operations of the National Coal Board (NCB) and enormous welfare costs which would have to be borne by the public purse. The accounts of the CEGB would show dramatically higher profits, and electricity charges to consumers might even go down. But this would be at considerable cost to the NCB and to other sectors of the national economy; these costs would probably exceed the apparent improvements in profit reported by the CEGB. The adoption of such an approach to public sector enterprises implies that state ownership is only justified on other grounds such as strategic considerations and that, if these do not exist, enterprises should properly be in the private sector. Other governments in power in the UK in the past have adopted an opposite view, that is that the public sector enterprises must operate to the advantage of the economy as a whole (Glyn 1984), with decisions taken on the basis of the total costs involved as explained earlier. Which view happens to be held by a government in power clearly has a major influence upon the decisions in any public sector industry, including location decisions, and must be taken into acount in examining the geography of public sector industries.

There are other reasons for the state having involved itself in industrial production other than to rescue important industries. As Chester (1976) has pointed out, the primary task of government in the historic past was as the provider of security, both internal and external, and of ensuring freedom for individuals and organizations to compete freely and fairly. To this regulatory function was first added the provision of goods and services more appropriately provided by a public sector agency and later the welfare function of ensuring social justice and a fairer distribution of national income and wealth. To achieve these goals there has been an increase in the direct intervention of the state in industrial activity in most Western industrial countries (Dicken and Lloyd 1981). In the UK this first occurred mainly through nationalization of major economic activities, beginning in the 1940s and continuing in the 1960s. The 1970s saw additional involvement through adoption of a National Economic Strategy which led to the establishment of the National Enterprise Board in England and the equivalent Scottish and Welsh Development Agencies in those countries. As part of their activities, the Board and the Agencies acquired interests in industrial companies through the purchase of minority or majority holdings. The effect of such action has been to produce changes which would not have occurred if there had been no public involvement.

There are two further relevant differences between public and private sector industries. The first is that the public sector industries can be expected to take account of total costs and of the social consequences of their actions and in the last resort cannot go bankrupt. Thus they are unlikely to have or to develop similar internal geographies to enterprises in the private sector. Secondly, there has always been difficulty in assessing the degree of economic efficiency achieved by the public sector. It has been suggested that their monopoly position in several instances in the UK has inhibited the achievement of the highest efficiency. Moreover, it is difficult to achieve agreement on the value of the social costs they are expected to take into account (Cable 1982). These can be seen as arguments for returning such industries to the private sector, on the basis that competition and accountability to shareholders would require them to improve their efficiency. Such arguments ignore the limited competition which occurs in several sectors of British industry because of the concentration of control into five or fewer large enterprises (Prais 1976; Watts 1980). There are, too, several examples of the economic inefficiency of private enterprises having led to their being rescued by takeover by the public sector, notably British Leyland and Rolls Royce. The arguments about efficiency will continue to rage. What needs to be made explicit is the fundamental point about public sector industry. Government ownership inevitably means the direct influence of government policy, resulting in the display of different distributions to those which would have been adopted if the industry was in the private sector. The more a public sector industry is required to operate as a private sector industry simply provided with financial support by the state, the more its geography will resemble that of private sector industry.

What has been argued so far is that it is useful to recognize, for geographical purposes, that public sector enterprises form a distinct group. But this is not the only justification for dealing with them separately. As in most Western countries, they are also large enough to have a major impact on the economic geography of the UK as a whole. Despite this they have received little attention in the work of economic geographers. What follows therefore, is largely exploratory. The size and importance of the public sector enterprises is described, their regional distribution and significance analysed and the geography of two example enterprises reviewed as the basis for further understanding.

The nature of public sector involvement

Public sector involvement in industry in the UK is complicated. Three major elements can be identified. The most obvious is direct state participation in industry through public corporations known as the nationalized industries and other Crown corporations (National Economic Development Office 1976). A second group can be identified as industries in which there is public sector involvement together with private investment. This is a more miscellaneous group, distinguished largely on the basis of the public sector involvement comprising less than 50 per cent of ownership. In this category is British Petroleum, which is the largest industrial enterprise in the UK and in which the government holds a 39 per cent share. This has been held since before the Second World War, but the government has never held a majority share. There are other industries which were formerly nationalized but into which private capital has been introduced since 1980. A good example is British Aerospace, which was nationalized from 1978 to 1981. The state has retained over 48 per cent of the shares, a further 2 per cent were sold to employees and the rest are held by other shareholders. With such a large share, the government effectively retains control. Another group of companies are those in which the British Technology Group (formed from the National Enterprise Board) has acquired a shareholding. Yet another category are industries in which there is public sector involvement at local government level rather than at the level of the national government. The best example is the West Midlands Enterprise Board established in the 1970s, which provides equity capital in the form of a minority shareholding in return for an agreement on the way a company operates, especially in relation to its employees.

Here, attention is confined to those public sector enterprises which are nationally owned and which are concerned with the supply of marketed goods. The nationally-owned criterion means the exclusion of enterprises wholly or partly owned by local government and those industries directly dependent upon selling to the public sector but not actually publicly owned (CIS 1982). The latter would include companies dependent upon the manufacturers of defence equipment for the UK armed services (*Statement on Defence Estimates* 1983; Pite 1980) as well as those supplying specialist products such as fire engines or refuse lorries essentially for the UK public sector market. The marketed goods criterion excludes publicly-owned social services such as health, local and national government administration and defence forces, together with services in the public sector which are marketed, most of which are in the fields of transport and communication. All these have considerable direct or indirect impact upon the geography of industry in the UK but need more attention than can be devoted to them here.

The principal public sector industries as defined above are the National Coal Board, British Gas, the Central Electricity Generating Board and the Area Electricity Boards for Great Britain and Northern Ireland, British Steel, BL, British Shipbuilders, Rolls Royce, British Nuclear Fuels, Short Brothers, the Royal Ordnance Factories, the Royal Mint, the Forestry Commission, the Milk Marketing Board and companies in which the British Technology Group and the Scottish and Welsh Development Agencies have a holding. Major public liability companies in which the government held substantial shares, and which are also included, are British Aerospace and British Petroleum. To these should be added government research and development, much of which is directly related to industrial production though not tied to any one specific industrial operation (Short 1981; Todd 1980; Pite 1980).

Measures of the size of public sector industry are fraught with difficulty. Over time, various governments have added new industries to the public sector and sold off all or parts of others. The situation was particularly dynamic after 1979 when a government came to power committed to a programme of privatization of public sector operations. There is also the difficulty of continual changes in the size of individual industries. Since 1979 such change has been largely through contraction as part of the industrial depression experienced in the UK. The Census of Employment does not produce separate data for the public sector industries by area and alternative sources are fragmentary and inconsistent. The statistics included below can, therefore, provide only a broad outline of the situation.

Size

The size of public sector industry as defined is difficult to determine because of the various ways in which relevant statistics are compiled. The nationalized industries, which include the transport and communications public corporations but which exclude such things as the Royal Mint and government research and development, in 1982 employed some 1.5 m. people and had a turnover of £44,000 m.; they invested over £5,000 m., 20 per cent of all fixed capital investment excluding housing (Central Office of Information 1982a) (Table 12.1).

The sectoral dominance of these industries varies widely. The public sector coal, gas and electricity industries dominate UK energy production and distribution, the only other major energy source being oil. Public sector involvement in domestic oil production is largely through the British National Oil Corporation which allocates exploration

Table 12.1 The major public sector enterprises in the UK

Enterprise	Turnover (£m.)	Capital investment (£m.)	Employment ('000)
British Aerospace	1,662	n.a.	80
British Gas Corporation	5,235	515	105
BL	2,877	n.a.	157
British National Oil Corporation	5,752	263	2
British Petroleum	25,347	n.a.	118
British Shipbuilders	1,026	37	66
British Steel Corporation	3,443	164	104
Electricity Council*	9,458	1,486	164
National Coal Board	4,728	736	282
Rolls Royce	1,258	n.a.	59
Royal Ordnance Factories	n.a.	n.a.	18
Short Brothers	110	n.a.	6
Total			1,161

* Including Central Electricity Generating Board, North of Scotland Electricity Board, South of Scotland Electricity Board, and the twelve area electricity boards in England and Wales.
Note: The figures are taken from the annual reports of the various bodies issued in the early 1980s and from other sources. They are not therefore directly comparable with each other. The table is intended to give an indication of the size of the activities.

licences and has the right to purchase up to 51 per cent of all UK oil production. Until 1983 it also undertook oil exploration and production itself, but these activities were separated off and sold to the private sector as Britoil. British Petroleum, in which the government has a 39 per cent shareholding, is one of the major oil companies in the UK and in the world: it owned 23 per cent of UK oil-refining capacity in 1983.

In the manufacturing sector, two major industries inherited from the early industrial revolution, steel and shipbuilding, have been nationalized, while BL emerged from the development of car making in Britain at the beginning of this century. British Steel is responsible for 85 per cent of British ingot steel production, British Shipbuilders account for 98 per cent of the UK output of merchant ships and for all slow-speed marine diesel engines. BL is the only domestically-owned major car manufacturer in the UK; it supplies less than 20 per cent of the British car market but accounts for over 30 per cent of the cars actually made in the UK. Most of the rest of the public sector manufacturing is concerned with modern high-technology industries. British Aerospace is responsible for the bulk of British airframe manufacture, most fixed-wing aircraft production, defence and telecommunications satellites, and most guided weapons systems output. In addition, the wholly publicly-owned Rolls Royce Company is responsible for almost all UK aero-engine production. Short Brothers Company of Belfast makes commuter airliners, airframes and missiles (Central Office of Information 1982b). The Royal Ordnance Factories manufacture weapons and armaments and have a much smaller share of the UK output of this industry. The Royal Mint is responsible for the manufacture of British currency and has an important export market which it shares with other UK currency manufacturers.

What this analysis reveals is that public sector industries provide a significant part of production industries employment in the UK and dominate certain sectors of UK production. Changes in their employment levels have a major effect in two ways. The first is through changing the geographical distribution of the industry concerned: this occurs through differential rates of change in different localities (Law 1981; Short 1981). The second geographical effect is that on particular places. In many instances the public sector industry operations at individual locations are on a large scale and tend to dominate the local labour market. Changes, whether growth or decline, thus have a major impact upon the employment characteristics and structure of particular places and, through those, on the society and economic health of those localities. A further important point is that because of their size and local importance,

Table 12.2 Employment in selected industries in the standard regions of Great Britain 1981 ('000)

Industry	SE	EA	SW	WM	EM	YH	NW	N	W	S	GB
Forestry	2	1	1	1				1	1	4	11
Deep coal mining	4			20	65	78	10	35	32	25	269
Nuclear fuel							9	6			16
Electricity	46	6	16	15	11	16	18	9	11	18	165
Gas	37	2	7	9	7	10	12	6	5	7	104
Water	18	2	6	7	5	6	7	4	5	4	63
Iron and steel	4	1		9	3	41	3	21	27	12	120
Shipbuilding and repair	28	3	19		1	3	5	34	1	29	123
Aerospace	52	1	31	15	23	8	36	1	7	12	185
Ordnance	2		2	4	3	2	10	3	1	3	30
Total	193	16	82	80	118	164	110	120	90	114	1,087
Percentage	18	2	8	7	11	15	10	11	8	11	100

Source: Employment Gazette
* The industries named are as defined in the 1980 Standard Industrial Classification; the numbers in brackets are the SIC class or group identifiers: Forestry (02), Deep coal mining (111), Nuclear fuel (15), Electricity (161), Gas (162), Water (17), Iron and steel (221), Shipbuilding and repair (361), Aerospace (364), Ordnance (329)

the decline of some of these industries has been a major factor in the past in influencing government regional policy and action. The most obvious examples are those of coal mining, steel making and shipbuilding, the decline of which has been a major element in decisions to provide regional assistance to ameliorate the local impact. Even in the mid-1980s, public sector industry-dominated labour markets in the UK exhibited above-average unemployment rates in many cases.

12.2 The regional pattern

The regional distribution of the public sector industries is indicated in Tables 12.2 and 12.3. The figures used are from the Census of Employment 1981 (*Employment Gazette* 1983a; 1983b) and thus include all employment in each of the industries shown and not just that in the public sector. This is significant for the present purpose in only three of them. The private companies involved in aerospace industries are almost all heavily dependent upon supplying aerospace products to the public sector. Westland, for example, makes helicopters for the UK armed services and Plessey and Marconi supply them with weaponry systems. The same is true of the ordnance industry, where the government-owned Royal Ordnance Factories supply armaments to the British armed services and are exporters in competition with private sector UK companies (*Statement on Defence Estimates* 1983). Shipbuilding and repair is different in that the public sector is almost exclusively concerned with merchant and naval ships. The private sector of this industry, which is also included in the figures, is almost exclusively concerned with pleasure boats and shipbreaking for scrap.

The percentage figures in Table 12.3 indicate the regional distribution of the public sector industries and the degree of dependence of each region upon them. The South East had the largest number of people employed, nearly 18 per cent of the UK total identified by this method. This was partly attributable to the large numbers employed in the public sector utilities in this region, in turn largely related to this being the region with the largest population in the UK. In addition, it also had almost half the UK employment in national government administration and so was heavily dependent for its prosperity upon the public sector. The mix of public sector industries in the South East, other than utilities, was strikingly different to that of most other regions. It had only minor representation of the older declining industries of

Table 12.3 The regional distribution of public sector industry employment* in Great Britain in 1981 and changes since 1971 ('000)

	SE	EA	SW	WM	EM	YH	NW	N	W	S	GB
Total employment*	198	16	83	87	118	172	116	126	92	113	1,122
Percentage of total in each region	18	1	8	8	11	15	10	11	8	10	100
Total as percentage of all production industry†	11	8	20	11	19	25	12	30	32	20	20
Percentage change in public sector industries 1971–81‡	−17	7	−6	−20	−11	−22	−20	−23	−35	−20	−20
Percentage change in all production industry‡	−32	−12	−12	−33	−15	−28	−34	−20	−29	−28	−28

Source: Employment Gazette

* The figures for total public sector employment are derived by summing totals for the following industries of the 1968 Standard Industrial Classification. The numbers in brackets are the SIC order of minimum list heading identifiers: Forestry (002), Coal mining (101), Iron and steel (311), Ordnance (342), Shipbuilding (X), Aerospace (383), Gas, electricity and water (XXI).
† Production industry is defined as all Index of Production Industries (Orders II–XXI of the 1968 Standard Industrial Classification) excluding Construction (X).
‡ 1981 minus 1971 expressed as a percentage of 1971.

coal, steel and shipbuilding, but it had by far the largest numbers employed in aerospace and defence industries.

Yorkshire and Humberside was the region with the second highest proportion of total public sector employment, which accounts for nearly one-quarter of its production industry employment. In contrast with the South East, the greater part of this employment was in the coal mining and steel industries with very little representation of the high-technology aerospace industries.

The percentage of total production industry employment which was in the public sector in each region showed a remarkably uneven distribution. The highest dependence was in Wales where it was over 30 per cent. As in Yorkshire and Humberside, over 50 per cent of this was accounted for by the coal and steel industries, but unlike Yorkshire and Humberside there was relatively little other production industry. The Northern region was close behind with almost 30 per cent of its production industry in the public sector. Even more so than Wales, it was vulnerable to economic decline by having three-quarters of this public sector industry in the declining industries of coal, steel and shipbuilding. At the other end of the scale, East Anglia had just over 8 per cent in the public sector. The South East had nearly 18 per cent of all public sector industry employment in the UK, but this represented only 11 per cent of the production industry labour force of the region. In addition, there were an estimated 25,000 employed in Ministry of Defence R and D establishments in the South East.

The changes between 1971 and 1981, shown in Table 12.3 using a different series of figures, show that the public sector industries had declined at a slower rate than index of production industries overall, a fall of 20 per cent compared with 28 per cent. The South East had a below-average fall in the public sector industries but a massive decline in index of production industries overall. As a result, the public sector contribution rose from just over 9 per cent to just over 11 per cent. This was not true in the West Midlands where there were above-average declines in both the public and the private sectors: the second highest in the UK in the former and the highest in the latter. Wales provides an intriguing and exceptional experience. It was the only region in the UK where decline in the public sector industries exceeded the fall in other production industry employment. As a result it was the only region in the UK where decline in the public sector enterprises contributed a smaller proportion of total production industry employment in 1981 than in 1971. The figures show that Wales, the Northern region and Yorkshire and Humberside were more successful than other regions in retaining private sector industries but suffered more from public sector industry declines.

12.3 Organization and ownership

In taking over industries previously privately owned, for example coal mining and steel making, or which had previously had a mixture of private and local government ownership, for example the utilities of gas, electricity and water, Britain adopted a state centralist model commonly known as nationalization. Other models of public ownership could have been adopted, such as transfer of ownership to separate independent regional corporations or to the county councils. Yet another alternative would have been to establish public corporations in which ownership of inter-related industries was combined, whether with the centralist structure of nationalization or with subnational control. Thus the gas, electricity and coal industries could have been combined into a single enterprise supplying energy in each region, as happens with utilities in other countries.

There were three important results of adopting the state centralist model. It was much easier for government to have a direct and relatively speedy influence upon the industries concerned. With control centralized and with chairmen who are government appointees, changes in government policy can be more readily implemented. The centralization of control also resulted in a well-ordered hierarchical structure of administration and organization, with a geography to match. The pattern and distribution of these functions would have been different if an alternative model had been used or if an industry had remained in private ownership. It also facilitated the spatial separation of the control functions of administration and R and D from production functions, to create a pattern again different to that likely to appear under disaggregated ownership. Lastly, the model is sectoral rather than spatial. Each public sector enterprise tends to aim at maximization of advantage attributable to itself, even if this is at the expense of another enterprise or to the detriment of the economy as a whole. The point is well illustrated by the example given earlier of the conflicts of interest which arise between the CEGB and the NCB. Invariably, such interindustry conflicts have a geographical impact. The location and distribution of the apparent or real benefits are usually different from those of the disbenefits (Fothergill and Gudgin 1984). Thus it is not only the fact of the change from private to public ownership, but the kind of public sector organization, that is of geographical significance.

Public sector ownership has also led to other important effects. The first is that public money has been made readily available to meet current operating costs and investment in a way which would have been at the very least unlikely in some of the heavy loss-making enterprises if they had remained in private ownership. Even in the profitable industries, there is a strong likelihood that the surpluses would have been transferred to other even more profitable investments in other industries and/or overseas had the industries remained in the private sector (*The Economist* 1981). As a corollary, the public sector industries are not allowed to become bankrupt. In some instances this has led to the accumulation of debts very much greater than would have accrued had bankruptcy in whole or in part been an available option. This again would have had a major differential effect on the geography of some of these industries. Beyond that, there has also been a considerable degree of cross-subsidization possible. Because of the state centralist structure, transfers from profitable parts of public sector industries to finance unprofitable parts have been facilitated. Finally, the nationalized industries have been considerably constrained in their ability to diversify operations overseas where more profits might have been obtained. Nor have they been allowed in some cases, for example BL, to purchase supplies from foreign producers which would have resulted in domestic production reductions; again with spatial results.

Some of these points are well illustrated in the following case studies of two selected public sector industries.

12.4 The National Coal Board

The National Coal Board (NCB) is essentially concerned with the mining and sale of coal and with some initial processing of it to yield products such as coke and smokeless fuel. Thus its activities are closely confined to the coal industry. Where coal mining elsewhere in the world is in the private sector, most of the major controlling companies have much wider interests. They are usually involved in a range of other energy industries including oil, natural gas and electricity production and/or in other industries which are large users of coal such as iron and steel making. Other than those which are local utility companies concerned mainly with the production and distribution of electricity and gas for local consumers, such companies, unlike the NCB, are usually multinational. This has allowed much greater flexibility in the use of their assets which can be more readily transferred between activities and locations. The absence of such flexibility and the restricted nature of the NCB in its activities and to the UK, of itself has produced a different geography. In its geographical organization, the NCB is strictly centralist and hierarchical. Overall policy and control is exercised from the headquarters located, not on any coalfield, but in central London. The coalfields are grouped into areas, each with an area office and director. The function of this second tier is essentially to implement policies laid down at the London headquarters. R and D establishments are maintained at Cheltenham and in Yorkshire. If the industry had remained in private ownership, there is little doubt that its organization would have been different. There would have been a number of companies involved, each linked with other interests such as electricity production or steel making: the nature and geography of organization would have developed to reflect this.

One of the most obvious differences which would have affected the geography of the industry if it had remained in the private sector is that it would have been much smaller. Coal mining in the UK emerged from the Second World War in a pitiful state. Investment to maintain existing mines, let alone to take advantage of best-practice technology and to replace exhausted capacity, had been kept to a minimum in the economic depression of the 1930s. The problems were exacerbated by the lack of resources and the need to maximize output during the war years (Humphrys 1973). Private owners would have been very unlikely to have found the massive amounts of capital needed to overcome these inherited problems in the immediate post-war years, despite the fact that the demand for coal remained high until the mid-1950s. The lesser amounts of capital that would have been available would undoubtedly have been invested in far fewer mines. In the event, the industry was nationalized in 1947 and the capital needed was provided by the state under a plan for coal which emphasized maximization of production rather than the minimization of costs. Thus the industry entered the 1960s much larger than it would otherwise have been. In the 1960s, competition from cheap imported oil caused the coal industry to lose markets and contract. The rate of decline was slowed by a government decision to reduce the competitiveness of imported oil by means of a differential tax. Even so, the industry continued to make substantial financial losses. In this situation it could be expected that private owners would not have been able to find the further large investments necessary to improve productivity and thus competitiveness. In addition, large consumers such as the electricity and steel industries would have turned to cheaper imported coal to a greater extent. In the 1970s, consumption of UK coal was further encouraged by government subsidization of coal burning in power stations and pressure on the electricity generating boards to add to coal-fired capacity. As a result, electricity generation took an increasing proportion of the coal produced even though the proportion of electricity generated by coal declined.

The demand for coal continued to fall after 1980 as a result of industrial depression in the UK. During this period, domestic production would have fallen much more rapidly if the government had allowed the import of the much lower-priced coal available from overseas. This would have affected the highest-cost British coalfields most and, since these are located in the north and west of Britain, this would have had an obvious geographical impact (Humphrys 1983). State ownership has had the further effect of limiting the amount of open-cast coal produced. Table 12.4 shows the much lower cost of producing such coal, which yields higher profits than any deep-mined coal and which private owners would undoubtedly have expanded. The result would have been more open-cast mines on the exposed parts of the British coalfields and fewer deep mines.

The major overall effects of public sector

Table 12.4 UK coal mining by area 1982/83

Deep coal mining areas	Saleable output (m. tonnes)	Wage-earners employed ('000)	Operating profit/loss		
			Total (£m.)	£/tonne	£/wage-earners
Scottish	6.6	16.9	−67	−10.16	−3.96
N.East	12.4	27.6	−67	−5.41	−2.43
Western	10.4	20.3	−8	−0.75	−0.39
S. Wales	6.9	23.4	−113	−16.38	−4.83
N. Yorks	8.4	14.4	−15	−1.83	−1.04
Doncaster	6.8	15.0	−14	−2.09	−0.93
Barnsley	8.1	15.0	−15	−1.80	−1.00
S. Yorks	7.3	15.5	−11	−1.43	−0.71
N. Derby	8.1	11.8	4	0.51	0.34
N. Notts	12.4	17.8	40	3.21	2.25
S. Notts	8.3	14.8	−19	−2.32	−1.28
S. Midlands	8.2	15.2	−27	−3.31	−1.78
Total	104.3	207.7	−312	−2.99	−1.50
Open-cast mining areas					
Scottish	2.8		37	13.19	
N. East	3.1		36	11.76	
N. West	1.0		7	6.70	
C. West	2.5		47	18.90	
C. East	3.2		43	13.59	
S. West	2.1		22	10.29	
Total	14.7		192	13.08	

Source: NCB Annual Report and Accounts 1982/83.

ownership, then, have been the creation of a particular geography of organization, a different mix of deep-mined and open-cast production and the maintenance of a much larger coal mining industry in Britain than could otherwise have been expected.

The regional distribution of coal mining in the UK can be explained only by reference to state centralist public ownership. This has been in part important in maintaining so much deep-mined coal production in the high-cost coalfields. The government has always taken account of the fact that reduction of coal output has regional social and economic effects of considerable significance. In 1983, government grants to help the coal industry amounted to £225 m., of which £15 m. was to promote sales of coal to the electricity generating boards. The government has also provided grants to stimulate the conversion of oil burning to coal burning by private industry (National Coal Board 1983). Although this benefited all the coalfields, it is the high-cost ones which benefited most. As Table 12.4 shows, these are geographically distributed: the highest losses were made by the peripheral coalfields and these have made the most persistent losses since nationalization of the industry. Even assuming current economic conditions, under private ownership a substantial proportion of the existing mines in these areas would have been closed by 1985. It is here, too, that the narrowly sectoral structure of the NCB plays a part. Much of the coal mined in the peripheral coalfields is consumed in electricity generating stations. By the 1990s, most of the power stations they sell their coal to will either be closed or will be peak load power stations burning little of their coal. An alternative form of ownership, private or public sector, in which coal and electricity production formed integrated regional or local supply utilities for example, would result in a much different pattern. Thus mines and generating stations would form single, local, integrated complexes, mutually interdependent. Instead, the NCB and the electricity generating boards pursue their own policies largely independently without even the benefit of a national energy policy.

12.5 The British Steel Corporation

The changes consequent upon the final nationalization of the bulk of the British steel industry in 1967 are excellent examples, of the differences between private and public sector ownership in their geographical effects. As Heal (1974: 276–7) has succinctly expressed it:

> Although strong economic and technological arguments favoured unifying the planning of a British industry whose total capacity was smaller than the United States Steel Corporation, the opportunity to do so originated in the political arena with the election of a large socialist majority to parliament in 1966. The 1967 Nationalization Act short-circuited all tentative moves towards a new corporate structure and imposed upon an unwilling industry a degree of internal coherence which it would otherwise not have obtained . . . By 1970 however, the central organization, located in London, had established complete control over its empire by means of a policy of divide and rule.

The new corporation was able to plan on a national, integrated and logical basis, creating the 'Heritage Programme'. Heal makes it clear that this would not have been possible if private ownership had remained (Heal 1974: 279). The programme began to be implemented in the first half of the 1970s. Capital was made available by the government to develop five integrated iron and steel making centres to achieve improved productivity through economies of scale. Without nationalization, it is unlikely either that these five centres would have been the ones chosen for development or that the capital necessary to implement the plan would have been forthcoming. The dramatic falls in the demand for steel after 1976 would thus have found the smaller steel-making units much more vulnerable and capacity would probably have been reduced even further than it was by the mid-1980s. Certainly the geography would have been different and would have been much more along ownership and company lines. Even within the new structure that was created, without government intervention the pattern would have been different because the Ravenscraig works in Scotland would have been closed. Writing about Ravenscraig, Warren (1984:148) suggested: 'It is still rather small and decidedly ill located; but its closure is socially and perhaps politically unacceptable.' The geographical distribution of the industry in the mid-1980s cannot be explained in terms of the environment external to the British Steel Corporation (BSC), of the cheapest coastal sites and proximity to markets, but only in terms of these 'given' external environment conditions being interpreted by the public sector rather than by the private sector. This is a conclusion reinforced by the comments on the significance of ownership for the geography of industry arising out of research into the electrical engineering industry (Massey and Meegan 1979). In the case of the steel industry, Pryke has been even more dogmatic in identifying the geographical results of the nationalization of the bulk steel industry in the UK. In a comment on the closure of the steelworks at Consett in County Durham, he criticizes the centralized structure of the BSC thus:

> it is questionable whether it is desirable to close Consett. . . . The reason why it is marked down for closure is that BSC has surplus capacity and that it has to shut something. This provokes the question of whether it is appropriate, when the decision between closing one works and another is finely balanced, to make a central decision? It might well be preferable for local initiative to be the determining factor (Pryke 1981: 209).

The quotation is included not to support the return of the UK steel industry to the private sector but to demonstrate the significance of nationalization for the geography of that industry.

12.6 Conclusions

This exploratory chapter has demonstrated the major significance of public sector industries in understanding the geography of industry in the UK. Industries in direct public ownership, plus those with substantial amounts of public sector shareholding, provided over 1.5 m. jobs in the UK in the early 1980s, nearly 20 per cent of index of production industry employment (excluding construction). Indirect employment, that is in enterprises directly dependent upon supplying these industries, would add a conservatively estimated 50 per cent to the total. The government has a responsibility for using the resources involved in the public sector industries to maximum economic benefit. It can, and to varying degrees does, influence decisions on the development of these industries, including their location. It can also ensure that the costs generated by decisions in any one of its industries are taken into account and attributed to that industry and not simply passed on to another part of the public accounts, so that the benefits of the decisions are real rather than merely cosmetic. To do otherwise is clearly immoral and an abuse of the responsibilities of public ownership. Moreover, the size of the public sector industries provides the government with considerable direct power to achieve the proper distribution of industry through location decisions of its own.

Such industries do not need to be persuaded to take location decisions to achieve desired spatial distributions by government incentives and planning instruments as do private sector industries. If these public sector industries are seen not to be pursuing government distribution of industry policy aims, it is hardly moral to try to persuade private sector industries to achieve them. Beyond that, it has been demonstrated that the location of public sector industries has been influenced by the fact of government ownership. This and the nature of state centralist nationalization of many of these industries, the absence of the bankruptcy option and the availability of public funds for finance, have all resulted in spatial patterns and distributions which are substantially different from those which would have arisen or survived under private ownership. The conclusion is that with the same external environment, different geographies will appear depending upon whether an industry is in the private or the public sector. It is in this area of the significance of the internal environment of an enterprise for location and spatial distributions that much current industrial geography research is being undertaken. As it proceeds in the field of public sector industry, greater understanding will be achieved, providing a more informed and practical basis for public policy and action.

References

Cable, J. R. 1982 The Nationalised industries. In Prest, A., Coppock, D. J. *The UK Economy*, 9th ed. Weidenfeld and Nicolson.

Central Office of Information 1982a *Nationalised Industries in Britain*. HMSO.

Central Office of Information 1982b *Britain's Aerospace Industry* HMSO.

Chester, T. E. 1976 The public sector, its dimensions and dynamics. *National Westminster Quarterly Review* Feb.: 31–44.

CIS 1982 The war lords: Report on the Arms industry. Anti Report 31, Counter Information Services, London.

Dicken, P., Lloyd, P. E. 1981 Modern western society. Harper and Row.

The Economist 1981 British Industry; are Britain's big companies leaner, fitter or just sadder. 4 July 1981: 73–4.

The Economist 1982 Non British Leyland? 14 August 1982:21.

Employment Gazette 1983a Census of Employment results for September 1981. **91**(5): Occasional Supplement 1.

Employment Gazette 1983b Further results from the 1981 Census of Employment. **91**(5): 61–5.

Forthergill, S., Gudgin, G. 1984 Is nuclear energy really necessary? *The Manchester Guardian* 11 April.

Glyn, A. 1984 When coal not dole is the best choice. *The Manchester Guardian* 28 May.

Hamilton, F. E. I. 1967 Models of industrial location. In Chorley, R. J., Haggett, P. (eds) *Models in Geography*. Methuen.

Heal, D. W. 1974 Ownership, control and location decisions. In Hamilton, F. E. I. (ed.) *Spatial perspectives on industrial organization and decision making*. Wiley, 265–84.

Humphrys, G. 1973 Power and the industrial structure. In House, J. W. (ed.) *The UK Space*, 1st edn. Weidenfeld and Nicolson, 224–7.

Humphrys, G. 1983 Power and the industrial structure. In House, J. W. (ed.) *The UK Space*, 3rd edn. Weidenfeld and Nicolson, 290–1.

Law, C. M. 1981 The defence sector in British regional development. University of Salford Discussion Papers in Geography, 17.

Lever, W. F. 1979 Industry and labour markets in Great Britain. In Hamilton, F. E. I., Linge, G. J. R. (eds) *Spatial analysis, Industry, and the Industrial Environment*. Wiley, 37–55.

Massey, D., Meegan, R. A. 1979 The geography of industrial reorganisation: The spatial effects of the restructuring of the electrical engineering sector under the Industrial Reorganisation Corporation. *Progress in Planning* **10**(3): 155–237.

National Coal Board 1983 *Annual Report and Accounts*.

National Economic Development Office 1976 *A study of UK Nationalised Industries*. HMSO.

Pite, C. 1980 Employment and defence. *Statistical News* S1.15.51: 15–19.

Prais, S. J. 1976 *The evolution of giant firms in Great Britain: A study of concentration in manufacturing industry in Britain 1909–1970*. Cambridge University Press.

Pryke, R. 1981 *The Nationalised industries: Policies and performance since 1968*. Martin Robinson.

Short, J. 1981 Defence spending in UK regions. *Regional Studies* **15**: 101–10.

Smith, D. M. 1979 Modelling industrial location: Towards a broader view of the space economy. In Hamilton, F. E. I., Linge, G. J. R. (eds) *Spatial analysis, Industry and the Industrial Environment*. Wiley, 37–55.

Statement on Defence Estimates 1983 Two vol. HMSO, Cmnd 8951.

The Nationalised Industries 1978 HMSO, Cmnd 7131.

Todd, D. 1980 The defence sector in regional development. *Area* **12**: 115–21.

Warren, K. 1984 Iron and steel. *Geography* **69**(2): 147–50.

Watts, H. D. 1980 *The large industrial enterprise: some spatial perspectives*. Croom Helm.

Introduction to Policies for Industry

Much of the analysis of locational decisions within industry has the unstated assumption that firms are free agents to select sites as and where they wish. In a world ruled by classical economics, such an assumption might produce a world or a country where labour markets adjusted to the preferences of industrial companies. If they all chose to locate in one area, then labour would be scarce, the price of it would be bid up and more would move in, re-establishing an equilibrium price for labour : conversely, in areas of labour surplus the price would come down and these low wages would attract firms into the area to utilize the cheaper labour. However, in countries such as Britain, these adjustment mechanisms are slow to operate – people have difficulty in moving to areas of lower unemployment, companies cannot relocate to adjust to wage differentials, and national wage agreements have the effect of evening out wage differentials. In consequence, there exist major differences in employment prospects from one area to another.

The classic response to these differences has been for the government to try to redistribute employment opportunities by inducing firms to establish branch plants or to move in entirety to areas with high unemployment. Britain's programme of regional policy (Ch. 14) is perhaps the most elaborate outside those countries with centrally-planned economies. In areas of low unemployment, restrictions were placed on growth; in areas of high unemployment all types of incentives and subsidies were offered – on land, on capital equipment and buildings, and on labour – to attract businesses. As long as the national economy was in reasonable condition, generating additional output and employment, it could be argued that this was a reasonable policy – it was equitable, even if the resulting distribution of industry was not the most efficient or what the companies would have chosen. Indeed, it was even possible that, nationally, output was higher as a result of easing bottlenecks such as high land prices, local shortages of labour and traffic congestion. By the 1970s, however, the advisability of such policies was being questioned. Concern was expressed over whether the national economic performance was being unduly harmed, and whether the regions qualifying for assistance were not too large and imprecisely defined to tackle the problem. In consequence, regional policy was reduced in power, and urban policy evolved in recognition of the concentrations of unemployment in certain towns and in the inner areas of the largest cities. Bill Lever describes in Chapter 15 how national programmes of urban aid developed out of a concern for the quality of housing, the environment, and social services and welfare, as it was recognized that without an adequate economic base such problems were likely to recur and recur, generation after generation. However, as urban policy increasingly addressed economic problems, the ability to create economic growth, anywhere, diminished as the world moved further into recession. Urban policy adjusted by looking to new sectors – the community, the informal sector, joint public – private ventures and worker-managed enterprises – to help in the programme of job creation. By looking in some detail at two contrasting conurbations, London and Glasgow,

Bill Lever is able to demonstrate the complex administrative and political structures which lie between urban policies for industry.

Overriding these policies is the government's concern for the national economy as a whole. Unemployment, so often used as the criterion for defining areas qualifying for special assistance, is only one facet of managing the national economy – others include inflation, the balance of payments, interest rates, sterling's exchange value and the growth in Gross National Product. If all these indices moved together, the economic role of national government would be simple. However, it is clear that they do not: interest rates may have to be kept high to help the value of sterling; reducing public sector expenditure to curb inflation may create unemployment. This presents governments with the need to identify priorities. The consequences, as Bill Lever shows in Chapter 13, for the manufacturing sector in Britain may not always be what the government would wish.

CHAPTER 13

National policy

William F. Lever

13.1 Introduction

The most frequently used approaches to the study of industrial change are those which direct attention to individual enterprises and their decisions on location, product type, investment and production technology, or to individual industrial sectors where the decisions are often inferred by statistical analysis of aggregate data on employment or, less commonly, on output. One of the exogenous forces which influences decisions at enterprise level or affects whole sectors is government intervention. Most emphasis has been placed, by geographers at least, on those policies which discriminate between different areas seeking to adjust the locational balance of industry and employment, usually with the objective of reducing spatial disparities in unemployment rates. By convention, these policies have operated at two distinct spatial scales – the regional and the urban – and these are dealt with in the two succeeding chapters. However, government intervention at the scale of the national economy has had powerful effects upon the industrial sector in Britain. Intervention at the national scale may have several objectives of which employment creation is only one. The other primary objectives of government in seeking to manage the national economy are the reduction of the rate of price inflation, the maintenance of a reasonable balance of trade and preferably one which generates a surplus in Britain's favour, assisting the overall rate of growth in output (without necessarily a commensurate rise in employment), and creating that exchange value for sterling which makes British goods and services competitive in international markets without causing inflation by raising the price of imported goods, especially necessary inputs to productive industries. It is clear from the attempts of most post-war governments that all these objectives are not simultaneously attainable and, indeed, some may be mutually incompatible. The range of policies which government may use to achieve these objectives is extensive but includes adjusting interest rates either to promote growth or to defend the exchange rate, spending public money in investment or directly in job creation, or in maintaining welfare levels, adjusting taxation rates, either to increase private expenditure or saving or to increase levels of government expenditure, and seeking to influence international trade flows either by creating tariff barriers or entering agreements to enhance trade links.

13.2 The aggregate economy

The total output of the British economy is measured in a number of ways: two of the most commonly used are Gross Domestic Product (GDP) at market prices and at factor cost. The former is defined as the total value of all sales of goods and services produced by UK residents. However, some of the values of sales are made up of taxes such as Value Added Tax (VAT), and, if these amounts are removed, the GDP at factor cost is generated. In absolute terms either definition of GDP for Britain is seen to rise. In 1972, GDP at market prices was £63.8 bn, in 1982 it was £274.2 bn; for the same years, GDP at factor cost was £55.7 bn and £232.6 bn respectively. In both cases, each successive year's

National policy

Table 13.1 Demand and output, 1978–82

	Percentage change from previous year				
	1978	1979	1980	1981	1982
Private consumption	5.8	4.9	−0.1	0.4	1.0
Government consumption	2.1	1.8	1.9	0.0	1.8
Fixed investment	3.5	0.9	−2.8	−8.2	2.3
Final domestic demand	4.6	3.6	−0.2	−1.2	1.3
Total domestic demand	4.4	4.4	−3.3	−1.6	2.0
GDP at market prices	3.8	2.0	−2.2	−2.2	0.8
GDP at market prices excl. N. Sea oil	3.2	1.1	−2.3	−2.6	0.5
Agricultural production	7.8	−0.1	9.6	−0.8	0.8
Mining	24.6	27.5	1.8	6.2	9.0
Manufacturing	0.6	0.1	−9.1	−6.4	−0.3
Construction, utilities	5.6	−0.5	−4.4	−7.5	−0.5
Services	3.0	1.9	−0.7	−0.7	0.5
GDP at factor cost	3.3	2.0	−2.9	−2.4	0.5

Source: Economic Trends, CSO

total was higher than that for the previous year. However, much of this apparent growth is due to inflation. If GDP is expressed in constant terms, it grew by 13.5 per cent in the seven years between 1972 and 1979, but between 1979 and 1981 it declined by 5.3 per cent so that in 1981 it was only little more than 1 per cent higher than it had been in 1973.

The sharp downturn in demand and output is shown in a number of ways in Table 13.1. Most components of demand fell very sharply between 1979 and 1981, but the 1982 data indicate a modest recovery which continued until mid-1984. Private consumption was relatively stable over the period: real take-home pay had been rising by about 4 per cent per annum between 1976 and 1980 but then fell sharply by 1.5 per cent in 1980–81. The impact of this fall, however, was somewhat alleviated by the rise in earnings from self-employment and by increased government transfers to households due to higher social security payments. Private consumption levels, however, were sustained as the savings ratio fell from 15.5 per cent in 1980 to 9.5 per cent in late 1982. Although government has tried to restrict expenditure, most of the savings and cutbacks have been made in investment rather than in consumption. Thus government investment fell from a peak of almost 6 per cent of GDP in 1973 to only a little over 1 per cent in 1982. Government consumption expenditure, however, remained fairly stable and only in 1981 was there no growth whatsoever. Private sector investment was consistently weaker after the 1979 peak, and this was particularly so in the manufacturing sector which by 1982 had returned to the levels of 1965. The fall in manufacturing investment reflects the sharp decline in output after 1980 and the consequent increase in utilized capacity accompanied by a severe squeeze in company profits. In 1981, pre-tax real rates of return had fallen to just over 2 per cent, the lowest on record and about half the rate in the previous, 1975, downturn. In fact, once the sharp rise in interest rates in 1980 and 1981 and higher local authority taxes are taken into account, then the rate of profit probably fell more than is suggested by the rate of return figures. Rates of return were also low in non-manufacturing, but not as low, and this therefore explains the shift of investment into industrial and commercial services. The high output growth figures in the mining sector in Table 13.1 are of course attributable to the rapid growth of North Sea oil output which moderated after 1979 (OECD 1983).

Table 13.2 shows how the recovery in the economy, first seen in 1982, continued and accelerated into 1983. GDP rose by 3.2 per cent, compared with only 0.8 per cent in 1982, and this increase in output reflected a rise in private consumption, in government consumption and in fixed investment. As total domestic demand grew by 4.6 per cent in 1983, compared with 2.0 per cent in 1982, some of this was met by higher levels of domestic output, but some of the increase in

Table 13.2 Demand, output and prices, 1982–84

	Percentage change from previous year		
	1982	1983	1984
Private consumption	1.0	4.3	1.8
Government consumption	1.8	2.6	1.0
Fixed investment	2.3	4.2	9.7
Final domestic demand	1.3	3.9	3.0
Total domestic demand	2.0	4.6	2.3
GNP at market prices	0.8	3.2	2.0
Consumer prices	8.3	5.1	5.0
Industrial production	0.1	2.5	2.2
Exports	0.9	1.0	5.7
Imports	3.9	5.6	9.0

demand was met by higher levels of imports. Imports rose by 5.6 per cent in 1983, compared with 3.9 per cent in 1982, whereas exports grew but only at the same rate in 1982 and 1983. This rise in output appears to have been sufficient to slow the rate of job loss, but increased productivity per worker meant that there was no commensurate rise in employment to match the 3.2 per cent rise in GDP.

By 1984, however, the growth in the economy faltered. The anticipated rise in GDP of 3.0 per cent was not realized, rising by only 2.0 per cent during the year. Some of this shortfall is attributable to the strike in the coal mining industry. During the year there was no growth in GDP in the first quarter, and a decline of 0.5 per cent in the second quarter, although there was some recovery in the last half of the year. Not all of the shortfall, however, is attributable to the loss of coal production. Personal consumption fell in the first half of the year, although some of this may be due to lower expenditure by miners and their families. The most obvious effect of the strike has been on industrial production which fell by about 3 per cent between the first and third quarters of 1984. On the expenditure side, there has been a clear effect on stocks, with coal stocks having been run down and some destocking elsewhere, and on import volumes which have risen as oil and coal have been imported to make good the coal production lost. Offsetting these developments, there has been a substantial rise in the amount of fixed capital investment which rose by 15 per cent, particularly in manufacturing where it rose considerably above the very low levels of 1981 (OECD 1984).

Thus 1983 and 1984 saw a rise in GDP although the figure of the latter year was depressed by the coal miners' strike. This growth in GDP is reflected in higher employment levels: employment grew by about 1 per cent between mid-1983 and mid-1984, having been static in the previous twelve months. This growth amounted to some 188,000 jobs but, significantly, unemployment continued to grow as the number of registered unemployed grew by 150,000 over the same period. What is happening within the national labour force, therefore, is that while declining sectors are continuing to release labour through redundancies, and some school-leavers are joining the registered unemployed, many of the jobs being created in expanding sectors are not being filled by registered unemployed. The weak recovery in Britain's traditional industrial sectors is reflected in the changing pattern of employment. Given that industrial production (excluding North Sea oil) by mid-1984 was still 10 per cent below its 1979 peak, it is not surprising that employment in this sector has continued to fall. Manufacturing lost some 103,000 jobs between mid-1983 and mid-1984 while services grew by 335,000 over the same period. Most parts of the service sector recorded increases in employment with particularly large gains in retailing and wholesaling, banking and finance, and business services. The effect of these sectoral shifts in the demand for labour meant that 181,000 of the 188,000 net new jobs were filled by females, the majority of whom had not hitherto been registered as unemployed. In addition, job creation has been overwhelmingly biased towards part-time work, with the number of females in full-time employment actually falling by 31,000.

Most studies of the relative importance of the manufacturing, primary, service and other sectors have tended to focus upon employment, and the decline in manufacturing employment as a proportion of all employment in Britain has already been commented upon (Chs 1 and 5). However, it is possible to examine the relative importance of these sectors in terms of their output value and their share of GDP. There is a problem in measuring the output of service sectors such as health, education and administration and government where there is no explicit charge and thus no value of total sales. Where this problem occurs, the value of output is measured as the total value of all the factors of input – labour, rents, capital costs and consumables – thereby making comparison possible. Table 13.3 shows firstly how the value of GDP has risen, from £55.7 bn in 1972 to £257.5 bn in 1983.

National policy

Table 13.3 Composition of GDP, 1972–83(%)

	1972	1973	1974	1975	1976	1977	1978	1979	1980	1981	1982	1983
Primary	7.8	7.7	7.2	7.8	8.6	9.5	9.5	10.3	11.9	12.9	13.7	13.7
Manuf'g.	30.7	30.4	29.0	28.0	27.5	28.4	28.6	27.0	25.5	23.8	23.1	22.6
Constr.	7.2	7.6	7.4	7.0	6.6	6.3	6.3	6.5	6.4	6.1	5.8	5.9
Services	54.3	54.3	56.4	57.2	57.3	55.8	55.6	56.2	56.2	57.2	57.4	57.8
GDP, factor cost (£ bn)	55.7	64.8	75.1	94.9	112.6	128.1	147.2	169.1	196.0	213.7	232.6	257.5

Source: Central Statistical Office: *Annual Abstract of Statistics 1985*

Within this total, the share made up by manufacturing industry has declined from 30.7 per cent in 1972 to 22.6 per cent in 1983; this fall has been almost continuous although there was a slight increase in the period 1976–78. Construction's share has over the same time period fallen from 7.2 per cent to 5.9 per cent, but there is clear evidence of cyclicality in its share of GDP, with peaks in 1973 and 1979 and a further rise in 1982–83. It is a common assumption that manufacturing's declining share of national output has been caused largely by the growth of the service sector (and there are employment figures to support this). The data in Table 13.3 show this to be untrue. It is true that the service sector's share of GDP grew from 54.3 per cent in 1972 to 57.8 per cent in 1983, but there has been little growth in its share since 1975 when it was 57.2 per cent. The more significant growth has been in the primary sector where, thanks to the exploitation, first of natural gas, and then much more importantly petroleum from the North Sea, the share of GDP has risen from 7.3 per cent in 1972 to 13.7 per cent in 1983. More strikingly, in absolute terms the value of output of oil and natural gas was as low as £69 m in 1972 (and as low as £12 m in 1975) but had risen to £16.2 bn by 1983.

This disparity between output change and employment change is obviously related to changes in output per head. Keeble (Ch 1) has already indicated that changes in productivity per worker explain the fact that while output in all manufacturing fell by 8.3 per cent between 1971 and 1983, employment fell by 32.2 per cent. This has occurred because in manufacturing industry (and in the primary sector) it has been possible to substitute capital equipment for labour: many of the routine, repetitive actions required by production processes have been mechanized. In consequence, fewer workers are required to sustain a given level of output and, of those, a higher proportion are employed in white-collar occupations and are not concerned directly with the production process. In services, the reverse is true: there has been a considerable increase in demand for services both personal and producer, but there has been less scope for the replacement of workers by machinery. While there has been considerable discussion about the introduction of (often computer-based) equipment in services such as banking, distribution and office activites, many of the major services such as health care, education, personal leisure services and social work services have proved extremely difficult to automate. The increasing use of part-time labour has further emphasized the employment growth within the service relative to levels of output.

13.3 Managing the economy

While measurement of the national economic performance in terms of GDP, employment and unemployment appears to treat the economy as though it were a single entity, it is of course made up of a large number of separate private enterprises plus the public sector which in its turn can be disaggregated into separate units such as the nationalized industries, public sector services, government and so on. These two sectors have rather different objectives: the private sector seeks to maximize long-run profits, the public sector seeks to achieve a desired level of output, whether of goods or services, at minimum cost. Increasingly, however, in the 1980s the public sector divisions such as nationalized industries, transport and utilities are being required to make profits by raising prices beyond production costs. Although the national economy is comprised of many separate units, government seeks to implement policies which affect it in its entirety. There are three broad apporaches to intervention in the national economy. One view, held by the current government, is that, ultimately, free market forces can be relied upon to equilibriate the national economy and, in consequence, intervention should be kept to a minimum. Thus parts of the public sector are sold into private ownership, and resources generally are channelled into the private sector and away from what is thought to be the non-productive public sector. Public sector expenditure is therefore cut and the government reduces the money supply in order to reduce the rate of inflation (Friedman 1962) which, as a policy objective, has a higher priority than reducing the rate of unemployment. The growth of unemployment is seen in the short term as being likely to moderate the level of wage demands and reduce the power of organized labour, which further reduces the rate of inflation. In the longer term, it is assumed that low inflation will enhance the competitiveness of British industry, export demand will increase, and some if not all of the lost jobs will be regained. It remains unclear, however, whether the sharp reduction in public investment, together with declining business confidence and a decline in consumer demand, do represent a reversible trend or whether they are likely to lead to mass unemployment, which in its turn may threaten public order (Medhurst 1984).

The second approach to the management of the national economy is that held by the Labour Party and often termed the Alternative Economic Strategy (AES), (Holland 1975). A part of this strategy involves the erection of tariff barriers around the British economy, for a limited period of time at least, thereby temporarily abandoning the tradition of free trade. An inevitable consequence of this would be taking Britain out of the EEC. This strategy also envisages an extensive system of planning agreements between government and the private industrial sector, giving the state much greater control over production and employment. Lastly, in contrast to the previous approach which had as one objective the diminution of the power of organized labour, the AES foresees a greatly enhanced role for worker participation in decision-making, arguing that long-term economic stability is likely only to be achieved by substantially increased levels of accountability. The AES has been criticized on a number of grounds. It is not clear, for example, how easy it would be for Britain to extricate itself from multinational agreements on trade. If, as has been suggested (Gudgin *et al.* 1982), the tariff barrier on goods and services were to rise steadily to a level of 50 per cent by 1990, then Britain's trading partners might well retaliate by discriminating against British goods, thereby cancelling out the employment effects of the protectionist policy. This retaliation, however, may only extend over two to three years and thereafter confer a net benefit on Britain. The impact of such a tariff barrier would, of course, be intensified by a decline in the value of sterling relative to the value of the currencies of Britain's major trading partners such as that which occurred in late 1984 and early 1985. A further criticism of the AES is that faced with increasing state control of industry and the prospect of greater decision-making power is in the hands of the workforce, capital would swiftly move out of Britain to countries where profits were higher and such restrictions did not exist. The AES therefore might require the reintroduction of exchange control regulations to restrict the export of capital.

The third approach to managing the economy rejects both these rather radical views and seeks to identify a much more centralist consensus. Authors such as Owen (1981) and Williams (1981) have drawn attention to the increasing volatility in political trends with the emergence of nationalist, ecological and social democratic parties which, they argue, reflects an increasing desenchantment with the inability of the two major parties to deal with long-run economic problems. Their solution is to seek a broader-based set of polices which rely on greater accountability and electoral reform (including proportional representation). For industry, their policies would involve greater worker representation on managerial boards, some form of prices and incomes policy, profit-sharing arrangements and a continuance of free trade within agencies such as the EEC (El-Agraa 1984).

These three approaches to management of the economy as a whole are obviously related to the major political parties, although the third, it is hoped, would draw support from sections of the Conservative and Labour Parties as well as the other parties, and form the basis of a centrist coalition. The history of government intervention into industry is markedly affected by the alternation of Conservative and Labour governments since the 1950s. The first major strategy for industry in Britain is often regarded as the National Plan, drawn up by the Labour government in 1965. This identified key sectors for development and laid down a plan of investment in infrastructure such as motorways and ports, and was to be administered by a new government department, the Department of Economic Affairs. The need to introduce counter-inflationary measures and to protect to exchange value of sterling in 1966 led, however, to the rapid abandonment of the National Plan (Opie 1972). The subsequent Conservative government did not develop a national strategy for industry, so that the next major initiative took place under the next Labour government in 1975 in the form of the National Economic Development Council (NEDC). The intention was to transform Britain into 'a high output, high wage economy . . . by improving our performance and productive potential' (Department of Industry 1985). The NEDC set up sector working parties whose responsibility was to make recommendations to raise productivity and competitiveness in forty key growth sectors, which by 1980 comprised 40 per cent of total manufacturing output. This was achieved by improving production technology, raising quality standards, providing funds for investment in new equipment, increasing specialist manpower training, improving marketing and providing assistance with exports. At the same time, the National Enterprise Board (NEB) was founded with £1,000 m to help productivity and sales. Although there were fears that these funds would be used to increase substantially the level of government ownership of Britain's manufacturing sector, by 1979 the bulk of the expenditure had gone to helping publicly-owned firms, even if these, as in the cases of Rolls-Royce and British Leyland, had in part come into public ownership because of very serious economic difficulties. Some of the NEB expenditure, however, was genuinely innovatory in high-technology sectors, such as the creation of a new micro-electronics company, Inmos, at a cost of £50 m.

With the return of a Conservative government in 1979, it was anticipated that government intervention in industry would be sharply reduced. To an extent, this prediction was realized. In 1981, the NEB and the National Research Development Corporation, which had been set up in 1948 to increase the rate of manufacturing innovation, were amalgamated into the British Technology Group which did, however, continue its funding to Rolls Royce and British Leyland. By 1982 and again in 1984, government's assistance to industry under regional policy was substantially reduced by reducing the areas which were eligible for assistance and by removing the concept of 'automatic' grants under the Regional Development Grant system and placing much more emphasis upon selectivity. Lastly, once the Conservative government was re-elected in 1983, an increased commitment to denationalization and privatization was reflected in the sale of a number of government assets of which the two largest were the state-owned oil exploration and production company, Britoil, and the telecommunications system, British Telecom (Clare Group 1982; NEDO 1982).

While we have stressed that government intervention in industry has often been strongly affected by which of the two major parties was in power, there are a number of common themes which have been developed in industrial policy almost irrespective of this. One of these is the concept of competition as a way of controlling prices and sharpening competitiveness. Arising out of the 1944 White Paper on employment, the 1948 Monopolies and Restrictive Practices Act sought to ensure that no product market became so dominated by a single producer that the producer was able to raise prices unnecessarily because of the lack of effective competition from others. Thus, proposed mergers between companies could be referred to the Monopolies Commission who might in some cases rule that a proposed merger was not in the consumers' interest. In a number of product markets, including tyres, detergents, automobile electrical systems and pet foods, the degree of producer concentration has been such as to arouse concern over the degree of price fixing which oligopolist structures might permit. Not all mergers, however, are viewed in this light, and the Industial Reorganization Corporation which existed in the late 1960s did seek to use mergers to create larger and more efficiently competitive units such as the linking of Associated Electrical Industries and GEC, and subsequently with English Electric, and the linking of the various car companies to create British Leyland (Dicken and Lloyd 1981: 91–2).

A second common element within industrial policies has been the encouragement of the small firm sector. In the early 1970s, the Bolton Committee drew attention to the capacity of small firms to create employment and to develop new products (Bolton 1971). The small firm sector has been dealt with in Chapter 9, but as an element in national policy for industry, government support has focused on tax relief for small businesses such as alleviation of corporation tax and the progressive raising of the minimum turnover threshold at which VAT becomes payable, and the creation of agencies offering them advice, on the grounds that they are unlikely to be able to afford to develop internally their own specialist advisers on finance, marketing, product development and so on.

Lastly, government policy for industry has placed increasing emphasis upon support for R and D, leading to both product and process innovation. The National Research Development Corporation was founded in 1948 to give assistance to research in universities and in industrial companies. Since that date, some 30 per cent of R and D expenditure in British industry has been funded by the government, and currently some 2.4 per cent of GDP is used by the government in this way, although just over half of this total goes on defence-related R and D. The Department of Industry initiated the Product and Process Development Scheme to encourage the introduction of new products and new production methods, and this was folowed in 1978 by the Microprocessor Application Project to deal specifically with the development of the silicon chip (see Ch 8).

13.4 Managing the labour market

It is clear from a number of earlier chapters (Chs 1 and 5) that, as there has been a sharp decline in demand for labour in the manufacturing sector and where service sectors have taken on additional labour, they have done so from those leaving education and from housewives, registered unemployment has risen sharply. The job decline problem, however, has been exacerbated by changes in the demographic structure of Britain. In 1961, the population of working age grew by about 250,000. The rate of increase subsequently declined to about 150,000 as the numbers reaching retirement age grew. However, the baby boom of the 1960s created substantial increase in the numbers of school-leavers, so that by 1981 the net addition to the population of working age rose to 340,000. This figure is then predicted to decline to 140,000 by 1991. Thus the government in the late 1970s was faced not only with a declining demand for labour but also an increased number of people of working age. By the peak year (measured in terms of the rate of increase), 1981, applying current participation rates, approximately 20,000 net new jobs would have had to be created merely to keep pace with the growth in the population of working age and to keep unemployment constant. This figure represents an addition, assuming no growth in jobs, of 0.8 per cent on the national unemployment in that year. In practice, the actual registered unemployment rate rose by 3 per cent, so approximately one-quarter of this rise can be attributed to demographic change and three-quarters to labour demand change.

The rapid worsening of the unemployment rate, and the growing concentration of unemployment upon certain groups – school-leavers, the unskilled, ethnic minorities – has led successive governments to intervene in the labour market by developing 'job creation' projects. The current sequence of job creation schemes began in the early 1970s with the inception of Community Industry targeted at disadvantaged young people. Job Creation was introduced in 1975 and in its three years of operation provided about 120,000 places, mainly for workers in the 16–24 age group. The intention was that the work undertaken by job creation schemes would either benefit the community through environmental or social projects or would benefit the workers through some aspect of training. There was therefore a clear intent that there should be no displacement effect allowing Job Creation schemes to displace existing businesses because of the government subsidy. Most of the places on Job Creation were provided by public authorities and voluntary and charitable organizations. In contrast, the Work Experience programme, started in 1976, was restricted to 16–18-year-olds and most of the 40,000 places (about 80 per cent) were provided by private industry. When Job Creation and the Work Experience Programme were abandoned in 1978, they were replaced by the Youth Opportunities Programme (YOP) and the Special Temporay Employment Programme (STEP). The YOP offered both courses of instruction and work experience to 16–18-year-olds with the latter comprising about 80 per cent of the places. STEP represented a recognition that unemployment was not just concentrated upon school-leavers, for it catered for those aged 19–25, and those over 25 who had been unemployed for more than one year. In 1980, the YOP scheme was expanded from 300,000 to 440,000 places and the budget was increased to £271 m., while the STEP was replaced by the Community Enterprise Programme (CEP) with 25,000 places and a budget of £122 m. Finally, in 1983, YOP was replaced by the Youth Training Scheme (YTS), available to all school-leavers without employment or a further education place, and the CEP was replaced by the Community Programme.

Several common themes run through this sequence of job creation schemes. Firstly, there is always a stronger emphasis upon school-leaver unemployed than the older age groups. The concern of government is that school-leavers who are unable to find work should not develop an attitude which disposes them never to seek employment and to regard unemployment as 'normal'. Secondly, there has been an increasing concern, seen most clearly in YTS, that training should be included. This is part of a wider discussion on the suitability of those leaving school for employment: criticism by employers of the educational system centres on a concern that school subjects, learnt in the classroom, do not fit their requirements. Thus schemes such as YTS are seen as essential training for work either in the workplace, in the community or in institutions of higher education. This is a rather different argument to that levelled at other manpower training schemes which states that as there is very little structural unemployment relative to the large volume of demand-deficient employment, there is little point in retraining redundant workers in new skills when these are already in excess supply. Thirdly, government has been careful to indicate that the job creation schemes will not create new enterprises in product markets or services which are already supplied by commercially-viable ones. The employment opportunities are therefore concentrated either in existing enterprises or new ventures supplying services to the community which would not be commercially viable without the subsidy. Fourthly, most of the jobs or places created have a maximum time duration, often one year. The hope in many instances, such as YTS places in industry, is that the employer will take on the employee on a permanent basis once the year has elapsed, but it is clear that many do not do so. Schemes such as YTS have been criticized on the grounds that they appear to assume that the problem of high unemployment is a temporary phenomenon and that such schemes are necessary to bridge the gap until economic recovery increases the real demand for labour, and that this assumption is misplaced.

In addition to job creation, the government has made a number of attempts to redistribute the existing stock of jobs in a different, and fairer, way. Two examples of such intervention into the national labour market are job-sharing and premature retirement. Premature retirement schemes are based upon government assistance – with lump sums or the maintenance of pension payments – to workers who can be induced to retire before the statutory age in order to create a vacancy which can thus be filled by a redundant worker. Work-sharing schemes depend upon two persons each taking part-time employment with government assistance to the employer to meet some of the duplicated non-wage fixed labour costs. The number of jobs created or released by such schemes, however, relative to the volume of unemployment is extremely small.

13.5 Britain's economy in an international context

One of the major factors in determining the shape of the British economy in the post-war period has been the very large changes in the patterns of international trade (Scammel 1980). In a period of relative economic growth up to the mid-1960s, the major trends were the growth of the industrial exports from Third World countries, the creation of multinational trading blocks such as the EEC and the European Free Trade Association (EFTA), the effect of political independence on many countries within the British Commonwealth and the emergence of the Oil Producing and Exporting Countries (OPEC). The major effect of this was to reduced dramatically Britain's role in the world trade in manufactured products. In 1950, the UK produced one-quarter of the total world trade in manufactured products compared with West Germany's 7.3 per cent and Japan's 3.4 per cent. By 1977, the UK's share had declined to 9.3 per cent, compared with West Germany's 20.8 per cent and Japan's 15.4 per cent (Dicken 1982).

Britain's competitiveness in world markets would appear to have diminished sharply over the post-war period. Competitiveness is measured as the outcome of three factors: firstly, the basic costs of production, usually reduced to the price of labour, sets the cost of goods; secondly, the exchange value of the currency influences the international price of goods; thirdly, the rate of inflation in different countries alters the dynamics of international trade flows. Where labour costs and inflation are low, then a country is likely to be competitive; similarly, where the value of a country's currency is declining in terms of international exchange rates, that country is also likely to be competitive. Competitive advantage defined in this way enables a country to sell in

Table 13.4 Visible trade of the UK (£m.)

	1972	1973	1974	1975	1976	1977	1978	1979	1980	1981	1982	1983	1984*
Exports (fob)	9,437	11,937	16,394	19,330	25,191	31,728	35,063	40,687	47,415	50,977	55,546	60,534	69,513
Imports (fob)	10,185	14,523	21,745	22,633	29,120	34,012	36,605	44,136	46,182	47,969	53,427	65,993	78,085
Visible balance	−748	−2,586	−5,351	−3,333	−3,929	−2,284	−1,542	−3,445	+1,233	+3,008	+2,119	−5,459	−8,572

Source: Department of Trade and Industry
* Annual estimate based on Jan.–Nov. data.

foreign markets more effectively but also to resist imports from other countries. For much of the period since the 1960s, Britain's competitive position has worsened relative to most countries. Relatively high labour costs, certainly when compared with the newly industrializing countries, and higher rates on inflation than most competitors amongst the industrialized countries, more than offset any decline in the value of sterling. This trend continued until 1980, since which time the slowing of the inflation rate together with a rapid decline in the value of sterling, especially against the American dollar, have partially reversed the stituation. For example, between 1978 and 1980, unit labour costs in Britain rose by 40 per cent, whereas the comparable figures for West Germany were 17 per cent, for the US 16 per cent and for France 23 per cent, while in Japan they fell by 10 per cent. The exchange value of sterling against a basket of currencies drawn from competitor industrial nations fell from an index of 100 in 1975 to 81.2 in 1977, recovered (thereby worsening Britain's competitive position) to 96.1 in 1980, and then fell to 84.5 in mid-1983 and to approximately 70 in early 1985. Against the American dollar, the decline was even more dramatic with the pound equal to $2.33 in 1980, $1.65 in late 1982 and $1.05 by March 1985 but recovering to $1.45 by mid 1986.

The consequences of these changes are shown in Table 13.4. For all visible trade, the UK was a net importer right through the 1970s, but the impact of North Sea oil in reducing imports in 1980, 1981 and 1982 is seen in the fact that in these three years the UK was a net exporter in terms of visible trade. By 1983 and 1984, the situation had deteriorated markedly with a growth in the import of manufactured goods and of fuel because of the miners' strike. Much of these deficits before 1979 and some of the deficit since 1983 was made up by the UK's earnings from invisibles such as banking, insurance, tourism and other services. More significantly, Table 13.5 shows the balance of trade on manufactured goods alone. The UK remained a net exporter of manufactured goods until as recently as 1982, the major deficits occurring in the other sectors such as fuel and agricultural products. Throughout the 1960s, the UK exported about 40 per cent more than she imported in manufactured goods; throughout most of the 1970s the surplus amounted to between 30 and 10 per cent. Only in 1983 did the UK become a net importer of manufactured goods and the shortfall grew very sharply between 1981 and 1985. Thus, despite the rapid decline in the value of sterling, British manufacturing industry seems to have been unable to compete both with Third World and Newly Industrializing countries and with developed industrial countries. The modest recovery in the national economy, and in world trade, in 1982–83 appears to have had the effect of sucking manufactured goods into Britain's markets.

Table 13.6 shows the balance of imports and exports by manufacturing sector for 1973, 1978 and 1983. In total, the proportion of home

Table 13.5 Manufacturing trade balance of the UK (£m.)

	1972	1973	1974	1975	1976	1977	1978	1979	1980	1981	1982	1983	1984*
Exports	8,114	10,067	13,306	16,063	20,701	25,824	28,029	30,870	34,811	34,898	37,316	39,919	46,115
Imports	5,983	8,697	11,693	12,605	16,578	20,702	24,421	25,689	31,177	31,993	37,083	44,905	52,588
Balance	+2,131	+1,370	+1,613	+3,458	+4,123	+5,122	+3,608	+1,181	+3,634	+2,905	+233	−4,986	−6,473
I : E ratio	136	116	114	127	125	125	115	104	112	109	101	89	88

Source: Department of Trade and Industry
* Annual estimate based on Jan.–Nov. data.

National policy

Table 13.6 Balance of imports and exports, by industrial sector (%)

	Imports/home demand			Exports/sales			Balance		
	1973	1978	1983	1973	1978	1983	1973	1978	1983
Food and drink	19	17	17	5	7	11	−14	−10	−6
Coal & petrol prod.	17	15	n.a.	14	14	n.a.	−3	−1	n.a.
Chemicals	22	29	35	28	38	41	+6	+9	+6
Metal manuf.	21	24	31	17	21	25	−4	−3	−6
Mech. eng.	26	33	33	36	46	41	+10	+13	+8
Instrum. eng.	46	56	56	50	56	52	+4	0	−4
Elect. eng.	27	36	43	25	40	38	−2	+4	−5
Shipbuilding	56	44	42	26	37	55	−30	−7	+13
Vehicles	23	36	51	36	43	37	+13	7	−14
Metal goods n.e.s.	10	13	17	13	17	18	+3	+4	+1
Textiles	21	31	41	23	29	28	+2	−2	−13
Leather	27	33	44	23	26	33	−4	−7	−11
Clothing	18	27	34	10	18	18	−8	−9	−16
Bricks etc.	7	9	10	11	15	11	+4	+6	+1
Timber, furniture	29	27	31	3	8	5	−26	−19	−26
Paper, etc.	19	20	20	8	11	10	−11	−9	−10
Other	15	18	23	17	20	18	+2	+2	−5
Total	21.4	24.8	31.4	19.6	25.3	27.0	−1.8	+0.5	−4.4

Source: Department of Trade and Industry

consumption made up by imports rose from 21.4 per cent to 24.8 per cent and subsequently to 31.4 per cent, while exports as a proportion of total sales rose from 19.6 per cent to 25.3 per cent and subsequently to 27.0 per cent. While the bases of the comparison are not the same – home demand against total sales – the balance does indicate an improvement between 1973 and 1978 and a marked deterioration between 1978 and 1983. Almost all sectors show an increase in the level of import penetration, the only exceptions being food and drink, and the shipbuilding sector where the massive decline in orders world-wide makes the figures difficult to interpret. The largest increases in import penetration are to be found in consumer goods, particularly vehicles, clothing and textiles, and in electrical engineering (which includes domestic electrical appliances). Rises in imports cannot really therefore be interpreted as industry stocking up with components or capital equipment preparatory to an upsurge in output: they are largely attributable to patterns of domestic consumption expenditure. Exports, as a percentage of all sales, rose in every sector between 1973 and 1978 but fell again in the majority of sectors, both capital (mechanical, instrument and electrical engineering) and consumer (vehicles, textiles, timber, furniture, paper) between 1978 and 1983.

By comparing the two ratios, the balance of trade in all sectors other than food and drink and shipbuilding worsened between 1978 and 1983. The most dramatic shift of all occurred in the vehicles sector which turned round from being a substantial net exporter in 1973 to a substantial net importer by 1983 with most of the damage being done in the period after 1978 (Wells and Imber 1977).

In addition to these major changes in the balance of trade flows across Britain's frontiers, there have been changes in the identity of our trading partners. Table 13.7 shows how membership of the EEC has affected the pattern of

Table 13.7 Export markets

	1972 (%)	1978 (%)	1983 (%)
EEC*	30.4	38.5	43.7
Rest of Western Europe	16.3	12.5	12.4
North America	16.7	12.0	15.6
Other developed countries	9.8	6.6	5.1
Oil exporters	6.7	13.1	10.0
Other developing countries	16.4	13.9	10.9
Centrally-planned economies	3.2	3.0	1.8

Source: Department of Trade and Industry
* Includes Greece, Denmark, Eire throughout.

Table 13.8 Import sources

	1972 (%)	1978 (%)	1983 (%)
EEC*	31.8	41.9	45.5
Rest of Western Europe	16.2	15.2	15.8
North America	16.3	13.4	13.7
Other developed countries	10.4	7.6	7.8
Oil exporters	9.3	8.4	4.2
Other developing countries	12.8	10.6	10.2
Centrally-planned economies	2.9	2.8	2.3

Source: Department of Trade and Industry
* Includes Greece, Denmark, Eire throughout.

Table 13.9 Balance of trade with Europe and the Commonwealth (£m.)

Europe	1972	1978	1983
Exports	2,935	13,621	26,509
Imports	3,525	16,547	30,104
Balance	−590	−2,926	−3,595
E : I	83	82	88
Commonwealth			
Exports	1,667	4,982	6,147
Imports	1,955	3,974	5,729
Balance	−288	+1,008	+418
E : I	85	125	107

Source: Department of Trade and Industry

Britain's trade. In 1972, only 30.4 per cent of all British (visible) exports went to the EEC (plus Greece, Denmark and Eire); by 1983 the figure was 43.7 per cent. The rest of Western Europe, East Europe and the 'other developed countries' (South Africa, Japan, Australia and New Zealand) all declined in importance. The need to recycle oil revenues forced up the share of Britain's exports going to oil-exporting countries between 1972 and 1978, but Britain's near self-sufficiency in oil by 1983 reduced their share of our export markets. For imports, again the major change was the effect of EEC membership, where the proportion of Britain's imports coming from the EEC rose from 31.8 per cent in 1972 to 45.4 per cent in 1983 (Table 13.8). All other import sources declined in importance over the same period and the drop in the oil-producing countries' share is particularly noticeable.

Lastly, Table 13.9 traces the changes in the balance of visible trade with two of Britain's major trading partners, the EEC and the Commonwealth. It has already been shown that trade with the EEC has grown in proportion from approximately 30 per cent of all trade by value in 1972 to 45 per cent in 1983. Britain has constantly experienced a net trading deficit with the EEC with exports by value amounting to only about 82 per cent of imports in 1972 and 1978, but this figure rose to 88 per cent in 1983. There has been a marked turnaround in Britain's patterns of trade with the Commonwealth. Exports to the Commonwealth comprised 17.3 per cent of all Britain's exports in 1972, but only 10.1 per cent in 1983. Imports from the Commonwealth made up 17.6 per cent of all imports in 1972, but only 8.7 per cent in 1983. Britain's balance of trade with the Commonwealth shifted from being a net importer in 1972 to a major net exporter in 1978, and to a slightly less emphatic net exporter in 1983. The main reason for the major boom in exports in 1978 was the success of sales to Nigeria which at that time was spending its massive oil revenues on imports. By 1983, exports from Britain to Nigeria were very substantially down.

13.6 Conclusions

The enterprises and sectors which form the traditional foci of interest in industrial geography are merely components in an aggregate national economy. A national economy such as Britain's is subject to external forces such as oil price rises and exchange value changes. The government, faced with a number of alternative and possibly conflicting goals, decides on its priorities and seeks to manage the economy as a whole towards the desired objectives. These external forces and national policies have their effects on the individual companies and sectors. To make any predictions as to the future direction of the manufacturing sector of Britain's industry, it is necessary to make certain assumptions about the government's objectives. In the short run, within the life of a government it is reasonable to assume little radical change in objectives. On this basis, models of the national economy indicate that fixed investment will be no higher in the short term with a fall in investment in housing, little change in public sector investment, a slow rise in consumer expenditure but a marked rise in exports, given the exchange value of sterling. On these assumptions, the NIESR model predicted a rise of 3.3 per cent in GDP in 1985 and of 1.4 per cent in 1986. These figures were not sufficient to bring down registered unemployment which remained around the figure of 3.2 m. through 1986 while the balance of payments returned to a modest surplus in 1986 (NIESR, 1984).

In the medium term, to the end of the decade, predictions on the national economy are faced with rather greater uncertainties. Gudgin et al. (1982) demostrate that if government continues with current policies, the level of unemployment by 1990 is likely to be 4.5 m. or 17.1 per cent nationally, whereas if conventional reflation is adopted the figures would be 3.3 m. or 12.3 per cent, or if the Alternative Economic Strategy were pursued the figures would be 1.9 m. or 7.0 per cent. The NIESR on an assumption of a continuance of existing policies sees real GDP rising by about 1.2 per cent annum and this rise in output being met be increased productivity so that no additional employment is created, and the rise in the population of working age forces up the numbers of registered unemployed to between 3.3 and 3.4 m. by 1989. This base case prediction is modified to take account of policies to create lower wages, or to allow for a faster rise in the level of world trade. In each case, however, the impact is only slighty different to that of the base case. Only by the more radical policies of the type described as the Alternative Economic Strategy would it appear possible that a substantial improvement might be created.

In the final analysis, however, geographers have tended to examine the internal decision-making processes within firms on such issues as locational choice, product type and production process. Where government policy is introduced, it is most commonly in the form of a study of the impact of regional or urban policy on locational choice. This chapter has sought to demonstrate that there are more national and global trends at work which influence many of the decisions at corporate level.

References

Bolton, J. E. 1971 *Report of the committee of enquiry on small firms*. Cmnd 4811. HMSO.

Central Statistical Office 1984 *Current economic trends*. HMSO.

Central Statistical Office 1985 *Annual abstract of statistics*. HMSO.

Clare Group 1982 Problems of industrial recovery. *Midland Bank Review* Spring: 9–16.

Department of Industry 1985 *An approach to industrial strategy*. HMSO, Cmnd 6315.

Dicken, P. 1982 The industrial structure and the geography of manufacturing. In Johnston, R. J., Doornkamp, J. C. (eds) *The changing geography of the United Kingdom*. Methuen.

Dicken, P., Lloyd, P. 1981 *Modern western society*. Harper and Row.

El-Agraa, A. M. 1984 *The Conservative Government's European policy 1979–84*. Discussion Paper Series 136, School of Economic Studies, University of Leeds.

Friedman, M. 1962 *Capitalism and freedom*. University of Chicago Press.

Gudgin, G., Moore, B., Rhodes, J. 1982 Employment problems in the cities and regions of the UK: the prospects for the 1980s. *Cambridge Economic Policy Review* **8**: 1–18.

Holland, S. 1975 *The socialist challenge*. Quartet.

Medhurst, K. 1984 Economic policy and political consensus in post-war Britain. In Hare, P. G., Kirby, M. W., (eds) *An introduction to British economic policy*. Wheatsheaf-Harvester.

NEDO 1982 *Industrial policy in the United Kingdom*. NEDO.

NIESR 1984 The home economy. *National Institute Economic Review* **110**: 6–26.

Opie, R. 1972 Economic planning and growth. In Beckerman, W., (ed.) *The Labour government's economic record: 1964–1970*. Duckworth.

OECD 1983 *Economic survey 1982–83: the United Kingdom*. OECD, Paris.

OECD 1984 *Economic trends 1984*. OECD, Paris.

Owen, D. 1981 *Face the future*. Cape.

Scammel, W. M. 1980 *The international economy since 1945*. Macmillan.

Wells, J. D., Imber, J. C., 1977 The home export performance of United Kingdom industries. *Economic Trends* **286**: 78–89.

Williams, S. 1981 *Politics is for people*. Penguin.

CHAPTER 14

Regional policy

Alan R. Townsend

14.1 Introduction

Many countries of the world have a recognizable 'regional policy' of one kind or another. In the EEC, for instance, regional development is one of the common objectives of member states, which typically involves attention to areas of declining industry as well as to remoter and poorer agricultural regions. The distinguishable feature of a regional policy is the definition of 'assisted areas'. Within the boundaries of a state, a need is seen by its government to demarcate certain areas for priority in new investment; along with this may go actual restrictions on the development of other areas. In the 'command economies' of the Eastern Bloc the control of individual enterprises for these and other ends is an integral part of national planning machinery: in the Western democracies the lesson is that the political need for regional policy must be strong and clear if it is to have a significant and sustained effect, but it may, in fact, provide very welcome subsidies to private industry.

14.2 Mainsprings of UK interest

The UK was a pioneer among Western European countries in developing a regional policy. Political objections to state involvement in private industry were sufficiently overcome to inspire a modest reaction to a major event, peak unemployment of 1931–32. Regional policy thus grew from small beginnings in pre-war conditions and generated a continuous stream of legislation, notably: Special Areas (Development and Improvement) Act, 1934; Distribution of Industry Act, 1945; Local Employment Act, 1960; Industrial Development Act, 1966; Industry Act 1972. As indicated by these titles, a concern with local or 'special' areas tended to alternate with wider interest in regional aspects of national industrial development. Regional policy in any guise was intimately involved with the subject of this volume, that of industrial change, in attempting to offset the effects of industrial decline in certain areas by diverting investment of mobile growth industries to those same areas. In terms of industrial change, it was often attempting to *replace* the oldest industries by the newest. Why, however, did the UK retain regional policy laws throughout the years between the deep recessions of 1929–32 and 1979–82? What were the mainsprings of attention to it?

1. The UK does have a wider distribution and variety of industrial coalfields than other countries. Whether for locational or structural reasons, several 'peripheral' fields were fairly persistently a problem. Their unemployment was a cause of great concern in the 1930s, and renewed redundancy in mining itself in the 1960s was the key feature in the revival of regional policy.
2. Several features combined to give policy a strong start just after the 1939–45 war: the concern of a Labour government for equity; the recommendations of a war-time report (Royal Commission on the Distribution of the Industrial Population 1940; the 'Barlow Report'); contemporary support and precedent for state involvement in the economy; and fear of a return to pre-war levels of regional

unemployment (see, for example, Daysh and Symonds 1953).
3. The *central control* of industrial development was given real teeth, unlike some other countries, through the land-use planning system; from 1947 to 1982, local planning permission for a factory required the issuing by central government of an 'industrial development certificate' (IDC): though not commanded to move to an 'assisted area', a firm could thus be encouraged to look at the possibilities.
4. Research on industrial movement from 1947 to 1951 (Luttrell 1962) showed that most industry (two-thirds, it was suggested) was genuinely mobile over the distances pertaining in Britain; thus industrial interest in moving in search of trainable labour was complementary to governments' interests in redressing pockets of unemployment.

The history of regional policy is presented in detail in McCrone (1969), Keeble (1976) and Maclennan and Parr (1979). It achieved respectability as a legitimate part of efficient national economic development in the early 1960s, notably in assuaging the inflationary effects of labour shortage in prosperous areas during a continuing period of full employment. The more abiding theme, however, emerges from the relationship of legislation to the economic cycle (Table 14.1) It is clear that the advancement of regional policy has generally occurred in reaction to cyclical unemployment, and with some eye to electoral consequences in areas affected. Only the Conservative government discovered, from its success in 1983, that a government could win an election with a reduced regional policy (see Section 14.8). Regional policy thus commonly represented a political reaction to past events, rather than necessarily the forward thinking of planning textbooks.

Policy changes and uncertainty following each general election have inhibited the plans of industrialists and contributed to the difficulties of local authority planning in making economic forecasts for their areas. It is as well at this stage to define the limits of regional policy in relation to other forms of spatial planning. In this chapter, we are concerned with powers given to the Department of Trade and Industry (and its predecessors) in all areas over IDCs, and in 'assisted areas' over buildings and inducements to factory investment and employment by new and existing firms, supplemented in later years by restricted assistance to tourism and certain service industries. These functions variously complemented and conflicted with the other interests of the Regional Economic Planning Councils and Boards 1966–79, established in an attempt to coordinate the affairs of the different national departments in each region, in transport, housing, industry etc. In the dispersal of industry from, say, London or Birmingham, the assisted areas were often competing as a destination both with each other and with overspill dispersal schemes including New Towns. Regional policy was later in conflict with urban policy for inner cities (see Ch. 15). Administrative arrangements and law generally provided an insufficient link between regional policy and local authorities, who gained direct interest in, and powers over, employment development only in the late 1970s (Townsend 1986).

14.3 The past identity of 'assisted areas'

Whitehall decisions over the identity of 'assisted areas' discriminated in effect both between and within regions and were changed with great frequency, especially between the 1958 and 1972 Acts. Three tendencies which were at work over time are illustrated in Fig. 14.1:

1. A marked extension of the proportion of the country affected (until 1977).
2. A great diversification in the type of area affected (cities and seaside resorts, for instance, were added to the original coalfield Special Areas).
3. Greater complexity was introduced through the evolution of a three-tier system; the conflicting attractions of local assistance and of a broader

Table 14.1 Unemployment peaks and regional policy

Peaks of national unemployment	Dates of regional policy legislation
1932	1934
1938	1937, 1945
1947	1950
1952	—
1959	1958, 1960
1963	1963
1967	1966, 1967
1971	1970, 1972
1977	—
1986(?)	1984

The past identity of 'assisted areas'

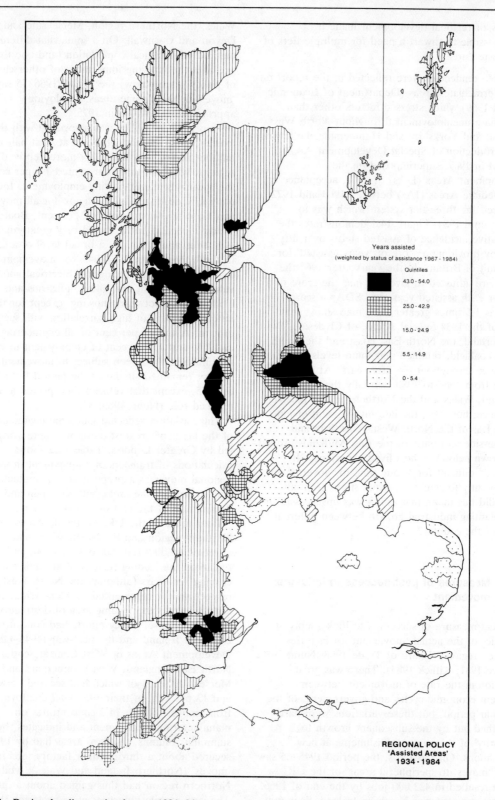

Fig. 14.1 Regional policy: assisted areas 1934–84

view of regional development made it impossible to thwart a need for multiple tiers of 'assisted area'.

All three tendencies were reflected in the report on the Intermediate Areas (Department of Economic Affairs 1969) which stressed factors other than immediate unemployment throughout North West England and Yorkshire and Humberside. Following the introduction of Special Development Areas (SDAs) in 1967, superimposed on the Development Areas (DAs) of 1966, acceptance of Intermediate Areas (IAs) between 1969 and 1972 produced the three-tier system which was to survive until 1984. Figure 14.1 demonstrates the cumulative incidence of assisted status over fifty years by providing a score of 'years assisted' for each unit of Britain. In the years after 1967 the scores are adjusted to give IAs *half* the score of a DA for each assisted year, and SDAs a score which is 1.5 times greater. The map shows that most of the 1934 Special Areas, of Clydeside, West Cumberland, the North-East coast and the South Wales coalfield, retained maximum levels of assistance throughout the fifty years. At one stage lasting from 1966 to 1980, virtually the whole of Scotland, Wales and the Northern region received assistance; however, the late involvement of most of the IAs of the North West, Yorkshire and Humberside and parts of the East Midlands proved (as shown below) a short-lived experience. It appeared difficult for a long time for government permanently to remove assisted status from an area; did this mean that policy was unsuccessful in transplanting industrial growth between different parts of the country?

14.4 Measures of past success in industrial movement

A direct measure of success is to look simply at statistics of the actual 'movement' of factories between regions (Board of Trade 1968; Nunn 1980; Pounce 1981; Killick 1983). There was great variation in the rate of movement between different economic cycles and governments of the post-war period, but these variations were partly smoothed out by the subsequent growth to 'maturity' of the new establishments at new destinations. On this basis, the period 1945–65 saw 1,057 'moves' to 'peripheral areas' of the UK, which resulted in 422,000 jobs by the end of 1966. These areas comprise Northern Ireland, Scotland, Wales, the Northern region, Merseyside and Devon and Cornwall. On a somewhat different method of statistical compilation (and one that excludes the growing importance of other channels of policy – see below) the period 1966–75 saw 940 'moves' to these same areas, employing approximately 130,000 in 1975.

Thus regional policy was associated with the existence, by the mid-1970s, of at least half a million factory jobs in the six more deprived regions, which in turn represented 7.6 per cent of the country's manufacturing employment. Including *other* UK destinations, the figure for all movement was close to 10 per cent (9.7 per cent, Board of Trade 1968: 25) but it varied very greatly between industries and areas. At a broad level of analysis, in the whole of the UK 1945–65, movement was dominated by engineering and electrical goods, vehicles, chemicals and allied industries and clothing and footwear, showing, except for the latter, a pattern of high correlation with sectoral growth. A full 15 per cent of all engineering jobs in 1966, and 13 per cent of employment in vehicle manufacture, had been subject to movement. The pattern remained similar in the period from 1966 to 1975, except that vehicles then played a much reduced role (Nunn 1980: 34).

This variation reflected industrial characteristics of the areas of greatest out-movement, which were led by Greater London. Table 14.2 shows calculations of transfers of employment in inter-regional moves as a proportion of manufacturing employment in the areas both of origin and destination, at the end of two study periods.

Four areas of the UK, Scotland, Wales, the Northern region and the North West, had each attracted 90,000–100,000 jobs by 1966, and remained the leading regions of attraction in the ensuing ten years (although the North West, in this respect mainly represented by Merseyside, clearly lost its role as the leading area of destination). To be more precise, policy efforts had concentrated on Northern Ireland and the post-war (1945–60) Development Areas of West Central Scotland, North East England, West Cumberland and Merseyside, each of which had secured between 9 and 13 per cent of their 1966 *total* employment from 'moves'. Table 14.2 concentrates on manufacturing employment and indicates, by summing columns, that two areas had by 1975 secured about a third of their factory jobs from 'moves' (Northern Ireland and Wales), while the Northern region had thus gained about a quarter. The new mobile industries were replacing the

Table 14.2 Employment in interregional moves, 1945–75

	Moves 1945–65*			Moves 1966–75*		
	Destination at 1966†		Origin at 1966	Destination at 1975†		Origin at 1975
Country/Region	Thousands	Percentage of M	Percentage of M	Thousands	Percentage of M	Percentage of M
N. Ireland	39.8	22.5	—	13.3	8.6	0.5
Scotland	94.7	12.6	0.6	28.5	4.5	0.2
Wales	93.7	28.3	2.6	32.8	10.3	0.4
Northern	89.6	19.2	1.1	35.5	7.8	0.5
North West	104.8	7.6	2.1	20.3	1.9	1.1
South West	36.9	8.9	2.7	19.4	4.5	0.7
Yorkshire & Humberside	31.3	3.5	4.8	9.8	1.3	1.0
E. Midlands	26.9	4.3	5.0	13.7	2.3	2.6
W. Midlands	8.7	0.7	7.3	1.6	0.2	2.1
East Anglia	16.7	8.8	5.7	17.4	8.8	1.3
South East	31.6	1.2	8.4	12.6	0.7	5.4

M represents total manufacturing employment
Source: Board of Trade (1968); Nunn (1980).
* There are differences in definition between the two periods, 1945–65 and 1966–75, both as regards the definition of a 'move' (see Nunn 1980: 39–43) and of certain English regions.
† Includes 108,500 jobs which were provided by moves from abroad, 1945–65 and 34,200 in the period 1966–75.

decline of coal mining (not included in these calculations for manufacturing) most notably in Scotland and Wales.

Where had these moves originated? The third and sixth columns of the table both show the South East to have 'lost' a greater proportion of its potential manufacturing jobs than others, although followed by areas of the Midlands and Yorkshire. If one takes these 'originating' amounts of mobile industry as separately given, it is possible to model the spread of 'moves' and accompanying employment to destination areas by the operational use of 'gravity models'. Keeble (1972) was able to allocate 'flows' of industry from the South East and West Midlands to the main assisted regions by reference to unemployment in those areas and the distance between origin and destination. Townsend and Gault (1972) similarly modelled flows between each of fifty recording areas of origin and destination, 1945–65, by reference to average numbers unemployed over the period and inversely to a function of distance; a correlation coefficient of 0.89 was achieved between the recorded and predicted values of 1966 employment from the sum of moves destined to the fifty areas.

That is to say that the attraction of new jobs to destinations was principally a function of labour availability, modified by factors of distance; on this interpretation it is distance, for instance, which explains how Wales captured more than its 'share' of mobile industry, that is due to its greater accessibility to London and the West Midlands relative to most other assisted areas; at a greater distance, Scotland's performance, 1945–65, only bears comparison with other major destinations because half (46,200 jobs) of the total shown in Table 14.2 arose from the strong attraction of plants from origins abroad (which were not included in gravity models). Unemployment is treated in these models as a surrogate for labour availability; a government survey (Expenditure Committee 1973: 573) confirmed that 'availability of labour at new location' was the *leading* factor determining the choice of location among 632 mobile establishments, mentioned by 72 per cent of them.

Neither politicians nor researchers paid much attention to the quality as opposed to the quantity of jobs provided, although some of the data were buried in the small print of the same report (Expenditure Committee 1973: 623); this showed that the typical mobile establishment employed between 40 and 59 per cent of its employees in semiskilled grades, whereas we may note that they represented only 25 per cent of the national manufacturing population in the 1971 Census.

Townsend et al. (1978) showed through interview data with workers how the high ratio of semiskilled workers carried fundamental implications for the nature and status of branch factories in the North East. Massey (1979) went much further in stressing the modern context of such establishments as part of national and international industrial corporations, who stood to gain from hiving off different stages of the manufacturing function to separate labour forces in separate labour market areas. This was a process helped by regional policy grants, while the assisted areas were as a result provided mainly with new jobs of a relatively low grade, more relevant to female workers than to members of craft unions in declining trades. 'The main impact of regional policy's success in increasing job opportunities in the Assisted Areas appears to have been to reduce the rate of net out-migration from the less prosperous regions and to raise their female activity rates rather than directly reducing the numbers of registered unemployed' (Regional Studies Association 1983, reporting Moore et al. 1981).

Conventional economic analysis, however, extended further than industrial movement, as discussed in this section. It attempted to embrace policy effects on the overall economy of assisted areas, including assistance to existing factories through the Regional Employment Premium (1967–77) and selective financial assistance (from 1972), together with all multiplier effects on the wider regional economy. 'It is clear that there is no methodology for assessing the effects of regional policy that is unequivocally defensible and robust' (Marquand 1980). The starting point for many analyses was the 'counter-factual' concept of what conditions would have been like in the absence of regional policy. These were commonly assessed by reference to a period of very low industrial movement, the 1950s, and the extent to which national industrial changes then accounted for departures from national rates of employment change in assisted areas. Applying the new data for national employment change after 1960, and subtracting from the actual performance of assisted areas, Moore et al. (1977; 1980) estimated that the main assisted areas received 280,000 jobs from regional policy 1960–76. Lower figures, however, are presented by Mackay and Thomson (1979) and Fothergill and Gudgin (1982: 138–52); the latter considered that some of the improvement in assisted areas would have occurred in the absence of regional policy, as part of the national statistical drift of industry from urban to more rural areas.

The Department of Trade and Industry (1983) made a fairly critical public assessment as a basis for further policy deliberations. They noted that academic research suggested that between 250,000 and 445,000 more manufacturing jobs were created in the assisted areas by 1979 than would have been the case had there been no grants or other forms of assistance. When allowance was made for local multiplier effects, the range was increased to between 350,000 and 650,000. The paper does note that 'policy was less effective in job creation terms in the late 1970s than in the late 1960s', but does not perhaps emphasize that the disappearance of jobs created by regional policy was also gathering force in this period, a point which will be emphasized in the rest of this chapter. We will have to conclude that the overall impact of regional policy (after including allowance for multiplier effects and a period of decline after 1976) was not in the end much different from the estimate we drew from Table 14.2, that it was responsible for providing of the order of half a million jobs in assisted areas.

14.5 Compromising factors of the 1970s

What was often missed at the time during the 1970s was that regional policy was declining in importance almost throughout the period, long before politicians admitted it, or in the end made a virtue of change. There was a number of inherent reasons why regional policy was increasingly compromised, within a very few years of the 1972 Industry Act. Most reasons were to do with the shape of the national economy, but there were other competing priorities requiring political judgement.

1. A short period of rapid investment was brought to an end in 1973; only a slow recovery in the economy had taken place before renewed conditions of recession set in in 1979. Many writers stress that the immediate cause of difficulty, the doubling of the price of oil in connection with the Middle East War (1973), was only the trigger which exposed more long-lasting difficulties in the British economy. Clearly, traditional regional policy could not be very effective without investment in *new* plants, but there were several other ways in which it was undermined.

2. A long period of full employment came gradually to an end. UK annual average unemployment rose from 2.9 per cent in 1974 to a

temporary peak of 6.2 per cent in 1977, inevitably involving many, but not all, of the traditionally prosperous areas. In economic terms, this tended to mean that a principal motivation of regional industrial movement among firms, labour shortage, was much less of a factor in prompting relocation.

3. Politically too, the operation of regional policy proved difficult in the absence of full employment. It seemed increasingly difficult for government to refuse IDCs in the relatively prosperous areas of the country (if industrialists were still planning expansion there). During the Labour government of 1974–79, the proportion of expansion in the South East and Midlands which was affected by IDC refusals fell from about 10 per cent to virtually nothing. In particular, some of the relatively prosperous regions included wide areas with unemployment rates as high as those used as criteria for regional assistance in the past.

4. The emergence of more problem areas as part of the difficult national conditions created competition for government attention and generated dilution and diversification of existing spatial policies. While areas of the Midlands were making claims to be added to the adjoining and now very large assisted areas, the inner areas of cities, wherever they were located, attracted justified concern. Due to longer-term trends in the cities, to a degree of success for regional policy in the preceding decade, and to particular characteristics of the 1973–75 recession, the assisted areas and those of the South East and Midlands had 'converged' in terms of their relative prosperity. Keeble (1977) and others therefore questioned whether spatial policy should be urban rather than regional. As shown later by Fothergill and Gudgin (1979), there were indeed basic statistical grounds for the view that the main direction of industrial change constituted a net drift from the cities. However, Townsend (1977) demonstrated that regional policy in practice was not necessarily in direct conflict with inner urban priorities. Indeed, it emerged later that spending on the latter remained much less. It was of particular significance, however, that urban problems (see Ch. 15) came to rival regional ones for the attention of politicians, researchers and the press.

5. Regional policy was no longer allied to an expanding part of the economy. Manufacturing employment declined by 11 per cent between 1971 and 1979, partly because investment was shifting towards the rationalization of production and reduction of labour costs. The view that this situation necessarily caused a reduction in movement deserves one qualification. Massey and Meegan (1982) stressed that in some circumstances expansion at new locations (often in new technical processes requiring less labour than the plants they replaced) was consistent with national net reductions in particular industries; these investments might be very significant in transferring work, implicitly or explicitly, between regions.

6. National government priorities themselves compromised regional policy. The Industry Act 1972 remained in place, but the regional employment premium was dropped in 1976–77, and clauses of the Act applicable to general industrial assistance (that is, irrespective of location) were increasingly used in Labour's industrial strategy (Cameron 1979). For instance, national industries centred on the West Midlands were picked out for government assistance, including BL (British Leyland, now Rover).

In retrospect, we can then see regional policy as one of a family afforded by full employment: yet even a Labour government was prepared to compromise the ideal of full employment when paying increasing attention to mounting inflation. In some ways, growing difficulties were obscured during the decade (Townsend 1983: 16–18), and the implications of new developments such as the UK's accession to the EEC in 1973 were unclear. On the one hand, Holland (1976) stressed how multinational companies might by-pass the regional objectives of different member states by 'playing off' them against each other; on the other hand, the European Regional Development Fund became available in UK assisted areas as such, principally serving to spread the costs of local authority capital investment. The six factors which we have described still fail to prepare one adequately for the scale of changes which occurred after 1976.

14.6 The decline of industrial movement after 1976

The resolution of the factors which we have just described produced a low rate of industrial movement, essentially reminding us both that regional policy can provide new plants for assisted areas only when there is relevant national investment upon which it can work, and that it is based on 'permissive legislation': governments did not, for instance, oblige themselves to apply IDC policy to a standard level of intensity at all times; unlike many kinds of fiscal or welfare policy, they could vary its application largely without public announcement. The facts of reduced spending in assisted areas were largely apparent in Annual Reports under the Industry Act (see section 14.7), but it was not explicit that this was increasingly concentrated on existing plants; furthermore, the monitoring of industrial movement itself was subject to long time-lags.

It was not until 1980 that government and academic research reported a decline in the impact of policy. Keeble (1980) reported that regional policy 'had ceased by 1976 to exert a measurable impact on the geography of manufacturing employment change in Britain'. Nunn (1980) found evidence of a long-term decline in mobility as from the mid-1960s (p. 10), and 'evidence of a marked change in the relative performance of assisted areas and non-assisted areas during the 1970s' (p. 8). Marquand (1980: 69) reviewed underlying causes in assessing the births and deaths of all manufacturing establishments as a proportion of the overall stock. What clearly was occurring at the end of her study, from 1973 to 1975, was a marked increase in the 'death rates' of establishments and a decrease of 'birth rates', leading to a growing 'natural decrease' in the population of establishments in most regions.

It was not until 1983 that official statistics reported manufacturing plant openings after 1976: 'During the period 1976–80 there were some 1700 openings of new manufacturing plants – an average of around 350 a year. This compares with an average of around 700 a year during the early 1970s, and around 1000 a year during the late 1960s . . . Not surprisingly, these openings accounted for only 1 or 2 per cent of all manufacturing jobs' (Killick 1983: 466). Observing that the pressures to move had been reduced, Killick stressed that there had been a reduction in the relative importance of longer distance moves and 'a continuation in the decline of inter-regional moves between different parts of the country; there were an average of 60 such moves per year during 1976–80, compared to around 180 during 1972–75 and over 200 during 1966–71' (p. 467). By the end of 1980, 299 interregional moves (1976–80) had generated only 14,800 jobs (of which 5,500 were in Wales), unequivocably the lowest rate since records began in 1945 (compare Table 14.2).

14.7 The decline of existing 'regional policy factories' after 1976

By 1980, manufacturing jobs in Great Britain were subject to an average of 33,000 redundancies a month, as part of the very large reduction of employment in conditions of national recession (Martin 1982; Townsend 1983). It is essential to ask not just whether these conditions inhibited the growth of new jobs in mobile plants (as described above, for instance), but whether factories established *previously* under regional policy were largely exempt (as might be expected from earlier recessions) or on the other hand sustained a proportionate (or even larger) share of this national decline.

The precise answer to this sensitive question can emerge only from government research, because only they define which establishments in assisted areas are counted as 'interregional moves'. However, an attempt to assess the probable incidence of closures, 1976–81, among regional policy factories has been undertaken from a data set prepared by Townsend and Peck (1985). This is an analysis of all job losses reported in the *Financial Times* by name of company and location. Each factory in the record was classified according to its location, including 1976 assisted area status, and by product, including cross-reference to a classification of industries by their mobility (Board of Trade 1968: 27). These enabled us to extract plant closures in assisted areas in industries which were more dependent than average on 'moves' in the period 1953–66. The resulting list of ninety-nine cases is considered to be closely congruous with the plants in our wider data base which originated from regional policy; it includes many industries which were unknown in the assisted areas before the post-war period, both in expanding industries such as electrical engineering and in some industries such as clothing which placed expansion in the areas despite contraction elsewhere.

These closures in mobile industries in assisted areas were led (in terms of workers affected) by Courtaulds, Plessey (see Peck and Townsend 1984) and British Leyland. Within individual corporations, the remarkable feature was that, in the overwhelming proportion of corporations affected, the closure of an assisted area plant *preceded* that of any in *other* parts of the country, appearing to confirm, for this period, the hypothesis that 'corporations close peripheral branch plants before others' (for further details see Townsend and Peck 1985). The timing of these closures, according to expected date of occurrence, is shown in Table 14.3.

The total of 45,500 represents the final loss, through closure, of nearly 10 per cent of the 'half

Table 14.3 The timing of major plant closures

	Cases	Employment before closure
1977	7	5,100
1978	20	10,600
1979	12	3,300
1980	29	11,100
1981	31	15,300
	99	45,400

million' jobs provided by regional policy (as in previous estimates above). It is, of course, a conservative estimate because it excludes other forms of redundancy, reported and unreported; many plants had reduced employment before closure, and others experienced major redundancies without closure. Including all kinds of job loss reported in this source over five years, Townsend (1983: 192) estimated the loss of between 90,000 and 100,000 jobs from 'regional policy factories'.

The rate of loss appears to be at least as great as that experienced by manufacturing in the nation as a whole. In the period 1979–81, the loss of jobs from government industrial estates in England occurred at exactly the overall rate for British manufacturing (Townsend 1983; 191). In 1980, the previously expanding mobile industries experienced job loss at a slightly faster rate than manufacturing as a whole (p. 179). Foreign-owned plants were fully involved in the withdrawal of investment from assisted areas (Townsend and Peck 1986), which was a new and fundamental feature for Strathclyde (Hood and Young 1982).

These broad inferences of failure in new plants must be set in the context of renewed structural decline in the assisted areas from 1976 to 1981. Tyler (1983) reports how the Development Areas showed their first negative 'differential shift' in performance since 1960 (one of 2.5 per cent). If we divide the period into two, then we find that the weaknesses of indigenous and new industries together were first registered in 1976–79 in the 'divergent' performance of the main assisted area regions in terms of redundancies (Townsend 1982), in unemployment (Frost and Spence 1983) and population change, 1977–80 (Champion 1983). The incidence of closures and redundancies in assisted areas, as for instance seen above for 1978, appeared partly to advance in time the impact of national recession in certain areas (particularly Tyneside, Merseyside and Clydeside), and thus in turn to moderate its impact compared with the national impact of fresh events from 1979 to 1981; these were felt most severely in the West Midlands, Wales and the North West. The impact of national recession is a complex matter in its own right, dealt with at different points in this volume and Townsend (1983). It convinced many geographers that consideration of regional policy was properly held subordinate to national economic trends, and that spatial problems were a wider matter than any simple definition of 'assisted areas' in peripheral regions. For one thing, the West Midlands conurbation was clearly the worst affected by recession among the UK's major urban areas, and it had always been a source of outward industrial movement rather than an 'assisted area'.

14.8 A straitened approach to assisted areas from 1979

Oblivious of the recession to follow, the Conservative government elected in June 1979 cut back assisted areas on the little-disguised basis of that month's unemployment figures (Townsend 1980). The map of assisted areas as it stood on election day (Fig. 14.2) had been comparatively little changed since 1972; although the Aberdeen area and much of North Yorkshire had been downgraded in a cautious review of 1977, the predominant feature had been the upgrading of areas such as Hull, Grimsby and Falmouth, meaning that 40 per cent of the British working population lay in assisted areas.

The Conservative government felt able to make a virtue of 'not spreading the butter too thinly' in difficult times, but all the indications are that they were acting on financial grounds in reducing the assisted areas to cover an intended 25 per cent of the working population. In doing so, they adopted a 'finer-grained' spatial approach than had been utilized for some years. It is appropriate at this stage to review the different possible methods of defining assisted areas in Britain, in general ascending order of scale (the Northern Ireland government has always offered financial inducements on an even basis over the whole province):

1. Site of redundant factory or mine.
2. Larger sites.
3. Parts of city/wards/parishes.
4. Overspill towns and New Town designated areas.
5. Employment Office Areas (EOAs – formerly Exchange Areas).

Regional policy

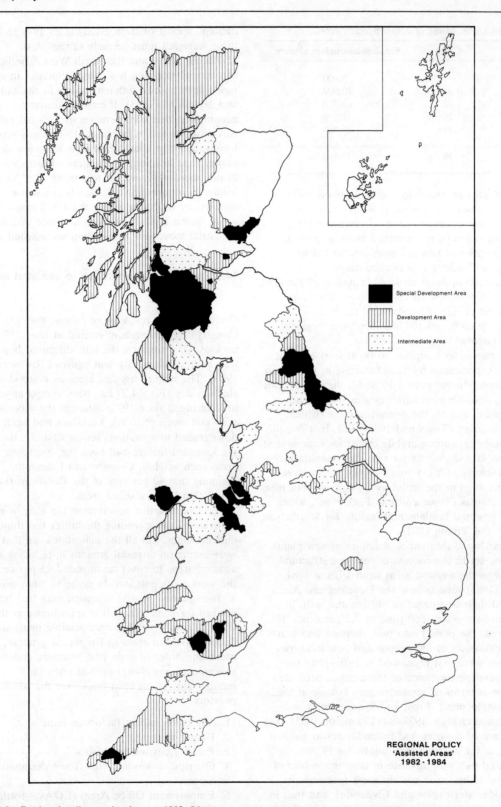

Fig. 14.2 Regional policy: assisted areas 1982–84

6. Local authority districts.
7. Travel-to-work areas (TTWAs – official groups of 5).
8. 'Growth areas'.
9. Contiguous subregional blocs of EOAs.
10. Regions.

Items (1) to (3) and (6) have never been utilized as such (up to the date of writing) in regional policy, although it is notable that urban policy has adopted level (3) in the definition of inner-city partnership areas, that rural policy (in the shape of RDAs of the Development Commission) is defined strictly in terms of parishes (level 3), and that the enterprise zones introduced by the 1979 Conservative government represent level (2), sometimes embracing larger redundant sites (level 1).

The use of the other units by past regional policy can be defined with the use of the above reference numbers as follows: 1934 (9); 1945–60 (9, with some use of 5, and 8 at Inverness); 1958–63 (5); 1963–66 (5), also (4) and (8). The complexity of this last period was wound up in 1966 through instituting primarily a regional approach, but this too was increasingly compromised through smaller excisions and superimposed units: 1966–72 (10) also (5); 1972–80 (10) and (7). Until 1980 then, the TTWA had been a means of assessing upgradings and downgradings within mainly regional assisted areas. As is shown from the shape of Figure 14.2, after the review of 1979 the TTWA (based on one or more EOAs) became the basic unit of assessment, decision and public definition. However, the simple use of unemployment data as the main statistic of assessment harked back directly to the previously discredited methods of 1958–66, including the Local Employment Act of 1960.

14.9 The instruments of regional industrial policy; the position in mid-1984

A principal purpose of assisting areas is for government to provide money or resources in support of their businesses. However, there are different ways in which this may be done, which have enjoyed different relative weight under different governments' philosophies for regional policy. Immediately after the war, for instance, the principal emphasis fell on the provision of government factories for rent on specially-provided industrial estates. This method was again fairly evident in 1984 in the shape of 'advance factory' building; this creation of standard factories ahead of the demands of a known tenant was attractive to government in these post-recession conditions, and had in several periods served to encourage a proportion of mobile investment to settle in assisted areas (Slowe 1981).

However, the main dilemma over the previous twenty years had lain in the balance of expenditure between 'standard grants' toward firms' new investment, the use of 'discretionary' or 'selective grants' toward their activities, and the use of 'labour subsidies'. Standard grants were first introduced in 1963 and offered the industrialist the certainty of a government contribution amounting to a fixed proportion of his new investment costs in buildings, plant and machinery; the predictability of the grant was held to be vital, both in 'selling' assisted areas to potential investors from outside and in underpinning the forward financial planning of existing and new firms within the areas; the firm did not normally have to prove that new jobs would materialize. 'Selective' grants, by contrast, did involve complex negotiations with government as they were a tailor-made alternative; prospects of viability and the need for assistance had to be proved, and the creation or safeguarding of jobs in assisted areas was a basic criterion.

The 'regional employment premium' was an explicit labour subsidy paid in respect of each manufacturing worker in 'assisted areas' from 1967 to 1977, but then abolished as the first of a number of financial economies made in the system. This trend exposed the balance of standard grants and regional selective assistance as an area of debate even before the 1979 general election. The growth of North Sea oil installations in Scotland was attracting extremely large 'standard grants' with very little return in jobs, while it was increasingly appreciated that the same grants when used for the modernization of process industry (steel and chemicals) in areas such as Teesside could even yield a net loss of jobs. Perhaps this money could be much better spent in the encouragement of smaller firms in the regions to take up 'tailor-made' selective assistance? Nonetheless, organizations such as the Confederation of British Industry naturally continued to defend the system of standard grants, and in 1979 they were withdrawn only from the lowest level of assisted areas (Intermediate Areas).

The *nature* of spending had not changed appreciably in the six years up to 1984, when the annual proportion of all regional industrial assistance spent on standard grants varied between 64 and 80 per cent. The more important feature was the absolute decline in *total levels* of spending on regional industrial assistance as a whole, which dropped by a full 30 per cent between 1974/75 and 1982/83, from £1,310 m. to £917 m. (including some back payments) at 1982/83 prices, and to an expected level of £643 m. in 1983/84. Clearly part of the reduction was due to industry's lower levels of investment during recession conditions. It is clear, however, that the efficacy of regional policy depends both on the amount of financial assistance which it channels to needy areas and on the quality of policy arrangements. In considering the latest changes in government plans, we have to distinguish carefully between these two aspects.

14.10 Regional policy decisions of 1984

In fact, changes made to regional policy with effect from November 1984 did involve further financial cuts, albeit within otherwise improved arrangements for spending regional industrial subsidies. These were presented in a 'white paper with green edges' of the previous December (Secretary of State for Trade and Industry 1983). Among the groups which sent in comments was the Industrial Activity and Area Development Study Group of the Institute of British Geographers (Wood 1984). The Group found itself part of an informed consensus of views which, within the departmental confines of the White Paper, could welcome some of its proposals. It *welcomed* the general principle of imposing a limit on regional development grants, finally set at a ceiling of £10,000 per job. It *welcomed* the plan to include offices and selected service industries (beyond their previous, almost nugatory levels of regional assistance) as a long-overdue reform. It was, however, concerned about some directions of ministerial thinking: the scope for geographical variations in wage rates had been overstated in the White Paper, and it depicted regional incentives as largely satisfying 'social' needs.

There was, however, general regret in many quarters, including the Confederation of British Industry, that the main thrust of the eventual government decisions of November 1984 was to save a further £300 m. per annum through a variety of means by 1987/88. Half of this reduction was through a re-definition of 'eligible projects' (to fall in line with EEC guidelines), but half of it arose from the abolition of the separate status of

Table 14.4 Changes in regional policy, 1984

	Special Development Areas	Development Areas	Intermediate Areas
Standard grants as percentage of factory investment:			
To 1980	22	20	20*
To 1984	22	15	'Selective assistance' only
From 1984	Status ended	15	
Working population included as percentage of GB:			
To 1984	13	9	6
From 1984	Status ended	15	20
Unemployed population (claimants) as percentage of GB, 1984:			
Before changes†	19	10	7
After changes‡	Status ended	23	25
Rate of unemployment, 1984 (%):			
Before changes†	19.0%	16.6%	16.1%
After changes‡	Status ended	19.6%	15.9%

* Grants for plant and machinery only, not buildings
† Data for September 1984
‡ Data for November 1984

SDA. As most of these cuts bore on 'standard grants' rather than 'selective' assistance, the measures represented a decisive shift in the balance of assistance away from 'standard grants', especially those for capital-intensive projects. 'Selective grants' remain the principal form of assistance available in IAs, but these now represent a much larger share of the 'assisted areas' than hitherto, as is shown in Table 14.4.

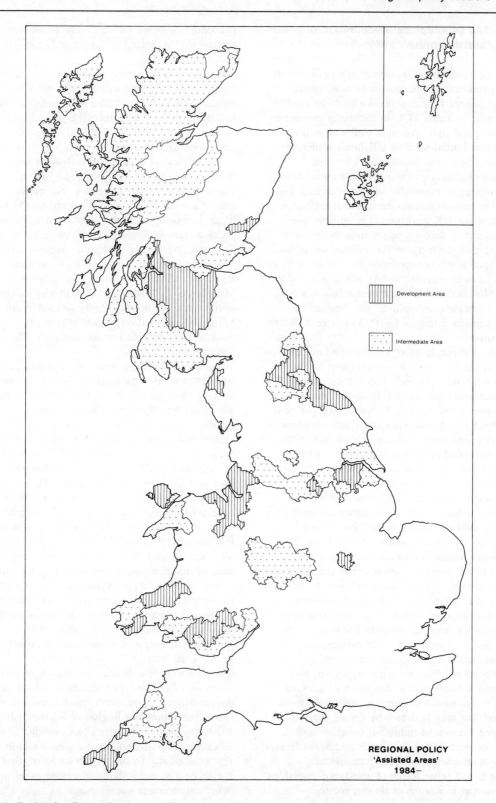

Fig. 14.3 Regional policy: assisted areas 1984

14.11 The geographical re-definition of 'assisted areas', 1984

Figure 14.3 defines the government's decisions of 1984 to reconstitute the 'assisted areas' under regional policy. This map must always be read in the context of Table 14.4, in particular bearing in mind the point that IAs must compete with each other for a limited total of UK funds under the heading of 'selective' assistance. They are, however, eligible for maximum assistance under the European Regional Development Fund. This attraction appears to have been the main factor in prompting the UK government to enlarge the 'assisted areas', through negotiating permission from the EEC to increase the assisted areas' share of the total working population to a level of 35 per cent. The large increase in IAs (demonstrated in Table 14.4) actually brought nearly half (48 per cent) of Britain's unemployed into the net of regional policy, compared with 36 per cent before the changes.

Rates of unemployment clearly remained the principal criterion used by government to distinguish DAs, IAs and others; their average unemployment rates in Britain at the time of designation stood at 19.6, 15.9 and 10.8 per cent respectively. However, compared with revisions of 1980, 1982 and 1983, the exercise embodied the use of improved statistical units, that is of new travel-to-work areas and a wider range of data for them. The basic units of decision-making in the composition of Fig. 14.3 are drawn from a set of 322 areas defined by the aggregation of ward areas from the 1981 Census of Population. These aggregations were computed by the Centre for Urban and Regional Development Studies, University of Newcastle upon Tyne, and described by Department of Employment (1984). Critics could rightly note that the degree of containment set is inevitably an arbitrary one, and produces travel-to-work areas of variable levels of containment for different social groups, for example males, females, car drivers, the unemployed etc. The government set out by proposing a multivariable definition of assisted areas: 'There would be merit in widening the basis on which the map is drawn by taking more measured account of industrial structure and peripherality, together with such additional factors as long-term unemployment, occupational structure, and other forms of assistance available. However the Government do not see the procedure as being entirely suitable for statistical treatment' (Secretary of State for Trade and Industry 1983: 17).

It is known that most of the above criteria were available and used in a composite index by officials. However, the final stages of decision-making rested on ministers 'taking each case on its merits', before having to defend marginal exclusions against criticisms from individual areas.

The net effect of successive changes since 1979 (i.e. the differences between Figs. 14.1 and 14.3) was to give the map more of a 'city region' and less of a traditional regional character. On the one hand, decisions effected in 1980 had withdrawn the previous comprehensive spatial coverage of Scotland, Wales and Northern England. On the other hand, the re-introduction of assistance throughout South Yorkshire and most of Greater Manchester in 1984, together with the historic decisions to include a broadly-defined West Midlands conurbation, meant that most conurbations outside London are now official 'assisted areas'.

Compared with the long-term average pattern from 1934 to 1984 (Fig. 14.1, above), the most striking changes of Fig. 14.3 lie in three regions which are heavily dependent on manufacturing activity, the North West, Yorkshire and Humberside, and the West Midlands. In the North West we see the coalescence of DAs around a broadly-defined Merseyside and extending as far into Wales as Rhyl and Wrexham. There is a continuous belt of assisted areas extending from Liverpool through Manchester and Sheffield to the ports of Humberside. The underlying explanation for these changes and those occurring in the West Midlands resides in the relatively greater growth here of unemployment, influenced by the structural factor of employment decline in steel, textiles, motor vehicles and engineering (Townsend 1983).

The new decisions were bound to arouse local criticism, and a map of 'upgraded' and 'downgraded' areas was provided by Hillier (1985). Perhaps the more telling lesson is that comparatively few of the 'assisted areas' of the 1930s or 1960s have escaped the need for present-day assistance. From the 1960s the main examples are in the Grampian Region of Scotland, from the 1930s the Whitehaven area of Cumbria. These cases are both attributable to special factors, the first case clearly to the impact of North Sea oil, the second to the stability of employment at the Windscale nuclear energy plant.

14.12 Conclusions; wider comment

Much of the literature on regional policy arose from the climate of the 1964–70 Labour governments, when the policy was arguably at the heart of a national strategy of thinking. The Conservative governments of 1979 and 1983 did not believe in a strong regional policy, and would have been unlikely to even if there had been more national resources for it. They retained regional policy as a domestic political necessity and, on the international scene, as a means of attracting foreign investment and EEC grants for 'assisted areas'. Many observers would judge that their cost-cutting measures were accompanied by perhaps a *welcome* re-emphasis on the social welfare role of regional policy, and a more cost-effective emphasis on job creation.

Some of the technical improvements achieved in the mechanics of channelling a modest supply of funds to needy areas may indeed count as methodological improvements in the long annals of regional policy. The new arrangements of 1984 were, however, immediately followed by a deferment in the payment of standard grants and concern over the number of areas competing for the set national total of selective assistance. It was hard to see the new prescription having much material impact on the large and growing spatial disparities in wealth and employment in the UK. Judged against the scale of the problem, which clearly reflects deep-seated social and economic upheavals, the measures cannot be regarded as more than a stopgap pending a re-assessment of economic policy at large.

The upheavals had themselves been associated with major changes in the departmental, and therefore the regional, pattern of public spending, all of which tended to dwarf the declining scale of regional policy spending itself, which remained less than 1 per cent of the public expenditure planning total. The latest available comprehensive data on regionally-relevant public expenditure (Short 1981a) had shown a bias in per capita spending towards the peripheral regions, arising not merely from regional policy but from spending on social services and social security. In the period 1979/80 to 1985/86, spending on social security alone was expected to have increased by 30 per cent and to represent no less than 30 per cent of total government spending at the end date (*The Guardian* 23 January 1985), with much of the increase spent in areas of greatest absolute additions to the unemployment register. On the other hand, defence spending is known to be biased towards southern regions (Short 1981b); this category expanded by 23 per cent, and over the same period and also in real terms, to make up 14 per cent of total spending, the second largest category in 1985/86. It is also one which is imparted increased significance for the processes of spatial development in Anglo-American literature.

No one imagined that a future government with different priorities could restore the active impact of regional policy to the past levels of 1963–73. Any doubts on this score were laid to rest by the widespread impression that 'world market factories' of multinational corporations (Froebel *et al.* 1980) were capable of finding their cheap labour in the Mediterranean and Third World rather than UK DAs; in actual fact, foreign corporations reduced their factory employment in the UK less than did UK corporations, 1977–81, and it was rare to associate factory closures in the UK directly with openings abroad (Townsend and Peck 1986), but the argument, while perhaps exaggerated, nonetheless remains a real one.

The interest of regional policy is thus to some extent a historical one, to do with factories which are now deeply embedded in the labour market geography of Britain rather than being expansive elements. Keeble's (1976) work on the 'dual population' of migrant firms or Massey's (1984) emphasis on the occupational characteristics of electronics or clothing branch plants relate very much to past 'layers' of investment activity, whose present character, in increasingly automated surviving plants, is *now* affected by a reduced emphasis on semiskilled female employment. The historical interest of regional policy factories requires more emphasis on the question of their selective survival in terms of industry, location and type of corporation. As noted earlier, Townsend and Peck (1985) reported a remarkably high incidence of redundancies and job losses in major corporations' plants in SDAs, 1977–79. These were overshadowed by the national pattern of redundancy in 1980–81, but may represent the underlying trend.

The period up to 1981 has already been the subject of research by Owen *et al.* (1986) and Townsend (1983), demonstrating that northern conurbations suffered worst from national recession, albeit in complex interplay of regional and urban decline. What has perhaps been missed is the scale of regionally-selective manufacturing

Table 14.5 Estimates of change in employees in employment, 1981–85 (September)

Country/region	Percentage change		Country/region		
	Manufacturing	Total		Manufacturing	Total
Scotland	−12.7	−1.2	E. Midlands	− 5.7	−2.0
Wales	−12.3	−2.5	W. Midlands	−11.7	−4.2
Northern	−14.6	−7.3	East Anglia	− 2.2	+5.8
North West	−14.6	−2.6	South East	− 6.7	+1.3
South West	− 6.7	+0.4	GLC	−15.6	−1.5
Yorkshire & Humberside	−11.7	−4.2	Rest	− 0.7	+4.1
			Great Britain	− 9.7	−1.2

Source: *Employment Gazette* May 1986.

decline between the 1981 Census of Employment and 1985, which is demonstrated in Table 14.5 in estimates from the Department of Employment.

This series has been subject to successive amendments including later ones reported by the 1984 Census of Employment. Taken as a statement of relative regional trends, it is consonant with data for unemployment and with a renewed increase in the share of British redundancies occurring in peripheral regions, the first four of Table 14.5. Manufacturing declined by 11 per cent or more in all those regions and in Yorkshire and Humberside, the West Midlands and London, while there were net gains of employment in East Anglia, the rest of the South East and the South West.

The fact of a north/south difference is the subject of widespread political comment, but the scale of *renewed divergence* has perhaps been little appreciated. Unfortunately it will not be until 1987 that data of sufficient detail will enable us to compare the employment performance of other conurbations with that of London. What is abundantly clear is that spatial issues are more complex than the regional policy map of the 1960s. They will remain a major challenge for many future governments, but are interwoven with urban policy issues, which are the subject of the next chapter.

References

Board of Trade 1968 *The movement of manufacturing industry in the UK*. HMSO.

Cameron, G. C. 1979 The National Industrial Strategy and Regional Policy. In Maclennan, D., Parr, J. B. (eds) *Regional policy; past experiences and new directions*. Robertson, 297–322.

Champion, A. G. 1983 Population trends in the 1970s. In Goddard, J. B., Champion, A. G. *The Urban and Rural Transformation of Britain*. Methuen, 187–214.

Daysh, G. H. J., Symonds, J. S. 1953 *West Durham: a problem area in North Eastern England*. Blackwell.

Department of Economic Affairs 1969 *The Intermediate Areas*. HMSO, Cmnd 3998.

Department of Employment 1984 Revised travel-to-work areas. *Employment Gazette*, Occasional Supplement No. 3.

Department of Trade and Industry 1983 *Regional industrial policy; some economic issues*. Department of Trade and Industry.

Expenditure Committee (1973) Trade and Industry sub-committee, Memorandum on the enquiry into location attitudes and experience. *Minutes of Evidence*. HMSO.

Fothergill, S., Gudgin, G. 1979 Regional employment change: a sub-regional explanation. *Progress in Planning* 12: 155–219.

Fothergill, S., Gudgin, G. 1982 *Unequal Growth: Urban and Regional Employment Change in the UK*. Heinemann.

Froebel, F., Heinrichs, J., Kreye, O. 1980 *The New International Division of Labour*. Cambridge University Press.

Frost, M. E., Spence, N. A. 1983 Unemployment Change. In Goddard, J. B. Champion, A. G. *The Urban and Rural Transformation of Britain*. Methuen, 239–59.

Hillier, J. 1985 Treasury wins on regional policy. *Planning* 11 January.

Holland, S. 1976 *Capital versus the Regions*. Macmillan.

Hood, N., Young, S. 1982 *Multinationals in Retreat. The Scottish Experience*. Edinburgh University Press.

Keeble, D. 1972 Industrial movement and regional development in the United Kingdom. *Town Planning Review* **43**: 3–25.

Keeble, D. 1976 *Industrial Location and Planning in the United Kingdom*. Methuen.

Keeble, D. 1977 Spatial policy in Britain: regional or urban? *Area* **9**: 3–8.

Keeble, D. 1980 Industrial decline, regional policy, and the urban–rural shift in the United Kingdom. *Environment and Planning A* **12**: 945–62.

Killick, T. 1983 Manufacturing plant openings, 1976–80. *British Business* 466–8.

Luttrell, W. F. 1962 *Factory Location and Industrial Movement*. National Institute of Economic and Social Research.

McCrone, G. 1969 *Regional policy in Britain*. Allen & Unwin.

Mackay, R. R., Thomson, L. 1979 Important trends in regional policy and regional employment: a modified interpretation. *Scottish Journal of Political Economy* **26**: 233–60.

Maclennan, D., Parr, J. B. (eds) 1979 *Regional policy; past experiences and new directions*. Robertson.

Marquand, J. M. 1980 Measuring the effects and costs of regional incentives. Working Paper 32, Government Economic Service (Civil Service College, London).

Martin, R. L. 1982 Job loss and the regional incidence of redundancies in the current recession. *Cambridge Journal of Economics* **6**: 375–95.

Massey, D. B. 1979 In what sense a regional problem? *Regional Studies* **13**: 106–25.

Massey, D. B. 1984 *Spatial Divisions of Labour*. Macmillan.

Massey, D. B., and Meegan, R. A. 1982 *The Anatomy of Job Loss: the How, Why and Where of Employment Decline*. Methuen.

Moore, B., Rhodes, J., Tyler, P. 1977 The Impact of Regional Policy in the 1970s. *CES Review* **1**: 67–77.

Moore, B., Rhodes, J., Tyler, P. 1980 New developments in the evaluation of regional policy. Paper at SSRC Conference on Urban and Regional Economics, Birmingham, May.

Moore, B., Rhodes, J., Tyler, P. 1981 The impact of regional policy on regional labour markets (mimeo.) Department of Applied Economics, University of Cambridge.

Nunn, S. 1980 The opening and closure of manufacturing units in the UK, 1966–75. Government Economic Service Working Paper No. 36, Regional Research Series No. 1.

Owen, D., Coombes, M., Gillespie, A. 1986 The differential performance of urban and rural areas in the recession. In Danson, M. (ed.) *Redundancy and Recession: Restructuring the Regions?* Elsevier.

Peck, F. and Townsend, A., 1984,' Contrasting experience of recession and spatial restructuring: British Shipbuilders, Plessey and Metal Box', *Regional Studies*, 18, 391–38.

Pounce, R. 1981 *Industrial movement in the United Kingdom, 1966–75*. HMSO.

Regional Studies Association 1983 *Report of an inquiry into regional problems in the United Kingdom*. Geo. Books.

Royal Commission on the Distribution of the Industrial Population, Report of 1940. HMSO, Cmnd 6153.

Secretary of State for Trade and Industry 1983 *Regional Industrial Development*. HMSO, Cmnd 9111.

Short, J. 1981a *Public Expenditure and Taxation in the UK Regions*. Gower.

Short, J. 1981b Defence spending in the U.K. regions. *Regional Studies* **15**: 101–10.

Slowe, P. M. 1981 *The advance factory in regional development*. Gower.

Townsend, A. R. 1977 The relationship of inner city problems to regional policy. *Regional Studies* **11**: 225–52.

Townsend, A. R. 1980 Unemployment geography and the new government's regional aid'. *Area*, **12**: 9–18.

Townsend, A. R. 1982 Recession and the regions in Great Britain, 1976–80: analyses of redundancy data. *Environment and Planning A* **14**: 1389–1404.

Townsend, A. R. 1986 Realpolitik as the distinctive *industry, employment and the regions*. Croom Helm.

Townsend, A. R. 1985 Realpolitik as the distinctive feature of industrial and employment policies. In Kivell, P. T., Coppock, J. T. (eds) *Geography. Planning and Policy Making*. Elsevier.

Townsend, A R., Gault, F D. 1972 A national model of factory movement and resulting employment. *Area* **4**: 92–8.

Townsend, A. R., Peck, F. W. 1985 The geography of mass-redundancy in named corporations. In Pacione, M. (ed.) *Progress in Industrial Geography*. Croom Helm.

Townsend, A. R., Peck, F. W. 1986 The role of foreign manufacturing in Britain's great recession. In Taylor, M. J., Thrift, N. J. *Multinationals and the Restructuring of the World Economy*. Croom Helm.

Townsend, A R., Smith, E., Johnson, M R D. 1978 Employees' experience of new factories in North-East England: survey evidence on some implications of British regional policy. *Environment and Planning A* **10**: 1345–62.

Tyler, P. 1983 The impact of regional policy on different types of industry and the implications for industrial restructuring. Paper to SSRC urban and regional economics group, Reading, July.

Wood, P. A. 1984 Regional industrial development. *Area* **16**: 281–9.

CHAPTER 15

Urban policy

William F. Lever

15.1 Introduction

Urban-based policies for industry in Britain are similar to regional policy in one respect: they offer differing types and degrees of assistance to existing or new enterprises at different locations. The amount of assistance can be related to the degree of severity of the areas' economic problems, although this may not necessarily be the case. However, there are several respects in which urban policy is quite different from regional development policy. Firstly, whereas regional policy is centrally managed by a single department of government, the Department of Industry, within a national context, urban policy is operated largely by the Department of the Environment which needs to work through local government. This introduces a number of complications as the local authorities may not always agree with central government on what is best for 'their' area. Secondly, a number of additional bodies have been created to assist in enterprise formation who have additional powers and expertise beyond those of the local authority sector. Thirdly, much of Britain's urban policy has developed out of concern for poor housing and environmental standards and low levels of public sector services such as education. During the 1970s it was accepted that many of these problems were incapable of separate solution without support for the local economic base in the form of job creation or maintenance. However, as urban economic development policy has emerged, it may still have environmental, community or social objectives linked to employment creation. Programmes such as the Community Programme and Community Business do at times tend to suffer from a confusion of objectives which lie on the interface between job creation and improving the quality of life in depressed urban environments. Lastly, regional policy, at least until 1982, tended to divide Britain into a 'North' and a 'South' – the benefits of any regional policy, it was assumed, would accrue to the recipient region. However, because the urban areas defined to receive priority treatment are so much smaller, and in many cases smaller than a self-contained local labour market, much of the benefit leaks out of the area.

15.2 The switch from regional to urban policy

During the 1970s there was a growing realization that the spatial structure of regional policy was having an inequitable effect. While there could be no doubt that regions such as the North and Scotland had higher rates of unemployment than those such as the West Midlands and the South East, *parts* of the North and Scotland were much more fortunate than parts of the West Midlands and the South East (Corkindale 1976). As Table 15.1 shows, in 1971 the worst local labour markets in the prosperous regions – both inner conurbation and free-standing towns – had worse unemployment rates (4.4–5.9 per cent) than the best local labour markets in the depressed regions (2.8–3.3 per cent). By 1981 the respective figures were 11.8–17.5 per cent and 6.4–10.1 per cent. Yet regional policy was still structured in such a way that investment might be expected to flow from the former to the latter. Regional policies were therefore being blamed, through the operation of Industrial Development Certificates and Regional Development Grants, for the decline

Table 15.1 Best and worst local unemployment rates, 1971 and 1981

	Greater London (%)	Inner London (%)	Best SE (%)	Worst SE (%)
1971	1.5	5.9	0.8 (Hereford)	4.4 (Oxford)
1981	7.8	11.8	4.7 (St. Albans)	13.2 (Chatham)
	Birmingham	Inner Birmingham	Best W Mid.	Worst W Mid.
1971	3.3	4.8	2.0 (Worcester)	5.3 (Coventry)
1981	13.8	15.2	7.4 (Stafford)	17.5 (Oakengates)
	Tyneside	Inner Tyneside	Best N	Worst N
1971	6.0	9.4	3.3 (Carlisle)	7.7 (Hartlepool)
1981	14.6	n.a.	10.1 (Carlisle)	26.0 (Consett)
	Glasgow	Inner Glasgow	Best Scot.	Worst Scot.
1971	6.5	9.0	2.8 (Perth)	14.3 (Bathgate)
1981	15.1	19.2	6.4 (Aberdeen)	21.8 (Irvine)

Source: Department of Employment Gazette

of the inner-city areas of London and Birmingham (Button 1978) and other labour markets in the prosperous regions (Townsend 1977). The consequences of this were an evaluation of regional policy to determine whether it might be spatially targeted more precisely, and to the growth of the belief that urban policy should have a more economic focus.

15.3 Urban problem definition

Defining urban problems in Britain has gone through a number of stages. In 1969 the Home Office launched a number of experiments in the form of Community Development Projects (CDPs) in which action teams were placed in areas of urban stress and their impact monitored. However, the areas selected experienced different problems, some economic such as high unemployment, others less so – immigrant communities, council housing with high vacancy rates, and rural outmigration, for example. These projects were founded on an assumption that there was a culture of poverty in which children born into such areas would suffer antisocial pathologies, underachieve, and then transmit the deprivation to another successive generation (Townsend 1974). The CDP projects were followed in the mid-1970s by the Inner Area Studies (IASs) of inner London (Lambeth), Birmingham and Liverpool, which indicated that the most fundamental cause of the urban crisis was the reduction in employment and income losses at both household and locality levels. To the Lambeth Inner Area Study, for example, 'job opportunities and the ability of residents to earn a living are fundamental' (Department of the Environment 1976).

The result of the CDP and IAS programmes was the 1978 Inner Urban Areas Act which defined a limited number of local authorities as having severe urban problems. The definition of these areas was based upon data from the 1971 Census of Population which used principal components analysis to define three components of urban decay – overcrowding, lack of household amenities, and unemployment and low income (Holtermann 1975). Those judged to have the worst deprivation, designated Partnership authorities, were provided within a fairly elaborate administrative machinery with power to set up joint programmes to tackle the problem, including substantial additional resources in the form of a 75 per cent government grant. There were seven such authorities – Liverpool, Birmingham, Lambeth, London's Docklands, Manchester/Salford, Newcastle/Gateshead and Hackney/Islington. A second tier of fifteen Programme authorities was set up with fewer powers and lower levels of government support. Lastly, Designated Districts were established with enhanced powers but no eligibility for government grants (Lawless 1981a; Bentham 1985). Key elements in the Inner Urban Areas Act were the power given to designated authorities to declare Industrial and Commercial Improvement Areas, facilitating the renewal of older areas by cooperation between the public and private sectors, to provide advanced factories and to assist small firms (Hausner 1982; Cameron 1983).

The granting of these powers to designated local authorities reflected the fact that central government programmes did not appear to be discriminating in favour of the inner areas of the larger cities. Analyses of Department of Industry grants since 1978 (Lawless 1981a) showed that in northern England the Merseyside Partnership area had done better than anticipated but that Manchester/Salford and Newcastle/Gateshead had done less well than anticipated, and 'most Programme administrations receive less grant than would be expected bearing in mind their share of manufacturing employment. Hull, Sheffield and the Wirral, however, are exceptions' (p. 3). Assistance from the National Enterprise Board (NEB), set up in 1974, has been only very marginal in the inner-city areas. Between 1974 and 1978, the NEB had provided financial assistance valued at £1,280 m. to fifty-five firms but the greater proportion had gone to British Leyland Ltd, Herbert Ltd, Ferranti Ltd, and Rolls Royce Ltd (£1,170 m. in total), and these companies, historically, had located outside the inner areas of the largest cities. Thus 'the picture which emerges is one of very limited investment in the Partnership and Programme authorities' (Lawless 1981a: 7). Only the Manpower Services Commission's programmes such as the Special Temporary Employment Programme (STEP) appear to have been 'bent' into the most deprived inner-city areas. The areas which received significantly more STEP funds in 1978–79 were Tyneside, Cleveland and Durham, Cumbria, South Yorkshire, Wolverhampton and Derbyshire, Nottingham and Merseyside. Major omissions from this list, however, include Greater London, Birmingham and Greater Manchester.

With the publication of the 1981 Census of Population, a re-analysis of the data on which designated authorities were defined took place (Bentham 1985). The number of Partnership authorities was expanded to nine, but only by separating the earlier joint schemes such as Newcastle/Gateshead, and converting the Docklands area into an Urban Development Corporation; the number of Programme authorities rose to sixteen and subsequently to twenty-three, with the addition of Blackburn, Brent, Coventry, Knowsley, Rochdale, Sandwell, Tower Hamlets and Wandsworth, and the number of Designated Districts rose to sixteen with the addition of Burnley, Langbaurgh and Walsall. Powers remained largely unchanged, however. In addition to the designated areas from the 1978 Inner Urban Areas Act and its subsequent revision (Department of the Environment 1981), there is a wider Urban Programme to cover projects in authorities with special social needs. This, however, received only £46.8 m. in 1982–83, compared with £223.2 m which went to the authorities designated under the 1978 Act.

15.4 Newer urban policies

After the urban riots of 1981, the Department of the Environment established the Financial Institutions Group comprising twenty-six executives, seconded from major banks, insurance companies and pension funds, to examine the financing of small businesses, employment and training. This resulted in two new urban policies. The Urban Development Grant Programme (UDG), modelled on the American Urban Development Action Grant Program, was devised to use a minimum of public sector funds to lever major private sector investment into the improvement of infrastructure and the provision of permanent jobs in inner-city areas. Some £70 m. was available under this scheme in 1983–84, consisting of 75 per cent central grants towards the local authorities' share of the project costs where it is hoped that a ratio of 1 : 4 between public and private funds will result. UDG is available only within the forty-three designated authorities. Inner City Enterprise (ICE) was set up as an independent property development company to develop viable inner-city projects for private sector financing. ICE is an attempt to draw the major financial institutions into close and financial participation in inner-city development, attempting to bridge the gap between the financial institutions' concentration upon projects of 'acceptable institutional quality' and the presumed existence of viable urban development projects which the institutions might not currently consider because of their novelty, size, complexity or the management needed to develop the project; £1 m. has been provided by the participating financial institutions for ICE's administrative costs.

Enterprise Zones (EZs) were introduced in 1980 with a first group of eleven (although only six were in inner-city areas) but the number had grown to

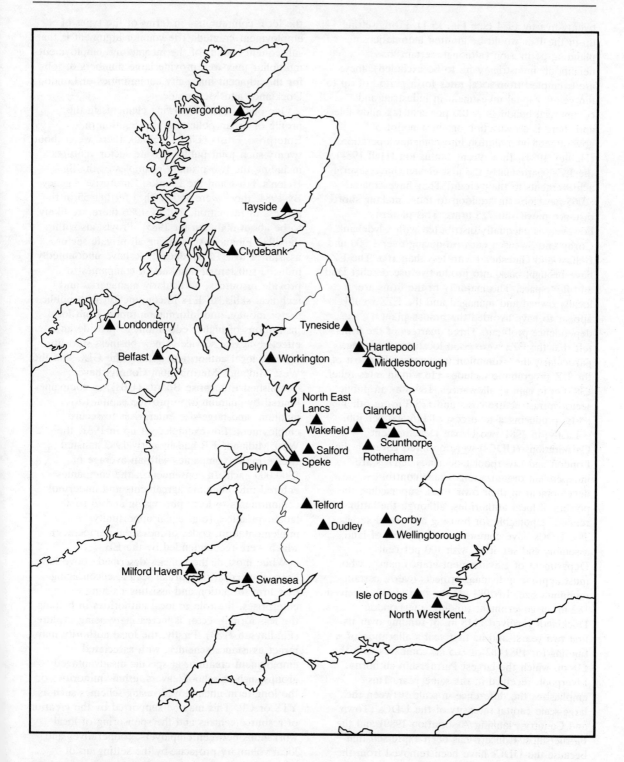

Fig. 15.1 Enterprise Zones 1985

twenty-five by 1984 (see Fig. 15.1). Firms setting up in the EZs would be granted automatic planning permission (although certain 'bad neighbour' industries were to be excluded), they are exempted from local rates for a period of up to ten years, capital investment in industrial and commercial buildings is 100 per cent tax allowable, and there is a restriction on the amount of government intervention into company operations (Hadley 1984). In a recent evaluation (Hall 1984), the EZs, particularly the first eleven, have several achievements to their credit. They have created 8,065 new jobs (in addition to those making short-distance moves) in 725 firms. This pattern, however, is unequally distributed with Clydebank, Corby and Swansea each producing over 1,000 and Belfast and Gateshead with less than 100. The EZs have brought back into productive use derelict land and floorspace. The majority of the firms are locally owned and managed and the EZs would appear to have avoided the branch plant dependence problem. Three-quarters of the firms felt that the EZs were good locations, stressing particularly the exemption from rates. The cost of the EZ programme includes £16.8 m in rates relief, £38.0 m in capital allowances, £39.8 m. in public sector infrastructure costs and £38.3 m. in other costs, producing a total cost of £16,500 per job.

Lastly, in 1981 two Urban Development Corporations (UDCs) were declared covering the London and Liverpool Docklands. UDCs are independent organizations which control development in their own areas, superseding the powers of local authorities, although the latter remain responsible for housing and social services. The UDCs have comprehensive powers of land assembly and servicing with 100 per cent Department of Environment grant finance, who must approve individual projects over a certain minimum size. In 1981–82 the two UDCs received £82 m. of government funding: the London Docklands received £118 m. of funding over its first two years, and its first year's allocation of funding for 1981–82 of £65 m. compares with the £18 m. which the largest Partnership authority, Liverpool, received in the same year. This emphasizes the difference in scale between the large-scale capital intensity of the UDCs (Town and Country Planning Association 1980) and the Partnerships. Concern has been expressed that because the UDCs have been removed from the local democratic process, much as the New Towns were, they may not be responsive to the needs of the local communities in terms of the types of employment created. The counter-argument is that subsequent rounds of the income and employment multiplier may well provide large numbers of jobs for the adjacent inner-city communities of London Docklands and Merseyside.

One of the fastest growing elements in the profile of urban policies is the creation of Enterprise Trusts (ETs). In 1980, there were about twenty such joint public–private sector ventures including the two primary examples of the St Helen's Trust and the London Enterprise Agency. By 1984, there were 160, with a further 55 in the process of formation, and by 1986 there are likely to be about 300 (Turner 1985). Provisions within the 1982 Finance Act making all private sector assistance to ETs tax deductible have undoubtedly induced substantial numbers of companies to provide resources, particularly managerial and technical skills, to ETs in conjunction with public sector money, and information inputs from local institutions of higher education, to provide an effective advice service to new business start-ups.

Some local authorities, notably the GLC and the West Midlands Metropolitan County, have established Enterprise Boards (EBs), as companies limited by guarantee, to provide capital for medium- and large-size enterprises to secure employment. For example, set up in 1981, the West Midlands EB had by early 1983 assisted twelve major companies with an average of £300,000 each. In consequence, the companies entered into planning agreements and undertook commitments to local purchasing and to some labour practices (e.g. equal opportunity implementation, codes of industrial democracy) which were recommended by the EB.

While most of the policies described above have been directed at the formal sector, encouraging new firm formation and assisting existing enterprises, the role of local authorities in helping the less formal sector is increasingly being explored (Findlayson 1983). Firstly, the local authority may target assistance schemes, with associated employment creation, at specific disadvantaged groups such as school-leavers, ethnic minorities or the long-term unemployed, using schemes such as YTS or CP. This may be supported by the creation of resource centres and the sponsoring of local workshops between employers, cooperatives and local voluntary projects, by the setting up of mutual help networks, and through enhancing basic adult education programmes.

15.5 Case study approaches

Unlike regional policy which is a single policy uniformly applied by central government, it is clear that urban policy is a mixture of national programmes implemented by local authorities and specialist local agencies. The better way to examine urban economic policy is via case studies. Accordingly, we select two very contrasting conurbations – London and Clydeside – to illustrate the full range of urban policies for industry.

15.6 The London case

The most significant feature in the London economy since 1950 has been the major decline of manufacturing employment. As Table 15.2 shows,

Table 15.2 Employment change in London ('000)

	1951	1961	1971	1981	1983
Primary & manuf.	1,448.9	1,453.0	1,082.2	650.0	583.0
Services	2,839.4	3,037.0	2,950.0	2,863.0	2,809.7
Total	4,288.3	4,490.0	4,032.2	3,513.0	3,392.7

total employment rose by 10 per cent between 1951 and 1961 and began to decline significantly only after 1966. Manufacturing employment held steady through the 1950s but then declined from 1,449,000 in 1961 to 583,000 in 1983. In the manufacturing sector, some industries grew during the 1950s but thereafter every major industrial category declined. In some the falls were quite startling: clothing, for example, declined from 177,000 employees in 1951 to 31,000 in 1981. By 1981, in only two sectors, electrical engineering and metal goods manufacture, had employment declined by less than 50 per cent of its 1951 total. In services, there has been growth, but this has been concentrated in sectors such as insurance and banking, and professional services, whereas there have been major job losses in sectors such as distribution and miscellaneous personal services. Analyses (e.g. Danson et al. 1980) demonstrate that London's economy has performed not only worse than that of the nation as a whole, but also worse than those of other major British conurbations such as Clydeside or the West Midlands. Explanations of this acute rate of job loss are based on the finding (Dennis 1980; Gripaios 1977; North and Leigh 1984) that most is due to closures and from contractions rather than outmigration. The firms and subsectors of industries which are located in London appear to be amongst the most backward and concentrate on the mature and declining stage of their respective product lifecycles. Fothergill et al. (1982a) have stressed the space constraints on inner London's industry, and its high cost as a cause of decline. During the period 1976–78 there was some evidence that industries such as furniture and electronics were handicapped by a lack of skilled labour but, given the deepening recession since 1979, this problem has abated although workers still tend to prefer employment in service/white-collar occupations. Despite findings by Wellbelove et al. (1981) that some inner London firms were capable of above-average profit rates, Fothergill et al. (1982b) showed that overall profit rates were lower in London than elsewhere.

The effect of these major losses is clearly seen in London's rising rate of unemployment, a rise exacerbated by the fact that where jobs are created in expanding sectors they are unlikely to be filled by workers released by the declining manufacturing industries. The male unemployment rate in London in 1971 was 2.3 per cent, in 1979 it was 4.4 per cent, and in 1982 it was 12.2 per cent. These figures are consistently *below* the national average where the corresponding figures were 4.3 per cent, 6.8 per cent, and 16.0 per cent respectively. Thus, while at the aggregate level for London as a whole, it is reasonable to treat the city as a relatively favoured location with little assistance to industry, the unemployment problem is heavily concentrated in the inner boroughs. Thus in Southwark, for example, the male unemployment rate rose from 3.6 per cent in 1971, below the national average, to 20.4 per cent in 1982, well above the national average. Even despite the continuing loss of residential population, male unemployment levels more than trebled in the period 1979–83. By 1984, these rates in the most depressed inner London boroughs such as Hackney and Tower Hamlets were close to 30 per cent, youth unemployment was 40 per cent and, by early 1985, 34 per cent of inner London's unemployed had been out of work for more than one year. Policy efforts have therefore been directed at enhancing job creation in the inner boroughs both using national programmes of assistance and devising programmes specific to those areas of inner London.

Using Boddy's (1984) definition of the degree of intervention, the 'market-based' policies in London are aimed at facilitating the growth of the private sector by relieving bottlenecks and problems of factor availability. Thus the policies of local authorities aimed at the provision of sites and premises can be seen as addressing the problem identified by Fothergill et al. (1982a). Local authorities have had a long-running interest in providing factory and warehouse space and developing industrial estates, not least because it enhances their stock of rate-paying property. It has proved a relatively uncontentious element in urban economic policy as it involves the creation of a lasting fixed capital asset which would survive the demise of occupant firms in a way that financial subsidy or rent/rates relief would not. For most London boroughs the discussion has centred on whether the borough itself should be the developer and landlord or whether a more effective course would be to assemble and service sites. Leigh et al. (1982) have emphasized the importance to local authorities of retaining control over at least a proportion of the stock of industrial premises, although central government policy has tended to move towards giving tax advantages to private sector industrial developers. Consequently, local authorities have increasingly tried to attract developers into their areas by the use of leaseback agreements which in effect guarantee rents to developers for a number of years. A second concern is that local authorities, in creating a reserve of industrial floorspace, are merely increasing employment mobility without significantly increasing employment or the rate of new firm creation. There remains a need for a stock of older (and therefore cheaper) industrial premises, but local authorities in London have shown a tendency to create new premises, often aimed at the new firm/high-technology sector, depending upon vacancy chains to release cheap old space at the end of the sequence and offering financial assistance towards the rents of such high-amenity premises (Valente and Leigh 1982). Such financial assistance, however, is likely to remain of limited utility as long as the pressure of demand for space from the commercial sector remains high and rents remain correspondingly high. A GLC report in 1984 concluded that 'there is however accumulating evidence that property-based methods of market intervention are not in themselves capable of making a major contribution to the task of maintaining even a baseline manufacturing employment presence in the inner city'. For example, a recent review by one Partnership Committee found that in the period 1977–81 its Industrial Development Programme had created 14,700 m^2 of floorspace, on which 332 jobs were located. However, of these, only 209 were either new jobs or jobs maintained, the rest being movers, and of these 75 per cent would have remained in the locality. Thus only 40–50 jobs might be attributed to the IDP programme.

The UDG programme also qualifies as a 'market-based' form of intervention in that its objective is to use public money to lever out private sector investment, much of it into redevelopment schemes refurbishing premises for industrial, commercial and other uses. Table 15.3 shows that for the whole of London in 1983–48,

Table 15.3 UDG allocations, London, 1983–84

	UDG award (£000)	Total cost (£000)	Gearing
Brent	1,822	11,000	1:9
Ealing	365	1,200	1:2
Greenwich	436	1,500	1:2
Hackney	2,532	13,574	1:4
Lambeth	1,035	6,114	1:5
Southwark	1,800	11,050	1:5
Wandsworth	3,017	20,096	1:6
Tower Hamlets	120	1,160	1:9
Lewisham	590	3,363	1:5
Islington	2,700	5,250	1:2
GLC non-borough	363	1,903	1:4
Total	14,780	76,210	1:4

some £14.8 m. of public money has succeeded in attracting a further £61.4 m. of private investment, a ratio of 1 : 4, close to the national average. Three-quarters of the schemes covered by the table involve the upgrading of industrial premises into workshop or factory unit space. Perhaps significantly, however, the schemes which have attracted the highest ratios of private sector involvement are those which produce shopping and office space such as the Clapham Junction scheme in Wandsworth and the Manor Park and Willesden Green projects in Brent where ratios of 1 : 6–1 : 9 are achieved. The industrial projects tend to attract lower levels of private investment: about a ratio of 1 : 3–1 : 4.

More thoroughgoing market-based strategies in London include the creation of the Isle of Dogs

The London case

Enterprise Zone and the London Docklands Development Corporation (LDDC). London's EZ was located in Tower Hamlets north of the Thames where there had been a massive reduction in port and port-related activities. Designated in May 1982, by late 1983 some 350 jobs were to be found in the EZ although half of these were in enterprises already committed to move to the area before the EZ was declared. By mid-1984 the LDDC claimed that there were 1,000 jobs in the EZ but this figure has been seriously challenged by local groups (e.g. AIC/JDAG 1984). Further criticisms of the EZ are that it merely stimulates short-distance moves by enterprises seeking rates relief and that in order to enhance its attractions, investment in other parts of Tower Hamlets has been deferred (Morphet 1984). From the perspective of the enterprises, savings on rates may be offset by rises in rents, and enterprises immediately outside the EZ are likely to complain of unfair competition.

The Isle of Dogs EZ lies within the boundaries of the LDDC, created in 1980. The LDDC's annual budget is about £40 m. of which, in 1983–84, the largest parts were allocated to improving the local infrastructure, transport services and land servicing (£13 m), environmental upgrading (£6 m) and land acquisition and reclamation (£4 m.)). A further £5 m. was spent on the provision of business services and advice. Thus much of the public sector money alleviates the costs normally borne by private sector enterprises and may therefore generate higher profits. In the short term, some 300 construction jobs have been created; in the medium term, about 2,000 transferred jobs in sectors such as newspaper publishing and 200 jobs in administration are to be provided, to replace an estimated loss of 1,500 jobs in a wide range of manufacturing and processing industries. In the longer term, the impression is of a site which, with enhanced facilities such as the proposed short-take-off-and-landing airport and

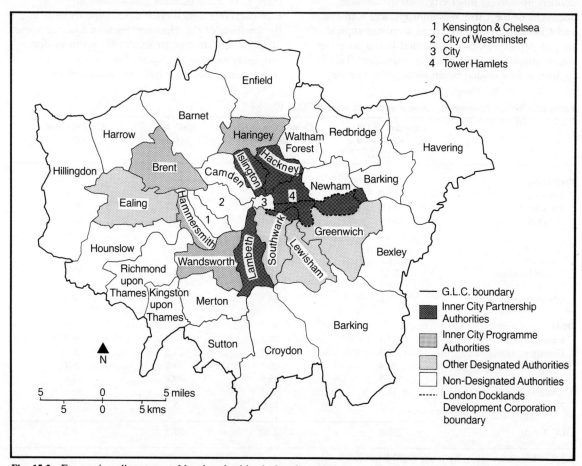

Fig. 15.2 Economic policy status of local authorities in London, 1983

good access to the City, will attract commercial office and related developments which in their turn will raise rents, thereby undermining the existing stock of local-employing small firms which will either close or leave.

The second approach of urban policy, termed 'neo-Keynesian' (Miller and Miller 1982; Lawless 1981b), is also concerned with improving efficiency and competitiveness in the private sector but does place more emphasis upon preserving employment and creating jobs for reasons of equity. Thus, firms in difficulty may be assisted with financial subsidies, on capital, labour or rents, or with business advice, particularly where they are significant employers of labour in depressed areas. The largest single element in this approach in London is the Urban Programme (UP) initiated in 1978. This created three tiers of a hierarchy in London, as Fig. 15.2 shows, in the form of Partnership, Programme and Other Designated District (ODD) authorities. London has been divided up into an inner city with the affluent boroughs of the City, Westminster, and Kensington and Chelsea, surrounded by an almost complete ring of depressed inner residential boroughs all of which qualify for some form of assistance. This in its turn is surrounded by an outer suburban ring where no assistance is available. The schemes have two elements – one is the increased use of financial assistance, the other is the development of a mechanism to coordinate intervention by a wide range of public sector agencies and departments of local government. The objectives of the Partnership areas can be exemplified by the plan for the Hackney–Islington scheme (August 1978) which defined three goals – halving the net rate of job decline by 1982, reducing local unemployment levels to those of the London average by 1982, and raising average household incomes to the London average by 1982. As Table 15.4 shows, between 1979–80 and 1983–84 the whole of the GLC area had received £296.2 m. at 1981 prices with the largest amounts going to the three Partnership boroughs, Islington, Hackney and Lambeth. Some elements of the UP in London have received approval but there has also been criticism. There have been significant improvements in environmental and infrastructural standards and there have been genuine innovations in manufacturing and service developments such as the creation of the Hackney Fashion Centre. Some of the administrative structures set up to evaluate proposals have been criticized as overly bureaucratic, but they have provided a very

Table 15.4 Urban Programme funding and RSG loss, 1979/80–1983/84

	Cumulative UP gain since 1979 (£m. 1981)	**Cumulative RSG loss since 1979 (£m. 1981)**	**RSG loss (%)**	**UP/RSG loss (%)**
Partnership				
Islington	43.0	84.7	52.6	50.8
Hackney	46.0	50.2	26.7	91.6
Lambeth	50.8	81.4	37.2	62.4
Programme				
Hammersmith	21.2	40.9	27.3	51.8
Tower Hamlets	20.3	63.0	44.7	32.2
Brent	4.1	55.4	27.3	7.4
Wandsworth	3.6	89.3	37.1	4.0
ODD				
Ealing	2.3	24.9	12.4	9.2
Haringey	3.3	48.3	28.3	6.8
Greenwich	30.4	66.5	49.7	45.7
Lewisham	11.5	83.5	39.8	13.8
Southwark	34.6	85.3	47.8	40.6
Newham	25.1	14.6	5.8	171.9
GLC Total	296.2	1532.6[a]	31.5	19.3

(a) Total includes losses by other GLC boroughs.

effective way of enabling voluntary agencies to acquire funding: some 800 groups dealing with community facilities, social services, health care, information and advice schemes have received assistance (GLC 1984).

The money received by the designated boroughs under the UP has to be placed in the context of local authority finance as a whole. As Table 15.4 shows, while these boroughs received almost £300 m. (in 1981 values) between 1979 and 1984 under UP, they lost over five times this amount in cumulative losses from the Rate Support Grant (RSG) from central government. Even in inner London where the need, as measured in terms of unemployment, was greatest, the acquisition of £260 m. under UP was offset by £865 m. of cumulative RSG loss. Only one borough, Newham, succeeded in acquiring more funds from the UP than it lost in RSG, not because it was particularly well funded under UP but because of its low rate of RSG loss. Much of the UP-borne expenditure has tended to be used to provide social and environmental services, rather than to assist new firm creations, offer business subsidies or to maintain industrial jobs, partly because these services were being eroded by the reduction in RSG and partly because most of the authorities concerned were Labour-controlled and therefore more likely to emphasize welfare than business support.

The UP in London has attracted a number of critical comments. Firstly, at the outset it was hoped that many projects initiated under it would eventually be taken on to mainline funding, thereby giving them some permanence. It is now clear that local authorities are unable to do this and many of the projects, with their associated employment, are coming to the end of their three-, four- or five-year lives. At the same time the Department of the Environment, as the funding body, is insisting on an increasingly high capital expenditure content and a reduction in the current expenditure content. This has the effect of forcing more schemes to deal with building refurbishment, land reclamation and environmental improvement which are 'one-off' projects, and fewer schemes to provide services with their proportionately higher employment components. A second criticism is that even where UP is directed at industry, the needs of the client group are not well served (North and Leigh 1984; Leigh *et al.* 1982). There is a tendency for local authorities to regard themselves as lenders of last resort: therefore enterprises receiving assistance may be economically the most marginally viable, unattractive to other lenders so that private sector leverage is slight, and suffer high rates of business failure. The final criticism is that of spatial spillover, whereby adjacent authorities tend to resort to competitive bidding for a fixed stock of mobile enterprises, thereby bidding up the cost of assistance without, net, increasing total employment, although to counter this charge some authorities have imposed a cost-per-job ceiling on UP assistance.

In order to estimate the employment effects of these types of subsidy, Damesick *et al.* (1982) and Turok (1984) used data for one borough, Southwark. Southwark devised a programme of providing premises, mainly small-sized, direct financial assistance, the declaration of three Industrial Improvement Areas, and enhancing local services and infrastructure. The objective between 1978 and 1985 was to create 5,000 new jobs and assist in the preservation of an unspecified number of others. Between 1978 and 1982, 119 firms with 4,098 workers were assisted but of these only fifty-six firms received assistance valued at more than £15,000. Of these, nineteen had closed by 1984 with a loss of 754 jobs and employment in the others declined from 1,694 in 1982 to 1,665 in 1984. Some firms did feel, however, that the provision of assistance had either been instrumental in keeping them in business or in helping them to expand, or had kept them in Southwark, with a joint net benefit of about 270 jobs. These figures suggest that it has proved extremely difficult for the local authorities to assist their economic bases.

The third approach to economic development has been termed 'socialist' in that it represents a much higher level of intervention and takes as its starting point the needs of the labour force rather than those of capital. It therefore uses resources released in the restructuring process for socially useful purposes and bases investment decisions upon social criteria defined by the workers and communities generally (Lever 1986). The major step in this policy element was the creation of the Greater London Enterprise Board (GLEB) in July 1982 funded from local rate income (the GLC could raise up to £40 m. from Section 137 of the Local Government Act of 1972 and a further £30 m. from its own superannuation/pension fund scheme). GLEB's priorities include stressing manufacturing over services, focusing on large firms rather than the small firm sector, encouraging producer co-operatives, using discretion in the placement of GLC contracts and developing technology networks and social infrastructure and social infrastructure (North and Leigh 1984). In the shorter run, GLEB has made a number of opportunistic deals with major companies on the verge of closure. For example, when Austin Furniture was placed in receivership, with 400 jobs, GLEB acquired the plant for £1.25 m, released it and provided a wage subsidy for 150 retained workers. GLEB also funded an employee buyout of Associated Automation to form a cooperative with 180 jobs in high-technology development.

15.7 The Clydeside case

If London can be described as an example of light to moderate urban economic policy, then the Clydeside conurbation is a case of full urban policy intervention. Table 15.5 shows that Clydeside's total employment remained fairly steady in the 1950s and then declined by 24 per cent between 1961 and 1984 at an accelerating rate. Decline in manufacturing employment was remarkably similar in London and Clydeside in 1951–83/84 (−59.8 per cent in London and −61.2 per cent in Clydeside).

Table 15.5 Employment change in Clydeside ('000)

	1951	1961	1971	1981	1984
Primary & manuf.	557.2	525.6	382.7	254.3	216.3
Services	287.0	315.3	406.8	431.2	423.7
Total	844.2	840.9	789.5	685.5	640.0

Service employment in Clydeside continued to grow much more recently than in London. The fact that London's unemployment rate has been lower than that of Clydeside therefore reflects the relative levels in 1951, the labour market processes within the two cities, and the relative rates of suburbanization of employment. A significant proportion of London's job loss is accounted for by short-distance movement outwards over the GLC's boundary; the Clydeside conurbation is a self-contained labour market in which job losses are attributable to closure, rundown and (rarely) long-distance outward movement. There also appears to have been a greater degree of mismatch between job losses and new job creations in Clydeside so that more workers made redundant have been unable to find re-employment and new jobs have been filled by workers not registered as unemployed.

The causes of Clydeside's job loss are in some respects different from those in London. It is true

(North and Leigh 1984). In the shorter run, GLEB has made a number of opportunistic deals with major companies on the verge of closure. For example, when Austin Furniture was placed in receivership, with 400 jobs, GLEB acquired the plant for £1.25 m, released it and provided a wage subsidy for 150 retained workers. GLEB also funded an employee buyout of Associated Automation to form a cooperative with 180 jobs in high-technology development.
and high-efficiency wages as contributing to the problems of firms in the conurbation, especially the inner city, and investment levels per worker are generally well below national levels on a sector-by-sector comparative basis. The Clydeside economy also has suffered particularly high losses of employment from the non-locally-owned branch plant sector: for example, in 1978–81 almost half the employment in foreign-owned manufacturing companies in the conurbation disappeared compared with only a quarter of the employment in British-owned companies. Lastly, the massive redevelopment programme in Clydeside, especially in Glasgow, initiated in the early 1960s, together with the uncertainties attaching to the implementation of a major urban motorway programme through the conurbation, created difficulties for enterprises, causing some to move, some to close and some to withhold investment and growth (Bull 1981).

The effect of these major job losses on Clydeside has been to substantially worsen unemployment levels. The male unemployment rate over the whole conurbation in 1971 was 10.3 per cent, by 1981 it was 18.8 per cent and by 1983 it was close to 20 per cent. Compared with British national rates of 3.2 per cent, 9.8 per cent and 12.3 per cent respectively, it is not difficult to see why Clydeside should have attracted high levels of urban policy concern. Just as in the case of London, at least as dramatic as the overall rise in unemployment are the wide spatial disparities. Figure 15.3 gives some indication at the District level of just how wide these disparities were. In 1981, for example, the three suburban dormitory areas of Eastwood, Bearsden and Milngavie, and Strathkelvin had an average rate of 7.3 per cent, the two New Town areas had an average rate of 12.6 per cent, the five Districts representing the older industrial periphery (Renfrew, Hamilton, Motherwell, Clydebank and Monklands) had an average rate of 18 per cent, and the City of Glasgow District had a rate of 23.5 per cent. At a much smaller spatial scale – that of individual

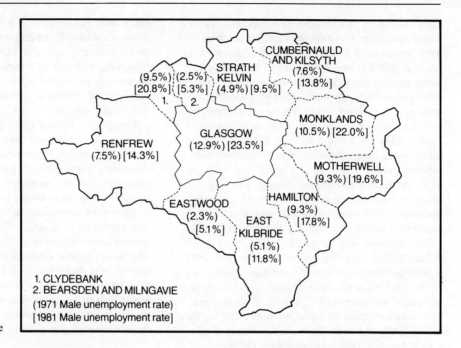

Fig. 15.3 District unemployment rates 1971 and 1981: Clydeside

1. CLYDEBANK
2. BEARSDEN AND MILNGAVIE
(1971 Male unemployment rate)
[1981 Male unemployment rate]

electoral wards – rates in excess of 40 per cent existed in 1981, many of them on the Glasgow peripheral housing estates. As with London, increasing unemployment rates have been accompanied by increasing duration of them. In Glasgow in 1979, for example, 33 per cent of the male unemployed had been out of work for more than one year; by 1985 the figure was 52 per cent, with 35 per cent out of work for more than two years.

Within the conurbation as a whole, it is possible to discern two types of urban policy: there are those which were developed before the mid-1970s in the context of a different national and local economic climate and which have continued to play a rather different role since then, and there are those which have developed since the major recession began in the late 1970s. The former tend to be operated by the long-established agencies such as local government, the latter by new agencies. In addition, we can identify a third type of urban policy – the area-based initiative – where attempts are made to integrate a range of plans and assistance within a given location: these would be the equivalent of the EZ and the LDDC in London.

While regional policy was dealt with in Chapter 14, it is relevant to examine the way in which it operates as an urban policy. Within Clydeside, the whole conurbation has qualified for SDA status since the 1960s. What is clear from an analysis of RDG allocations in the period 1979–83 is that regional policy is anti-urban. Despite the fact that Glasgow's unemployment rates have been persistently higher than those of the surrounding area, with 55 per cent of the manufacturing employment in the conurbation, Glasgow received only 31 per cent of the assistance. As RDG is linked, statutorily, with expansion, this imbalance may reflect the different economic structures of Glasgow and the outer conurbation. However, a sector-by-sector analysis shows that in only three sectors out of nineteen did grant per capita figures for Glasgow exceed those for the outer city. In the four-year period 1979–83, it has been estimated (Lever 1985) that some 3,000 jobs have been created or maintained by RDG in Clydeside but less than 1,000 have been located in Glasgow. Regional Selective Assistance appears to have been even more biased away from the inner city which received only 22 per cent of the funds granted. Of regional policy assistance, only the Office and Service Industries Scheme (OSIS) was bent into the inner city, unsurprisingly because of the dominance of Glasgow's CBD within the conurbation, and OSIS provided only £7.2 m. compared with £137.1 m. provided by RDG and RSA for manufacturing in 1979–83.

As in London, local authorities have evolved a programme of providing space and associated infrastructure. In this respect, however, they have been outbid, not by the private sector as in

London, but by the Scottish Development Agency which took over the role of industrial landlord from the Scottish Industrial Estates (Corporation). Local authorities are now faced with the role of filling gaps in the market, concentrating on the small premises sector and, with more success, on managed workshops where the supply of premises is linked to jointly provided business services and advice. This has left the SDA with the job of refurbishing older industrial premises (often with subdivision into many smaller units) and with the more intractable problem of disposing of large premises built during the 1950s and let, in many instances, to incoming branch plants of national and international companies which arrived in the 1960s and have now departed.

Perhaps the best example of a policy, devised in a different economic climate and now confronting a much more rigorous environment, is the attraction of inward investment to Clydeside. This became such a key element in economic policy for Scotland in the 1960s and early 1970s, in conjunction with regional policy assistance, that Firn (1975) estimated that by 1973, whilst Scottish-owned manufacturing companies made up 72 per cent of all plants in Scotland, 59 per cent of all manufacturing employment lay outside Scottish control, and 80 per cent of all new manufacturing jobs created between 1958 and 1973 were in companies whose ultimate ownership lay outside Scotland. Henderson (1980) demonstrated that the attraction of inward investment was particularly difficult for large urban areas such as Glasgow which between 1945 and 1970 attracted some 18,350 jobs in incoming firms when on a 'fair-shares' basis the city might have expected 52,400 such jobs. By contrast, the remainder of the conurbation attracted 48,800 jobs, some 32,400 more than might have been expected: of these, some 13,000 are attributable solely to the 'New Towns effect' as East Kilbride and Cumbernauld pulled investment away from the inner city of Glasgow. By the late 1970s, local authorities were having increasing difficulty in mounting effective campaigns to attract inward investment, and in Scotland the process was centralized and unified within the Scottish Development Agency's Locate in Scotland (LIS) programme (Young 1984). LIS claims to have brought 9,000 jobs in new companies and 10,600 additional jobs in existing companies to Scotland in 1981–84, but because LIS, and indeed the SDA as a whole, operate on market principles, LIS has a very strong tendency to offer incoming firms the most attractive sites and these tend not to be in cities such as Glasgow, nor in the areas of greatest need as defined by unemployment rates. Were LIS to discriminate in favour of these areas it would, it is argued, lose out in competition with other national development agencies such as that of the Republic of Ireland in attracting Japanese or North American capital.

These conventional older approaches to economic development have generally not worked to the advantage of inner-city problem areas. Accordingly new policies have been developed in recognition of the deepening recession and the need to bring other forms of enterprise and employment into being and to bring new agencies into existence to augment the local and national governments' attempts at economic regeneration. The most obvious element has been to increase the rate of new firm formation which is generally low in Britain (see Ch. 9) and lower still in Clydeside. Small firm promotion, in conjunction with environmental improvement and the provision of small industrial premises in the urban cores, have

Table 15.6 Small firm financial support in Clydeside/Scotland

1. Financial assistance

UK Small Firms Loan Guarantee	£43 m.	1,300 applicants (Scot.)
SDA Small Business Div. loans	£6 m.	344 projects (Scot.)
Glasgow District Council, grants, loans	£2 m.	– (Clydes.)
EIB/ECSC loans	£300,000	– (Scot.)
IDS Better Business Services	—	121 projects (Scot.)
Other ERDF non-quota support	—	157 projects (Scot.)
Glasgow District Council, rent free period	—	63 projects (Clydes.)

2. Advisory services

IDS Better Business Services		
Glasgow Opportunity	£400,000	120 projects (Glasgow)
Scottish Co-operative Development		
Lanarkshire Industrial Field Executive		
British Steel (Industry)		
EEC non-quota support for consultancy services		

been the main elements in the inner-city revitalization programme (Hood et al. 1982). Small firm development in Clydeside has an unusually wide range of support agencies with differing forms of advisory and financial support, from national programmes of the Department of Industry, through the SDA, through local authorities and increasingly through specialist agencies linking the public and private sectors. Table 15.6 gives a summary indication of the levels of activity of the various agencies involved in small firm promotion and estimates of their take-up in Scotland and/or Clydeside. A broad range of assistance is covered comprising financial support, advisory services, management and labour training, and premises. This diversity raises the potential problem of overlap and confusion although this danger does appear to have been minimized by cooperation between agencies which have developed interlocking rather than overlapping functions. For example, the Small Business Division of the SDA provides only specialist technical and investment advice, leaving the provision of general advice on setting up in business to Glasgow Opportunity and other local Enterprise Trusts. Similarly, while much of the finance for small firms is provided by the Small Business Division (SBD) of the SDA, this is complemented by the local authorities' power to offer loan finance to enterprises which are considered too marginal or risky for the SBD to contemplate.

The second new direction is the encouragement of high-technology industry. This has largely been the preserve of the Scottish Development Agency, as local authorities are rarely conversant with the special needs of these sectors. At the Scottish level, the most important sector in the high-technology area has been micro-electronics, where 25,000 people are employed. Significantly, 80 per cent of this labour force is in non-Scottish-owned companies and concern has been expressed about the vulnerability of this sector (Hood and Young 1982). Historical factors have directed the electronics industry to eastern Scotland with some major employers drawn to relatively isolated labour markets such as Greenock (IBM, National Semiconductors). Few of these firms have been drawn to Glasgow; only East Kilbride with its very different environment has attracted major electronics companies. Once again, the SDA has demonstrated that its prime economic concern lies at the level of the Scottish economy and it is not prepared to risk a promise of inward investment by pushing relatively unattractive inner-city environments. Within the high-technology sector, however, other developments have been drawn to Clydeside by the SDA. The Motherwell Food Park has been established to provide joint services in the food preparation and testing industry at a point with good access to central Scotland, and the West of Scotland Science Park has been established in west Glasgow.

A third innovation, paralleled in London, is the concern of urban policy to involve the private sector, both for its expertise and for its financial resources. The prime example of this joint public–private venture is the creation of Enterprise Trusts, of which the most prominent is Glasgow Opportunity, launched by a number of major companies in 1983 with £300,000 of private sector support. The SDA contributed a further £105,000 to start-up and operating costs, and several major companies including IBM, Coats Paton and Marks and Spencer seconded managers to act as advisers. Glasgow Opportunity provides a range of support services ranging from informal advice to financial planning and market advice; by early 1985, it assisted some 120 projects with about 500 jobs. Other smaller Enterprise Trusts have been established in Clydeside in Motherwell, Monklands and Cumbernauld, and significantly the SDA has chosen this format to integrate their newest area-based initiative in Greenock–Port Glasgow to the west of the conurbation.

Recognizing that formal enterprises are unlikely to provide a complete solution to the unemployment, a number of ways of involving the community and voluntary sectors have been devised as part of the urban programme. In Clydeside a new wave of enterprise has arisen alongside the 'small firm' sector, comprising worker-managed enterprises and community businesses. The latter are trading organizations owned and controlled by the local community, aiming to provide self-supporting, viable employment for local people (Community Business Scotland 1981). There have been difficulties in reconciling the two objectives of local accountability and economic viability, although some contributions to employment have been made. In Clydeside the number of jobs is quite small, with approximately 600 generated by late 1984 (plus some 300 homeworkers on piece rates). For the individuals concerned, the benefits may be significant and there will be multiplier effects. However, in terms of the scale of the unemployment problem in the conurbation, the overall impact is small (McArthur 1984).

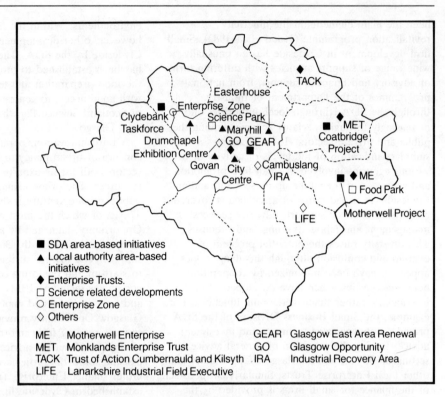

Fig. 15.4 Area-based employment initiatives 1985: Clydeside

ME Motherwell Enterprise
MET Monklands Enterprise Trust
TACK Trust of Action Cumbernauld and Kilsyth
LIFE Lanarkshire Industrial Field Executive
GEAR Glasgow East Area Renewal
GO Glasgow Opportunity
IRA Industrial Recovery Area

Lastly, in both London and Clydeside, as we have shown, the proportion of long-term unemployed has risen sharply. Thus policies have had to be devised to confront a situation in which an assumption of a cyclical return to full employment no longer holds. Some of these are national training programmes such as the Training Oportunities Scheme (which ceased in 1984) and the YTS, currently being augmented: some are job creation schemes such as the Community Programme and its predecessors such as the Special Temporary Employment Programme and the Community Enterprise Programme (McArthur and McGregor 1985). One alternative is self-employment, and the Enterprise Allowance Scheme (EAS) was established in 1983 to assist the unemployed to set up in business on their own by providing a wage subsidy. However, in an area such as Clydeside with a long tradition of a few major industrial employers, it would appear that there is less entrepreneurial skill to be tapped. The uptake of EAS has been especially low in areas where the need is greatest. Clydeside appears to be particularly handicapped by its low level of home ownership which prevents large numbers of potential users of EAS using a second mortgage as capital, and house repairs, which are a promising market for many self-employed builders, tend to be undertaken within the public sector Direct Works departments. The lack of premises further disadvantages the peripheral public sector housing estates, relative to the inner-city zones, despite their greater need. Thus a potentially useful element in urban policy may be bent away from the area of greatest need.

While much of what we have described as urban policy is either a national programme with varying rates of uptake, such as EAS, or is a conurbation-wide programme such as regional aid status, the Clydeside conurbation has the most diverse set of area-based programmes of any area in Britain. Figure 15.4 gives an impression of the number and types of area-specific programmes. Some, such as the Glasgow East Area Renewal (GEAR) project and the Motherwell Project are close to the London Docklands Development Corporation, except that they are coordinated by the Scottish Development Agency rather than an Urban Development Corporation; others such as the Clydebank Enterprise Zone or Glasgow Opportunity offer a limited range of incentives; and others again such as the West of Scotland Science Park and the Motherwell Food Park are sector-specific as well as spatially very restricted. It has long been recognized that Clydeside – like London – is an open labour market so that any area-based strategy which creates employment may not just confer a benefit on the local unemployed. GEAR, for example, the largest of all the projects, is a very open labour market in which 75 per cent of jobs employ people resident outside the area, and 75 per cent of the resident workforce are in jobs located outside the area. The rationale for area-based programmes therefore is that the areas selected are ones of potential and that trickle-down will allocate the employment and income benefits spatially. As Fig. 15.4 indicates, there are several types of area-based initiative. GEAR, the most comprehensive, is jointly managed by the SDA and two tiers of local government. It links a massive environmental programme to many forms of industrial assistance – premises, services, loans and grants, advice and training. Some £300 m. of capital expenditure is committed to GEAR between 1977 and 1987, although only £50 m. is a net extra resource above previously existing investment programmes including Glasgow District housing (£41 m.), Scottish Special Housing Association (£39 m), Strathclyde Transport (£17 m.) and health care (£4 m.). By 1984 some 3,300 jobs were attributed to the GEAR programme although a proportion of these were in pre-existing companies.

The Clydebank Taskforce project was created out of the closure of a major industrial employer – Singer's sewing machines – with the loss of 3,000 jobs in a town whose other manufacturing employment base – shipbuilding – was also uncertain. The Taskforce, operated by the SDA, was largely concerned with clearing the Singer site and providing new premises. However, the Taskforce initiative was overtaken by the EZ programme. Given the headstart created by the SDA's Taskforce, the Clydebank EZ emerged as Britain's most successful (Roger Tym and Partners 1983). By 1984, some 1,800 jobs were located on the site, of which 5 per cent were in firms already previously located in the designated area, 29 per cent were in new start-ups and 66 per cent were in firms which made short-distance transfers into the EZ. The Taskforce area now has provision for approximately 2,600 jobs, generated by an SDA expenditure of £21 m. which has brought in a private sector investment of £18 m. in industrial and commercial property and £24 m. in residential property. The inclusion of private sector residential development in areas such as GEAR and Clydebank exemplifies the integrated nature of these schemes and the extent to which massive public sector expenditure can, when concentrated into particular areas, engender sufficient confidence in private sector enterprises which have hitherto been reluctant to enter rather risky inner-city areas. The public sector investment, however, will only achieve this if the scheme for the area is genuinely integrated so that planning policies are adjusted (many Clydeside authorities having had a long-run aversion to new private sector housing development, believing that all land zoned for housing would be required by the public sector, so acute were the housing problems and the need for renewal) and the supplies of public services, particularly education and health, enhanced.

The SDA's Integrated Area Projects are basically contracts between the SDA and the relevant local authority by which the SDA agrees to deliver a number of specific programmes within an agreed time period on the basis of an independent assessment of an area's needs. Most of these are infrastructural projects and, unlike GEAR, no social or community programmes are included. The Motherwell Project will run for the period 1982–87, will involve £57 m. of SDA resources and has a target of 3,000 jobs of which 840 were in place by mid-1984. It is intended to attract £62 m. of private industrial and commercial investment and £37 m. of private residential development. The Coatbridge Project will run for the period 1983–86, will involve £15 m. of SDA money, linked to £2.5 m. of private industrial development and £2.8 m. of private residential development. It has a target of 1,300 jobs of which 230 were in place by mid-1984.

The increasing number of SDA-initiated area-based projects has inevitably stimulated local authorities into similar but less adequately resourced ventures. The SDA has argued that the sites selected for its projects are areas with commercial potential to be realized and that they are not rescue operations after major industrial collapses, although the Clydebank scheme (and an earlier scheme at Glengarnock in the south of the conurbation, where a steel plant closed) does appear to be an attempt to retrieve the economy of a threatened town. Similarly, local authorities have argued that potential is a key determinant in their selection of areas for treatment. Glasgow District's earliest choices of Maryhill and Govan, it could be argued, were both areas of great need, but also of considerable potential, close to the city centre with ample land and good motorway links. The more recent area projects in Glasgow on the peripheral housing estates at Drumchapel and Easterbouse look significantly more like cases of need than of potential. This formulation has also been used by Glasgow District Council to cover major new developments in the Central Business District and in conjunction with the creation of the Scottish Exhibition Centre.

In order to complete the diverse range of area-based programmes for employment, it is necessary to include the Enterprise Trusts in the conurbation described above, and the two science-related developments of the West of Scotland Science Park in west Glasgow and the high-technology food park in Motherwell. Both have the backing of the SDA, the local authorities and the local universities.

15.8 Conclusion

Urban policy is much more confused than regional policy: it is implemented by a very large range of agencies, some of which (e.g. national government and some local authorities) may be in conflict over political objectives and finances. It also suffers from confusion of objectives, some of which are strictly economic (improved efficiency and output) and some of which are social (community improvement, enhancing the quality of life) but which have employment effects. The deepening recession since 1979 has meant that policies devised focusing on mobile investment, new firm formation and an infrastructural basis to development are inadequate. New policies (and agencies) have been created dealing with the informal sector, working with the community, and examining more closely the relationship between the public and private sectors. Whereas regional policy developed slowly, and generally by increasing the areas qualifying for assistance on the grounds that the spatial patterns of prosperity and depression had altered little throughout the post-war period, urban policies will require to be more dynamic. Some areas, such as the New Towns, once thought economic successes in their respective regional contexts, now require large amounts of help; within the large cities, the rate of inner-city decline appears to be slowing, while the peripheral estates will require to be the next foci of attention.

References

Association of Island Communities/Joint Docklands Action Group 1984 *Stifling the island's enterprise*. AIC/JDAG.

Bentham, C. G. 1985 Which areas have the worst urban problems? *Urban Studies* **22**: 119–31.

Boddy, M. 1984 Local economic and employment strategies. In Boddy, M., and Fodge, C. *Local Socialism?* Macmillan.

Bull, P. 1981 Redevelopment schemes and manufacturing activity in Glasgow. *Environment and Planning* **13**: 991–1000.

Button, K. J. 1978 Employment and industrial decline in the inner areas of British cities – the experience of 1962–77. *Journal of Industrial Affairs* **6**.

Cameron, S. 1983 Economic development – the changing public/private sector relationship. *Planning Outlook* **26**: 75–8.

Community Business Scotland 1981 *The development of Community Business in Scotland*. CBS.

Corkindale, J. T. 1976 Employment in the conurbations. CES Inner City Employment Conference, University of York.

Damesick, P., Howick, C., Key, T. 1982 Economic regeneration of the inner city: manufacturing industry and office development in inner London. *Progress in Planning* **18**(3): 133–266.

Danson, M. W., Lever, W. F., Malcolm, J. F. 1980 The inner city employment problem in Great Britain, 1952–76: a shift–share approach. *Urban Studies* **17**: 193–210.

Dennis, R. 1980 The decline of manufacturing employment in Greater London, 1966–74. In Evans, A., Eversley, D. (eds) *The inner city*. Heinemann, 45–64.

Department of the Environment 1976 *Lambeth inner area study, IAS/IA/II: London's inner area problems and possibilities*. DoE, para. 2.1.

Department of the Environment 1981 *Policy for the inner cities*. HMSO, Cmnd 6845.

Findlayson, D. 1983 Some alternative responses. *Local Government Policy Making* **10**: 37–41.

Firn, J. R. 1975 External control and regional policy. In Brown, G. (ed). *The Red Paper on Scotland*. EUSPB.

Fothergill, S., Kitson, M., Monk, S. 1982a *The role of capital investment in the urban–rural shift in manufacturing industry*. Industrial Location Research Project Working Paper 1. University of Cambridge.

Fothergill, S., Kitson, M., Monk, S. 1982b *The profitability of manufacturing industry in the UK conurbations*. Industrial Location Research Project Working Paper 2. University of Cambridge.

GLC 1984 *Government policy and the inner cities*. GLC Inner City Conference, 5–6 April 1964.

Gripaios, P. 1977 Industrial decline in London: an examination of its causes. *Urban Studies* **14**: 181–90.

Hadley, G. 1984 Enterprise Zones in Britain. *Planning Outlook* **27**: 34–8.

Hall, P. 1984 Enterprises of great pith and moment. *Town and Country Planning* **53**: 296–7.

Hausner, V. A. 1982 *Urban development in the United Kingdom and Europe*. PSI.

Henderson, R. I. 1980 The location of immigrant industry within a UK Assisted Area: the Scottish experience. *Progress in Planning* **14**(2): 193–226.

Holtermann, S. 1975 Areas of urban deprivation in Great Britain: an analysis of the 1971 Census data. *Social Trends* **6**: 31–47.

Hood, N., Milner, M., Young, S. 1982 *Growth and development in small successful manufacturing firms in Scotland*. Report prepared for the Scottish Economic Planning Department.

Hood, N., Young, S. 1982 *Multinationals in retreat: the Scottish experience*. Edinburgh University Press.

Lawless, P. 1981a The role of some central government agencies in urban economic regeneration. *Regional Studies* **15**: 1–14.

Lawless, P. 1981b *Britain's inner cities: problems and policies*. Harper and Row.

Leigh, R., North, D., Gough, J., Escott, K. 1982 *Monitoring manufacturing employment change in London 1976–81*. Report to the Department of the Environment.

Lever, W. F. 1982 Urban scale as a determinant of employment growth or decline. In Collins, L. (ed.) *Industrial decline and regeneration*. Edinburgh University Press.

Lever, W. F. 1985 Old policies in a new role. In Lever, W. F., Moore, C. (eds) *The city in transition*. Oxford University Press.

Lever, W. F. 1986 Economic development policy in London. In Ewers, H. J. (ed.) *The future of the metropolis*. de Gruyter.

Lever, W. F., Mather, F. 1985 The changing structure of business and employment in the Clydeside conurbation. In Lever, W. F., Moore, C. (eds) *The city in transition*. Oxford University Press.

McArthur, A. A. 1984 *The Community Business movement in Scotland: contributions, public sector responses and possibilities*. Discussion Paper 17, Centre for Urban and Regional Research, University of Glasgow.

McArthur, A. A., McGregor, A. 1985 Policies for the disadvantaged in the labour market. In Lever, W. F., Moore, C. (eds) *The city in transition*. Oxford University Press.

Miller, C., Miller, D. 1982 Local authorities and the local economy. *Town and Country Planning* **51**: 153–4.

Morphet, J. 1984 Area based employment initiatives. In *Planning and employment in London*. London Branch RTPI Working Party on Employment.

North, D., Leigh, R. 1984 Alternative approaches to urban economic policy: the case of London. In Barr, B., Waters, N. (eds) *Regional diversification and structural change*. Tantalus, Vancouver.

Town and Country Planning Association 1980 Urban Development Corporations. *Town and Country Planning* **49**: 26–7.

Townsend, A. 1977 The relationship of inner city problems to regional policy. *Regional Studies* **11**: 225–51.

Townsend, P. 1974 *The cycle of deprivation*. Paper presented to the British Association of Social Workers Conference, University of Manchester.

Turner, D. 1985 Enterprise Agencies reach maturity. *Town and Country Planning* **54**: 196.

Turok, I. 1984 *The impact of urban economic policy in a context of decline*. Paper presented to the British Section of the Regional Science Association Conference, 5–7 September 1984. University of Kent at Canterbury.

Tym, R. and Partners 1983 *Monitoring Enterprise Zones – year two report*. Department of the Environment.

Valente, J. Leigh, R. 1982 Local authority advance factory units – a framework for evaluation. *Planning Outlook* **24**: 27–35.

Wellbelove, D., Woods, A., Zafiris, N. 1981 Survival and success of the inner city economy: the performance of manufacturing and services in Islington. *Urban Studies* **18**: 301–14.

Young, S. 1984 The foreign-owned manufacturing sector. In Hood, N., Young, S. (eds) *Industry, policy and the Scottish economy*. Edinburgh University Press.

Index

activity location, 158–9
Acts of Parliament *see* legislation
aerospace industry *see* British Aerospace
AES *see* Alternative Economic Strategy
agglomeration economies and linkages, 110
aggregate economy, 211–14
aggregate patterns in behavioural approaches, 42, 43–4
aggregate travel model, 24
Aglietta, M., 60
Alchian, A. A., 38
Alderman, N., 104
Alternative Economic Strategy, 215, 222
applied research, 98
Armstrong, H., 77
Asia, 168, 187, 193
  see also Japan
assisted areas, 224, 228, 229, 237
  innovation in, 102
  labour markets, 77
  maps of, 225, 232, 235
  multiplants in, 163
  past identity, 224–6
  redefinition, 236
  straitened approach to, 231–3
Austria, 128
automation, 3, 173
autonomy and linkages, 111–12
Averitt, R. T., 60
Aydalot, P., 9, 17

Babcock International, 171, 175, 177, 179
balance of payments, IMF-OECD and, 177, 182
Bannock, G., 127, 130
Barlow Report, 223
Barrett, S., 89
Bass, 154, 171, 176
Bassett, K. A., 77
BAT *see* British American Tobacco
Beecham Group, 171–2, 173, 175, 177

Beesley, M., 134
behavioural approaches to industrial location studies, 38–55, 57
  in Britain, 42–51
    aggregate patterns, 42, 43–4
    decision-making, 44–51
      external relationships, 48–51
      location, 45–6, 51
  elements of, 39–40
  firm-environment characteristics, 40–2
  origins of, 38–9
Belgium, 9, 183
Benson, J. K., 64
Bentham, C. G., 241, 242
Better Business Services, 252
BICC, 154, 171, 175, 178
Binks, M., 128–9, 131, 132, 136
Birch, D. L., 134
birth rates, 18
BL *see* British Leyland
'black economy', 128
Blackaby, F., 2, 112
Blackburn, A., 182, 188
BOC International, 171, 173, 175, 177, 178
Boddy, M., 22, 89, 246
  on structural approaches to location, 56–66
Bolland, A., 132
Bolton, J. and Committee, 126–30, 132, 137, 144, 216
Bosanquet, N., 83
Bowater, 154, 171, 176, 177
BP *see* British Petroleum
'branch plant economy', 46
  see also plants
Brech. M., 7
Brechling, F., 77
Breheny, M., 7, 17
brewing industries, 132, 176
  multinationals, 170–2
  multiplants, 152, 154–5, 156–8

British Aerospace, 3, 171–2, 173, 198, 199, 200
British American Tobacco, 160, 164, 171, 173–7 *passim*
British Gas, 199
British Leyland, 7, 114, 170–1, 173
   employment loss, 154
   as public sector industry, 198–200, 216, 229–30, 242
   robotization, 3
   trade, 178, 179, 191–2
British National Oil Corporation, 199, 200, 216
British Nuclear Fuels, 199
British Petroleum, 169, 170–1, 173, 174. 177, 198–200
British Rail, 160
British Shipbuilders, 171, 173, 199, 200
British Steel Corporation, 2, 16, 135, 160, 171, 173, 179, 199, 200, 205–6, 252
British Technology Group, 198, 216
British Telecom, 216
British Venture Capital Association, 141
Britoil, 199, 200, 216
Broadbent, T. A., 82
Brooke Bond, 171
Browett, J., 59
BS *see* British Shipbuilders
BSC *see* British Steel Corporation
   buildings, industrial, 14, 86–95, 246, 249
   employment change and, 91–2
   floorspace supply, 87–9
   implications for national economy, 94–5
   locational change, 43–4, 46
   locational consequences, 92–3
   stock, quality of, 89–91
Bull, P., 250
Burmah Oil, 170–1, 173, 177
Business Expansion Scheme, 10, 133–4
'Business Link', 164
business services linkages, 116–19
Business Statistics Office, 149
Buswell, R. J., 47, 104
Button, K. J., 241

Cable, J. R., 198
Cadbury Schweppes, 171, 173, 175, 177
   location of plants, 152
Cadbury, Sir A., 145
Cairncross, A., 2
Cameron, G. C., 5, 229
Cameron, S., 242
Campbell, M., 174
Canada, 128, 182, 187
capital, 79–82
   location theory and, 70–2
   small and new firms and, 133–4, 140, 141, 144
   venture, 141, 144
   versus labour, 82–3
   see also investment; restructuring
capitalism, structural approach to, 57–8
Carney, J., 56

Carr, M., 38, 40, 45
catering, small and new firms in, 126–7, 132
   see also food
Cathcart, D. G., 131, 137, 139, 140
CBI (Confederation of British Industry), 233, 234
CDPs (Community Development Projects), 52, 241
CEGB *see* Central Electricity Generating Board
Census of Employment, 238
Census of Population, 236, 241, 242
Census of Production 70, 71, 72, 80, 150
Central Electricity Generating Board, 197, 199, 202
Centre for Advanced Land Use Studies (CALUS), 89
Champion, A. G., 57, 231
change *see* industrial change
chaotic conception, 63, 64
Chaplin, P., 132
Chappell, H., 132
chemicals and rubber industries, 3, 4–5, 226
   labour and capital, 5, 71, 79–82
   multinationals, 169–72, 173, 174, 175, 177, 183
   multiplants, 150, 157, 164
Cheshire, P., 17, 71
Chester, T. E., 197
Child, J., 40, 42
Chisholm, M., 2
Chrysler, 186, 191
circulating commodity capital, 71
Clark, D., 139, 150, 153–4, 163
Clark, J., 3
climate, 17
cloning, 62
closures, 2, 44, 45–6, 136, 163, 230–1
clothing and textiles industries, 32, 226
   labour and capital, 5, 70, 77, 79–82
   multinationals in, 169–72, 176–7
   multiplants in, 150, 156, 157, 160
   output, 4–5, 16
   small and new firms in, 139
Clydeside urban policy, 250–6
coal mining *see* National Coal Board
Coatbridge Project, 255
Coats Patons, 172–4, 176, 177, 254
commodity capital, 71
Commonwealth, 7, 179, 187, 218, 221
communications, 17
   see also linkages
Community Business, 240
Community Development Projects, 52, 241
Community Enterprise Programme, 217–18, 254
Community Industry, 217
Community Programme, 218, 240, 244, 254
competition/competitiveness, 27, 216, 218, 248
   decline in, 96, 103
   small firms and, 137, 142
components of change analysis, 63
Confederation of British Industry, 233, 234
Consolidated Goldfields, 171
construction industries, 5, 154, 159, 161

multinationals, 171–2, 176, 177
  small and new firms, 126–7, 129, 130, 131
control and material linkages, 110–12
Cooke, P., 2, 56, 58, 60
Coombes, M. G., 14, 104
Cooper, M. J. M., 45
co-operatives, worker, 132
core and periphery
  employment and floorspace change, 43–4
  labour and capital, 71
  regional development policy and, 33–4
  small firms and, 141–3, 145
  structural approach to, 60–1
  see also rural; urban
Corkindale, J. T., 240
corporate
  environment see environment and firm
  organization
    changes in, 6–9
    linkages and, 116–19
  see also enterprises
cost of living 71–2
costs
  labour, 24, 25, 30–1
  land, 25, 93
  least-cost see neoclassical location theory
  production, 14, 28
  public sector, 197, 215
  research, 6
  spatial interaction with revenue, 29–30
  surfaces, 31
  transport, 69, 72, 110
    labour and, 72
    location and, 24–6, 28, 31, 33
  variable, model of, 25–6
  wages, 70, 72, 83
  see also public expenditure
Counahan, M., 190
Courtaulds, 153–4, 170–1, 176, 177, 178, 230
Coyne, J., 128–9
CP see Community Programme
Crompton, D., 114
Cross, M., 50, 132, 139, 140, 163
Crum, R. E., 79
Curran, J., 128, 133, 136
Cyert, R. M., 39

Dalgety-Spillers, 171, 177
Damesick, P., 249
Danson, M. W., 244
Darnell, A., 131
DAs see Development Areas
Davies, J. R., 137
Davy Corporation, 175, 179
Dawson, J., 127
Daysh, G. H. J., 224
De Genova, D., 17
decentralization, 72
  see also urban-rural shift

decision-makers see behavioural approaches
'de-industrialization', 1–2, 127, 169, 172
  see also recession
Delta, 173, 175
demand
  curve and market areas, 27–8
  deficient unemployment, 74–5
  and output, 212–13
Denmark, 183, 186
Dennis, R., 44, 245
Department of Employment, 75, 134, 236, 238
Department of Environment, 87, 240, 242, 249
Department of Trade and Industry, 95, 133–4, 153, 156
  and buildings, 87
  and location, 69
  and multinationals, 186, 188–91, 193
  and national policy, 216, 217, 219–21
  and new firms, 139
  and regional policy, 45, 224, 234, 236
  and urban policy, 240, 242, 253
deskilling, 84
Dessler. G., 40
Development Agencies, 87, 89, 164, 199
Development Areas, 44, 69
  multinationals and, 186, 188–90
  national policy and, 226, 230, 234, 236, 237
  technological change, 99–104
Development Commission, 89, 233
Dicken, P.
  on competition, 137
  on decision-makers, 39–40
  on least-cost location, 36
  on multinationals, 182, 183
  on multiplants, 150–1
  on ownership, 7, 46, 47
  on public sector, 197, 216
  on small firms, 137
  on trade, 218
  on unemployment, 81, 84
  on urban firms, 44, 81, 150
Diffusion Survey, 102, 104, 106
distance costs see transport costs
Distillers, 172, 178
Distribution of Industries Act (1945), 223
diversification, 136–7
Doeringer, P. B., 83
Dunford, M., 60
Dunlop, 169, 170–1, 175, 177, 191, 192
Dunnett, P. S., 114
Dunning, J. H., 47, 167–9, 172–3, 177, 182, 183

East Anglia
  birth rate, 18
  employment, 11–13, 14–15, 70, 227, 238
  innovations, 16, 99–102
  motor vehicles industry, 115
  multinationals, 185–6
  multiplants, 47

new firms, 18
overseas investment, 47
public sector industries, 200–2
regional policy, 225, 227
small firms, 10, 18, 130, 135, 136, 138–43
unemployment, 76, 78, 109
East Midlands, 18, 32
employment, 11–14, 15, 70, 227, 238
innovation, 99–102
motor vehicle industries, 115
multinationals,185–6
multiplants, 150, 152–4, 157, 159,161–2
public sector industries, 200–1, 204
regional policy, 226–7
small firms, 18, 130, 131, 135, 137,138–43
unemployment, 76, 77, 78
urban policy,242–3, 244
Eastern Europe, 168
EBs (Enterprise Boards), 244, 249
economic impact of small firms, 134–8
Economic Planning Region, 116, 119, 158
economies
agglomeration and linkages, 110
of scale lacking, 145
education level of small firm entrepreneurs, 132, 140, 144
Edwards, R. C., 60
El-Agraa, A. M., 215
electricity *see* energy industries
Electricity Council, 199
electronic and electrical products industries, 226
innovations, 100
labour and capital, 5, 30–1, 70, 71
multinationals, 170–2, 175–8, 183, 186, 188, 189, 191
multiplants, 150, 154, 156, 162
output, 3–5, 16–17
small and new firms, 132, 136, 137, 144
empirical analysis, need for, 63
employment
change, 5, 9–13, 213–14
buildings and, 43–4, 91–3
innovations and, 105–6
multinationals and, 172–3, 174, 176
multiplants and, 8, 155, 161
density, fall in, 91
encouraged *see* regional policy; urban policy
linkages, 111–12
national policy and, 217–18
peak, 43, 108, 135
public sector, 201–2, 204, 206
and small and new firms, 126–9, 134–5, 138, 139
trends, 4–6
see also labour; unemployment
Employment Office Areas, 231–2
Employment Protection Act, 133
energy industries, 71, 199–200, 201–2
see also National Coal Board
engineering industries, 156, 226

labour and capital in, 5, 70, 71, 77, 79–82
multinationals in, 170–2, 173, 175–81, 192
multiplants in, 150, 154, 156
output, 3, 4–5, 16
English Industrial Estates (EIE), 87,89
Enterprise Allowance Scheme, 73, 134, 154
Enterprise Boards, 244, 249
enterprise, industrial *see* firm; multinational; multiplants; new firms; public sector; small firms
Enterprise Trusts, 244, 253
Enterprise Zones, 242–4, 247, 254–5
entry modes of multinationals, 188–9
environment and firm, 39, 402, 45
EOAs (Employment Office Areas), 231–2
ERDF (European Regional Development Fund), 10, 252
establishments *see* plants
Estle, E. F., 70
ethnic origin and unemployment, 75
ETs (Enterprise Trusts), 244, 253
Europe
cheap labour in Mediterranean area, 237
investment in, 168
motor vehicles, 114, 180
multinationals, 173, 179–80, 183, 184, 186, 189–91, 193–4
small firms, 128
trade, 220–1
see also European Economic Community *and individual countries*
European Economic Community, 1, 9, 215, 218, 229
Common Agricultural Policy, 184
investment and, 7, 168–9
medium-sized firms, 47
motor vehicles, 114, 180
multinationals, 182, 183, 184, 188, 189, 190
regional policy, 10, 223, 234, 236, 237, 252
Social Fund, 10
trade, 5, 220–1
urban policy, 252
see also individual countries
European Free Trade Association, 183, 218
European Regional Development Fund, 229, 236
Evans, A. W., 47
external relationships and decision-making, 48–51
EZs (Enterprise Zones), 242–4, 247, 254–5

factors of production *see under* production
fdi (foreign direct investments) *see* multinationals
finance *see* capital
financial institutions, committee on *see* Wilson
Financial Institutions Group, 242
Findlayson, D., 244
fireworks industry, location of, 32
firms: defined, 149
see also corporate; enterprise
Firn, J. R., 44, 46, 252
floorspace *see* buildings

food and drink industries, 4
  labour and capital, 5, 79–82
  multinationals, 170–3, 175, 176, 177–8
  multiplants, 150, 152, 157
  small and new firms, 132, 139
Ford, 3, 7, 153–4, 169, 190, 191
'Fordism', 60
foreign see international; overseas
foreign direct investment see multinationals
forestry industry, 199, 200
  see also timber
Forsyth, D. J. C., 182, 186
Fortune 500, 168–70, 172
Fothergill, S., 67, 68
  on assisted areas, 228–9
  on buildings, 88
  on industrial buildings, 86–95
  on London, 245–6
  on multiplants, 162
  on new firms, 130, 135, 137–41
  on organization, 203
  on profitability, 80
  on urban-rural shift, 11–15, 43–4, 64, 104, 161
Fowler, R. F., 77
France, 9, 180, 219
  multinationals, 109, 182, 183, 186–7
franchising, 132–3
Fredricksson, C. G., 112
Freeman, C., 3, 17, 97
frictional unemployment, 74
Friedman, A., 60, 62
Friedman, M., 215
Froebel, F., 59, 237
Frost, M. E., 76, 77, 231
Fuchs, V. R., 71
functional specialization, 8, 47–8, 62

Galbraith, J., 40
Gallaher, 171
Ganguly, P., 8, 128–30, 138–9
gas see energy industries
Gault, F. D., 227
GEAR (Glasgow East Area Renewal), 254–5
General Electric Company (GEC), 114, 153–4, 171, 173, 174, 175, 177, 178, 216
General Motors, 7, 170, 190, 191
geography see location; spatial; regional
Germany see West Germany
Gibb, A. A., 113
Gillespie, A. E., 12, 14, 76
Gilmore, J. M., 110
Gilmour, J., 71
GKN see Guest, Keen and Nettlefolds
Glasgow East Area Renewal, 254–5
Glasgow Opportunity, 252, 253, 254
Glaxo Holdings, 172, 173, 175, 177, 178
GLEB (Greater London Enterprise Board), 249–50
Glyn, A., 197

Goddard, J. B., 3, 8, 17, 47, 48, 57, 67–8, 116, 152
  on technological change, 96–107
Goodman, J. F. B., 71
Goodyear, 183, 191
Gordon, D. M., 60
Gould, A., 7, 9, 12, 17, 18, 130, 135, 136, 139–40
government, 9
  multinationals and, 167
  small and new firms and, 125, 133–4
  see also national policy; public sector; regional policy; urban policy
Grant, R., 8, 9
Greater London Enterprise Board, 249–50
Green, H., 50, 126
'greenfield' developments, 7
  see also urban-rural shift
Greenhow, F., 133
Gregory, D., 25
Gripaios, P., 245
Gross Domestic Product, 12, 211–14, 217, 221
Gross National Product, 172
Gudgin, G., 9, 79, 136
  on assisted areas, 228–9
  on buildings, 88
  on capital, 79
  on migration, 86
  on multiplants, 162
  on national policy, 215, 222
  on new firms, 130, 135, 137–41
  on organization, 203
  profitability, 80
  on urban-rural shift, 11–15, 43–4, 64, 104, 161
Guest, Keen and Nettlefolds, 153–4, 170–1
  as leader of engineering multinationals, 175–81

Haddad, P. R., 31
Hadley, G., 244
Haggett, P., 77
Haig, R. M., 8
Hall, P., 17, 244
Hamilton, F. E. I., 39, 41, 45–6, 124, 158, 196
  on multinationals, 167–95
Hardill, I., 150
Harrison, R. T., 45, 131
Hart, M., 131
Haug, P., 99, 189
Hausner, V. A., 242
Hawker Siddeley, 171, 175, 177, 178, 179
Hayter, R., 6, 45, 151
head offices, location of, 47–8, 50, 62, 152–3, 162
  see also South East 'core' functions
Heal, D. W., 205
Healey, M., 44, 46, 51, 123, 139
  on multiplants 149–66
Henderson, R. A., 163, 252
Heritage Programme, 205
Herron, F., 156
Hewlett Packard, 17, 186, 189

hierarchy, locational, 152–3, 162
Hill, C., 38
Hillier, J., 236
Hoare, A. G., 110, 111
Hodge, M., 193
Holland, S., 192, 215, 229
Holtermann, S., 241
Honeywell, 17, 183
Hood, N., 182, 231, 253
Hoover, E. M., 24, 183
Hotelling, H., 26
Hough, J., 133
Howard, R. S., 45
Howdle, J., 134
Howells, J. R. L., 6, 17, 47, 104, 152
Humphrys, G., 124
  on public sector industries, 196–207
Hymer, S., 58, 192

IAs see Intermediate Areas
IAS (Inner Area Studies), 241
IBM, 7, 17, 189, 192, 253
ICE (Inner City Enterprise), 242
ICI see Imperial Chemical Industries
ICL, 154, 172, 173, 175, 177
IDC see Industrial Development Certificates
Imber, J. C., 220
IMI, 172, 175, 176, 178
Imperial Chemical Industries, 154, 164, 169, 170–1, 173–8 passim, 196
'In Business', 164
Industrial Activity and Area Development Study Group, 234
industrial change in UK, 1–20
  changing environment, structure and organization, 1–10
    corporate organization, changes in, 6–9
    institutional changes, 9–10
    sectoral and employment trends, 4–6
    technological change, 2–4
  changing location, 10–18
    major trends, 10–14
    regional reversal, 14–18
    urban-rural shift, 14
  see also factors of production; industrial enterprises; location theory; policies
Industrial and Commercial Improvement Areas, 242
Industrial Development Act (1966), 223
Industrial Development Certificates, 15, 44, 224, 229, 240
Industrial Development Programme, 246
Industrial Recovery Area, 254
industrial relations, 83, 136
Industrial Reorganization Corporation, 216
Industry Act (1972), 223, 228, 229–30
Industry Training Board, 130
information-processing, location of, 47–8
information technology, 97

Inland Revenue see taxation
Inner Area Studies, 241
Inner City Enterprise, 242
Inner Urban Areas Act (1978), 241, 242
innovation, 61–2, 82, 98
  multiplants, 159
  process, 98, 101–2
  product, 100–1
  significant, 98–9
  small firms and, 132, 137, 145
  see also technological change
Innovation Survey, 100, 104
inputs and linkages, 111
institutional change, 9–10
Integrated Area Projects, 255
intensification, 61, 82, 159
interest rates, increase of, 79, 212
Intermediate Areas, 226, 233, 234, 236
  innovations in, 99–100
international
  comparisons
    in innovation adoption, 102
    of small firms, 128
  context
    of location, 59
    of national policy, 218–21
  trade, 2, 9, 218–21
    see also multinationals
  see also individual countries; overseas
International Monetary Fund-OECD Common Reporting System on Balance of Payments, 177, 182
investment, 45, 61, 159
  overseas, 7, 81–2
    see also multinationals
  Public sector, 196–7
  recession and, 79–80
  small firms, 133
  see also capital; foreign direct: regional policy; urban policy
Ireland, 9, 174, 183, 187
Isard, W., 23
isodapane map, 24, 26, 32
Italy, 9, 180, 182

James, V. Z., 116
Japan
  car industry, competition, 114
  industry restructured, 172
  investment in UK, 7, 9, 167, 168, 169, 182, 183, 186
  labour costs, 219
  multinationals, 173, 179, 183, 186, 188–91, 193–4
  robotization, 3
  small firms, 128
  trade, 218, 221
Jennings, A., 131, 132, 136
Jepson, D., 114

Job Creation, 217
job loss *see* unemployment
job-sharing, 218
John Brown (firm), 172, 178, 179
Johnson, P., 125, 129, 131, 137, 140
Jones, D. T., 114
journey-to-work and unemployment, 70–7
    see also transport

Kaldor, N., 192, 193
Keeble, D.
  on agglomeration, 110
  on buildings, 88
  on industrial change, 1–20, 43, 44, 45, 57
  on interest rates, 79
  on multiplants, 160
  on new firms, 130, 135, 136, 139–40
  on output, 214
  on regional policy, 224, 227, 229–30, 237
Kelly, M., 137
Kelly, T., 3–4, 16, 17
Kennelly, R. A., 31
Killick, T., 156, 226, 230
Kirwan, R. M., 5
Kitson, M., 12, 14–15, 44, 67
  on industrial buildings, 86–95
Kondratieff waves, 1, 3, 4, 17, 97
Krumme, G., 38

labour, 72–8
  costs
    in electronics industry, 30–1
    location and, 24, 25
  importance of, 59
  location theory and, 70–2
  national policy and, 217–18
  new spatial division of, 162–4
  process and relations of production, 60–1
  versus capital, 82–3
  see also employment; unemployment
Labour Force Survey, 75
Lambooy, J. G., 97
land, 69
  costs, 25, 93
  use planning system, 223
  see also buildings; location
Langridge, R., 17
large firms, 196
  behavioural analysis, 39–40
  capital, 81, 83
  and small firms, 129, 131–2
  see also multinationals; multiplants
Law, C. M., 47, 159, 160, 182, 186, 200
Lawless, P., 241, 242, 248
Layard, R., 77
LDDC *see* London Docklands
Le Heron, R. B., 109
least-cost location *see* neoclassical location

legislation
  employment, 133, 223, 233
  industry, 223, 228, 229–30
  inner cities, 241, 242
  local government, 249
  monopolies, 216
  nationalization, 205
  special areas, 223
Leigh, R., 48, 157, 161, 245, 246, 249
Lever, Sir Harold (now Lord), 125
Lever, W. F., 14, 67, 68, 111, 112, 120, 209–10
  on labour and capital, 69–85
  on national policy, 211–22
  on urban policy, 240–57
Lewis, E. W., 47, 104
Lewis, P. W., 31
Lindberg, O., 31
Lindmark, L. G., 112
Linge, G. J. R., 41, 172, 180, 187
linkages, regional development and industrial
  change, 48–51, 71, 108–21
  analysis
    problems in, 109–10
    traditional approach to, 110–12
  corporate organization and business service, 116–19
  dependence and industrial change, 112–13
  industrial decline and, 113–16
  industrial economy, 108–9
  material, 110–12, 113–16
liquidation, 130
LIS (Locate in Scotland), 252
Little, A. D. Ltd, 252
Lloyd, P. E.
  on competition, 137
  on innovation, 137
  on multinationals, 182
  on multiplants, 150, 164
  on new firms, 44, 81, 136, 140–1
  on ownership, 47
  on public sector, 197, 216
  on small firms, 18, 137, 139
  on unemployment, 81, 84
  on urban firms, 44, 81, 136, 150
Loan Guarantee Scheme, 10, 133–4
local
  labour markets, 76–8
  manufacturing units *see* plants
  markets and small firms, 137, 140, 141
  material linkages, 110
  unemployment, 77–8
Local Employment Act (1969), 223, 233
Local Government Act (1972), 249
Locate in Scotland, 252
location
  activity, 158–9
  adjustment of multiplants, 154–5, 157–61
  buildings and, 92–3
  change, 10–18

quotients (LQ), 70
theory: labour and capital, 70–2
    see also behavioural; neoclassical; structural
see also regional; spatial
London
  capital markets, 79, 80
  employment loss, 162
  innovation, 104–5
  offices, 117, 118
  public sector industries, 203
  regional policy, 226
  small firms, 142–3, 145
  urban policy, 241–50
  see also South East
London Brick Company, 154, 159, 161
London Docklands Development Corporation, 247, 254
London Enterprise Agency, 244
Lösch, A., 27, 29, 37
Loveridge, R., 60
Lovering, J., 56, 62, 163
Lucas Industries, 154, 170–1, 175, 178, 179
Luttrell, W. F., 38, 224

McAleese, D., 190
McArthur, A. A., 253–4
McCrone, G., 224
McDermott, P., 6, 31, 40–2, 46, 49, 109–10, 111, 112, 182
Macey, R. D., 150
McGreevy, T. E., 45
McGregor, A., 5, 254
McGuire, J., 125
'machinofacture', 60
Mackay, D. I., 71
Mackay, R. R., 228
Maclennan, D., 224
McPhail, C. I., 78
Makenham, P., 75
Malecki, E. J., 45
management
  buyouts, 132
  information, 97
  innovations, 98
  -labour relations, 60–1, 83, 186
Mandel, E., 3, 36
Manners, G., 76
Manpower Services Commission, 75, 78
manufacturing industries, 200, 213
  decline *see* 'de-industrialization'
  encouraged *see* regional policy; urban policy
  small and new firms, 126–30, 134–5, 136–7, 144
  see also employment; *individual industries*;
    multinationals; multiplants; unemployment
March, J. G., 39
Market Research Society, 117
markets
  areas and demand curve, 27–8
  labour *see* labour
  local and small firms, 137, 140, 141
Marquand, J. M., 228, 230
Marsh, P., 17
Marshall, J. N., 40, 48, 49–50, 68, 163
  on industrial change, linkages and regional development, 108–21
Martin, R. L., 2, 13
Marxism, 36, 57, 58–9
Mason, C., 18, 44, 50, 123, 182
  on small firms, 125–48
Massey, D.
  on assisted areas, 2
  on deskilling, 5
  on electrical engineers, 156, 162, 206
  on employment loss, 52, 56, 59, 61, 155
  on investment, 82, 237
  on multiplants, 8, 14
  on regional policy, 228, 229
  on restructuring, 56–8, 61–3, 82–3, 116, 159, 162
  on transport costs, 161
material linkages, 110–12, 113–16
maternity hospital, location of, 33
Mather, F., 250
Medhurst, K., 215
medium-sized firms and locational change, 45, 46
Meegan, R. A.
  on deskilling, 5, 14
  on electrical engineers, 153, 156, 162, 206
  on employment loss, 56, 59, 61–2, 155
  on investment, 82
  on regional policy, 229
  on restructuring, 56, 61, 63, 82–3, 116, 159, 162
  on transport costs, 161
Mensch, G., 3, 97
mergers *see* takeovers
Metal Box, 157–8, 171, 176
metals and metal-working industries 4–5
  decline, 109
  innovations in, 100, 101–2
  labour and capital in, 5, 70, 77, 79–82
  multinationals in, 170–2, 173, 175, 176
  multiplants in, 150, 163
  new and small firms in, 136
Microprocessor Applications Project, 217
Middle East, 179, 228
Midlands *see* East Midlands; West Midlands
Milk Marketing Board, 199
Miller, C., 248
Miller, D., 114, 248
Minimum List Heading, 16, 98
mining
  small and new firms in, 126–7
  see also metals; National Coal Board
MNE *see* multinational enterprises
mobility, industrial, 226–8
  decline, 229–31
  see also rural-urban shift

Mok, A. C., 60
Monk, S., 12, 14–15, 44, 67
    on industrial buildings, 86–95
monopolies, 198, 216
Moore, B., 10, 14, 15, 80, 228
Moore, C. W., 109
Morgan, K., 16, 56, 60, 61, 62, 106
Morphet, J., 247
Moseley, M. G., 109
Moses, L. N., 23
Motherwell Project, 254, 255–6
motor vehicles industries, 226
    decline, 16, 109, 114–15
    labour and capital, 5, 71, 77, 79–82
    multinationals, 7, 169–73, 178, 186, 190, 191–2
    multiplants, 150, 159
    public sector, 199
    small and new firms, 126–7, 130
multinational enterprises, 7–8, 167–95, 196
    changing international position of UK 167–8
    foreign in UK, 7, 182–4
        character of, 190–1
        investment, origins of, 183
        investment, quality of, 184
        regional impact of, 184–90
        size and importance, 182–3
    linkages and, 50, 51
    overseas shift, 7–8, 174–81
    relative decline of, 168–74
multiplants, 2, 6, 8
    geography of, 151–3
    linkages, 48–51
    locational adjustment, 154–5, 157–61
    locational change, 46
    numbers and distinctiveness of, 149–51
    ownership, 46–8
    pattern of changes, 153–6
    spatial growth, 154–5, 156–7
    urban and regional system and, 161–4
Murgatroyd, L., 56
Muth, R. F., 72

Nash, P. A., 45
National Coal Board, 5, 171, 177, 199, 202, 203–5, 213, 223
national context of location, 59
National Economic Development Council, 216
national economy and industrial buildings, 94–5
National Enterprise Board, 216, 242
National Plan, 215–16
national policy, 211–22
    aggregate economy, 211–14
    international context, 218–21
    managing economy, 214–17
    managing labour market, 217–18
National Research Development Corporation, 216, 217
National Semiconductor, 189, 253
Nationalization Act (1967), 205

nationalized industries *see* public sector
NCB *see* National Coal Board
NEB *see* National Enterprise Board
NEDC (National Economic Development Council), 216
neoclassical location theory, 23–37
    practical applications, 30–1
    revenue as spatial variable, 26–9
    spatial interaction of cost and revenue, 26–9
    variable-cost model, generalized, 25–6
    Weberian antecedents, 23–5
    welfare formulations, 32–5
Netherlands, multinationals in, 182, 183, 186–7
    see also Royal Dutch Shell
new firms, 8, 9, 244
    economic impact of, 134–8, 145
    employment and, 135, 139
    factors in creation, 131–4
    numbers of, 129–30
    regional distribution, 138–40, 141
    unemployment and, 131, 135
    see also small firms
new plants, 44, 45–6
New Towns, 87, 89, 224, 231, 244
    in Clydeside, 250, 252, 256
Newly Industrializing Countries, 219
    see also Third World
NIESR, 221–2
Nigeria, 221
Nissan, 7, 188
Norcliffe, G., 41
North, D., 45, 48, 157, 161, 245, 249
North and North East
    birth rates, 18
    capital, 79
    employment, 11–13, 14, 15, 70, 83, 163, 227, 238
    GKN in, 180–1
    innovation, 99–106
    linkages, 111–12, 113
    locality studies, 56
    motor vehicles industries, 115
    multinationals, 185, 187
    multiplants, 150, 153, 162, 163, 164
    offices, 116, 118–19, 120
    ownership in, 49
    as 'periphery', 43
    public sector industries, 200–2, 204, 206
    regional policy, 225–7, 232, 236
    revenue, 28
    small firms, 131, 135, 138–44
    takeovers, 48
    unemployment, 76, 77, 109, 135
    urban policy, 240–3, 244
    'working class', 48
    see also Yorkshire and Humberside
North West
    birth rates, 18
    capital, 81

employment, 11–13, 14, 69, 70, 164, 227, 238
GKN, 180–1
innovation, 99–102, 103
linkages, 110
motor vehicles industries, 115
multinationals, 182, 185–7, 189
multiplants, 150, 153, 157, 159, 160, 162, 164
offices, 116, 117, 118–19, 120
as 'periphery', 43
public sector industries, 200–1, 204
regional policy, 225–7, 231–2, 236
small firms, 131, 136, 137, 138–43
unemployment, 76, 77–8, 109, 162
urban policy, 241–3, 244
'working class', 48
Northern Engineering Industries, 172, 175, 178, 179
Northern Ireland
behavioural approaches, 38
Development Agency, 87
employment, 11–13, 227, 238
innovation, 99–102
linkages, 111–12
multinationals, 185–7
regional policy, 226–7
small firms, 138–44
unemployment, 75–6
urban policy, 243, 244
Nunn, S., 156, 226–7, 230

Oakey, R. P., 45, 137, 142, 144, 158
ODD (Other Designated District), 248
OECD, 80, 192, 212–13
balance of payments and, 177, 182
Office and Service Industries Scheme, 251
office machinery industry, 144, 150, 169, 170–2
office market, provincial, 117–19
oil
multinationals, 169, 170, 173, 174, 177, 198–200
North Sea, 79
price rise, 74, 108, 110, 169–70, 228
public sector, 199, 217
OPEC, 218, 220–1
Opie, R., 216
organization
corporate
changes in, 6–9
linkages and, 116–19
geography of, 40
linkages, 112
of public sector, 196, 202–3
small firms, 132–3
OSIS see Office and Service
Other Designated District, 248
output see production
'overmanning', 172
overseas
investment, 7, 81–2
see also multinationals

origin of innovations, 99–101
ownership, 46–8, 51
shift of UK multinationals, 174–81
see also international
Owen, D., 12, 14, 76, 215, 237
Owens, P. L., 11, 47, 174
ownership
changes, 45
decision-making and, 46–8, 51
material linkages and, 110–12
overseas, 46–8, 51
of public sector, 196, 202–3

Paine, S., 8
Palander, T., 24
paper and printing industries, 4–5, 31, 136
labour and capital, 5, 71, 79–82
multinationals, 170–2, 176–7
multiplants, 150, 154
Parr, J. B., 224
Parsons, G., 47
Partnership Programme, 246, 248
part-process structures, 62, 63
see also plants
Pearce, R. D., 172–3, 177
Pearson, 172, 176
Peck, F., 56, 62, 157–8, 160, 230, 231, 237
periphery see core and periphery
Perrons, D., 56
petroleum see oil
Peugeot-Citroen, 186, 190
pharmaceutical industry, 4, 6, 152
multinationals, 170–2, 173, 175, 176–8
Pilkington Brothers, 170–2, 176, 177, 178
Pires, A. da R., 2
Pite, C., 199
plants
defined, 149
location of, 41–2, 44, 45–6, 62, 63
see also multiplants
plastic industries see chemicals
Plessey, 17, 154, 171, 173, 175, 177, 201, 230
policy see government
political parties
management of economy and, 10, 215–17
regional policy and, 224, 231
Pounce, R., 156, 226
Povey, D., 128–30, 138–9
practical applications of neoclassical location theory, 30–1
Prais, S. J., 6, 8, 149, 174, 198
Pred, A. R., 39
Predöhl, A., 72
prices and demand and output, 212–13
see also competition
privatization of public sector, 198, 215, 216
process
innovations, 98, 101–2
production, 58

Product and Process Development Scheme, 217
product change, 100–1, 158
production/output
　costs, 14, 28
　decline, 2–5
　　see also 'de-industrialization'
　factors of see buildings; capital; labour; technological change
　geography of, 61–3
　labour process and relations of, 60–1
　linkages, 111
　multinationals, 167–8, 173
　national policy and, 211–14
　process, 58
　public sector, 204, 206
　small and new firms and, 129
　as social process, 58
　social structure and organization of, 57–8, 62
profits
　decline, 2, 79–80
　maximization, 29, 31, 33, 197
Pryke, R., 206
public
　facilities, location of, 33
　expenditure, 10
　　see also regional policy; urban policy
　infrastructure, 97
　sector, 196–207, 214–15
　　British Steel Corporation, 205–6
　　buildings supply, 89, 246, 249
　　different from private, 196–201
　　National Coal Board, 203–5
　　organization and ownership, 196, 202–3
　　size, 199–201, 201–2
　utilities see energy industries

R and D see research and development
Rabey, G. F., 113
Racal Electronics, 169, 173, 175, 177
Rainnie, A., 136
Randolph, W., 12
Rank Hovis McDougall, 171, 175, 178
Rate Support Grant, 249
rational abstraction, 63
rationalization, 82, 83, 160, 229
　unemployment and, 61–2, 63–4
Ravallion, M., 72
Rawstron, E. M., 29–30
RDAs (Regional Development Areas), 233
Reati, A., 80
recession, 2, 15–16, 223
　buildings and, 87–8
　investment and, 79–80
　multiplants and, 154, 160
　oil price rise and, 74
　small firms and, 132, 136
　see also 'de-industrialization'
Reckitt and Colman, 172, 176, 177
redundancies, 44, 230

multiplants, 160
small and new firms, 125, 131
see also unemployment
Reed International, 154, 171, 176, 177
Rees, J., 102
Reeve, D. E., 44, 164
regional
　differences
　　in technology, 97
　　in unemployment, 75–6
　distribution of small firms, 141–4
　impact of multinationals, 184–9
　policy, 15, 223–39
　　behavioural response, 38
　　core and periphery and, 33–4
　　labour and capital, 69, 76
　　location and, 44–5
　　measures of past successes, 226–8
　　in 1970s, 228–33
　　in 1980s, 233–6
　　structuralist attitude to, 64
　　to urban policy, 240–1
　　see also assisted areas
　reversal, 14–18
　specialization, 33
　systems, multiplants and, 161–4
　see also location; spatial
Regional Development Areas, 233
Regional Development Grants, 188, 216, 224, 240, 251
Regional Economic Planning Council and Boards, 224
Regional Employment Premium, 228
Regional Selective Assistance, 251
regions see East Anglia; East Midlands; London; North; North West; Northern Ireland; Scotland; South East; South West; Wales; West Midlands; Yorkshire and Humberside
relocation, 94
　see also urban-rural shift
research and development, 6, 16, 97–8, 103
　government and, 217
　multinationals, 172–3
　multiplants, 152–3, 162
　small and new firms, 137, 141
restructuring, 2, 14, 61–2, 82, 162
retailing, small and new firms in, 126–7, 130, 131
retirement, premature, 218
revenue as spatial variable, 26–30
Rhodes, J., 10, 14, 15
Riddell, P., 125
Robert, S., 12
Robinson, P. A., 31
Rodger, J., 131
Rolls Royce, 171–2, 173, 174, 176, 178, 198, 199, 200, 216, 242
Rothman's International, 171, 177
Rothwell, R., 45, 137, 141, 145
Rowntree Mackintosh, 172, 175, 177, 178

Royal Commission on Distribution of Industrial Population (1940), 223
Royal Dutch Shell, 169, 170–1, 173, 174, 177
Royal Mint, 199, 200
Royal Ordnance Factories, 199, 200, 201
RSA (Regional Selective Assistance), 251
rubber industries *see* chemicals
Rugman, A. M., 191
rural areas
  buildings in, 88–91, 93
  cost of living, 71–2
  shift to *see* urban-rural shift
  small and new firms in, 140

St Helen's Trust, 244
sales patterns in small firms, 137–8
Sant, M. E. C., 44, 45
Sargent, V., 132
Sayer, A., 56, 57, 60, 61, 62, 63, 106
Scammel, W. M., 218
Schmidt, C. G., 109
Schmookler, J. A., 97
Schofield, J. A., 44
Schumpeter, J. A., 96
Schwartzman, J., 31
Science Policy Research Unit (SPRU), 98
scientific instruments, 100
  see also electronics
Scotland
  birth rates, 18
  capital, 72, 81
  employment, 11–13, 69, 70, 71, 72, 163, 227, 238
  GKN in, 180–1
  innovation, 99–102, 106
  linkages, 111–12
  motor vehicle industries, 115
  multinationals, 182, 183, 185–8, 189
  multiplants, 153, 159, 163, 164
  public sector industries, 200–1, 204, 205
  regional policy, 225–7, 231–2, 233, 236
  small firms, 138–44
  technology, 99–102, 106
  unemployment, 76, 77
  urban policy, 240–1, 243, 244, 250–6
Scott, A. J., 71
Scott, M., 136, 137
Scott-Ward, J., 114, 115
Scottish Co-operative Development, 252
Scottish Development Agency, 87, 89, 164, 199, 251–6
Scottish Industrial Estates, 252
Scottish Special Housing Association, 255
SDAs *see* Special Development Areas
Searjeant, G., 5
secondary sector unemployment, 75–6, 83–4
Seddon Atkinson, 179, 186
segmentation of economy, 50–1
  see also large firms; small firms

self-employment, 73, 128–9, 131, 132
Seneschall, M., 6, 7, 8, 9
sensitivity of local unemployment, 77
service industries, 52
  employment, 213
  growth, 73
  ignored in research, 62
  small and new firms in, 126–7, 129–31, 144–5
  women in, 71, 73
  see also linkages and regional development
Sharp, M., 7
shift-share analysis, 63, 64
shipbuilding, 171, 173, 199, 200
  decline, 2, 4–5, 16, 113–14
Short Brothers, 199, 200
Short, J., 199, 200, 237
Shutt, J., 131, 164
significant innovations, 98–9
Silverman, D., 40
Singh, A., 192
size of firms *see* large; small
skills, 97
  small and new firms and, 135–6, 140
Slowe, P. M., 233
Small Business Division, 253
Small Engineering Firms Investment Scheme, 10, 134
Small Firm Counselling Scheme, 133, 134
Small Firm Information Centres, 133
small firms, 125–48, 196
  capital, 89
  definition, 125–7
  economic impact, 134–8
  geography of, 46, 71, 138–44
  linkages, 49–51
  multiplants and, 149
  recent trends, 127–9
  revival of, 8–9, 18, 129–31
    factors in, 131–4
Small Firms Loan Guarantee, 252
Smart, M. W., 76
Smith, D. M., 20, 57, 76, 197
  on neoclassical location theory, 23–37
Smith, I. J., 6, 7, 47, 48, 49, 152, 160, 161, 163, 164, 186, 189
Smyth, D. J., 173–4
social process, production as, 58
social structure and organization of production, 57–8, 62
Soete, L., 3
South East, 31
  capital, 79
  climate, 17
  'core' functions (R & D, head offices, etc.), 8, 17, 47–8, 57, 103, 104, 152, 189, 202–
  employment, 11–13, 14, 15, 69, 70, 71, 162–3, 227, 238
  GKN in, 180–1

'information rich' see 'core' above; innovation below
innovation, 9, 16, 17, 56, 99–106
locality studies, 56, 65
M4 corridor, 56, 186, 189
motor vehicle industries, 114–15
multinationals, 182, 185–7, 189, 190
multiplants, 47, 152–3, 157, 160, 162, 163
new firms, 18
offices, 117–18
overseas investment, 47
public sector industries, 200–2, 203
regional policy, 227
revenue, 28
small firms, 10, 18, 130, 131, 136, 138–44
takeovers and mergers, 48
unemployment, 75–6, 77, 78, 109
urban policy, 240–3
see also London
South West
climate, 17
employment, 11–13, 14, 15, 70, 227, 238
GKN in, 180–1
innovation, 16, 99–102
locality studies, 50
motor vehicles industry, 115
multinationals, 185–6
multiplants, 153, 157
public sector industries, 200–1, 204
regional policy, 225–7, 232
small firms, 10, 18, 138–43
unemployment, 76, 77–8
Spain, 7, 128
multinationals, 186–7, 190
spatial
change and behavioural analysis, 40
distribution
of multiplants, 151–3
of small firms, 138–44
division of labour, new, 162–4
growth of multiplants, 154–5, 156–7
interaction of cost and revenue, 29–30
variable, revenue as, 26–9
see also location; regional
Special Areas Act (1934), 223
Special Development Areas, 226, 234, 237
Special Temporary Employment Programme, 217, 254
specialization
functional, 8, 47–8, 62
regional, 33
Spence, N. A., 76, 77, 231
Sraffa, P., 71
Standard Industrial Classification, 102
Stanworth, J., 128, 133, 136
state see government
statistical definition of small firms, 126–7
Steed, G. P. F., 17, 38, 109, 115
steel industry see British Steel

STEP see Special Temporary Employment
Stoney, P. J. M., 114
Storey, D., 6, 18, 50, 129–30, 131, 134–5, 139–41
Storper, M., 56, 57, 59
structural approaches to industrial location, 36, 52, 56–66
labour, 59
process and relations of production, 60–1
methodological issues, 63–4
national and international context, 59
policy considerations, 64–5
production, 58
geography of, 61–3
labour and, 60–1
social process, production as, 58
social structure and organization of production, 57–8, 62
structure and outcome, 59
structural unemployment, 74–5
subsidiaries, small firms as, 129, 132
substantial innovations, 98–9
substitution, 72
'successful' small firms, 141–4
sunk capital, 71
Support for Innovation Scheme, 10
Swales, J. K., 44
Sweden, 38, 128
multinationals, 182, 183, 187, 193
Switzerland: multinationals, 182, 183, 187
Symonds, J. S., 224
systems approach, 57

takeovers and mergers, 40, 46, 48
tariffs
barriers, 215
eliminated see European Economic Community
taxation, 87
avoidance, 128
incentives and small firms, 133–4
restrictive, 108
VAT, 129–30, 133, 138–9, 211, 216
Taylor, J., 32, 77
Taylor, M., 6
on environment of firms, 40–2, 48, 49
on labour and capital, 60, 71
on linkages, 109–11
on location, 31, 52, 57
on multiplants, 151, 160, 161, 163
on segmented economy, 49, 63
on small firms, 49–50, 149
technological change, 2–4, 16–17, 69, 96–107
definition of, 97–8
employment and, 105–6
research and development, 103
urban and regional contrast, 104–5
see also electronics; innovations
technological definition of success, 141–2
technology
information, 97

linkages, 111–12
transfer, 189–90
see also innovations
textiles see clothing
Third World, 8, 237
trade, 5, 218, 219–21
Thirlwall, A. P., 2, 77
Thomson, A. W. J., 45
Thompson, C., 11
Thomson, L., 228
Thorn-EMI, 154, 171, 173, 175, 177, 178
Thrift, N., 52
on labour, 60
on linkages, 109, 112, 115
on location, 52, 57
on multiplants, 151
on segmented economy, 49, 63
on small firms, 49–50, 149
Thwaites, A. T., 3, 17, 45, 67–8, 142, 144
on technological change, 96–107
timber and furniture industries, 136
labour and capital, 5, 79–82
multinationals, 170–2
multiplants, 150
tobacco industry see British American
Todd, D., 199
Tomkins, C., 163
Tootal, 176, 177
Tornqvist, G., 31
Townroe, P., 15, 38, 44, 45, 46, 71, 109, 110, 158
Townsend, A., 14
on regional policy, 223–39
Townsend, J., 56
on innovation, 3, 98–9
on linkages, 113
on multiplants, 154, 157–8, 160, 163
on production change, 62
on recession and unemployment, 13, 44, 62, 160, 163
on regional differences, 13–14
on urban policy, 76, 241
trade see under international
Training Opportunities Scheme, 254
transport costs, 69
labour and, 72
linkages and, 110
location and, 24–6, 28, 31, 33
transport industries, 126–7, 150, 199
see also motor vehicles
travel-to-work areas, 233
'triangle, locational', 23–5, 36
Tube Investments, 172, 175, 176, 178, 179
Turner, D., 244
turnover in small and new firms, 126, 127, 129
Turok, I., 249
Tyler, P., 10, 14, 15, 80, 231

UDCs see Urban Development Corporations
UDG see Urban Development Grant

Ultramar, 170–1, 173, 177
unemployment, 109
growth, 213, 217
increase in, 2, 72–3, 75, 83
local, 77–8
multinationals, 167
multiplants, 153–4, 155, 158
regional policy and, 223–4, 227–9, 231, 234, 236
remedies, 217–18
secondary sector, 75–6, 83–4
small and new firms and, 131, 132, 135, 139
structural approach to, 56, 60, 61–4
types of, 74–5
urban policy and, 245–6, 248, 250–1
urban size and, 71
vacancies and, 74–5
see also employment; labour
Unilever, 154, 164, 169, 171, 174, 177
United Biscuits, 171, 175
United States
industry restructured, 172
innovations, 100, 102
investment in, 8
investment in UK, 47, 167, 168, 169, 184
labour, 70, 71
linkages, 109
motor vehicle industries, 180
multinationals, 7, 173, 179–80, 182, 184, 186–91, 193–4
small firms, 128, 134, 137
trade 220–1
Unlisted Securities Market (USM), 133, 143
UP (Urban Programme), 248–9
urban areas
buildings, 43–4, 45–6, 88–91, 93–4
cost of living, 71–2
innovation, 104–5
labour markets, 71
multiplants in, 161–4
unemployment, 71–2, 76
Urban Development Corporations, 242, 244
Urban Development Grant Programme, 242, 246
urban policy, 240–57
Clydeside, 250–6
labour and capital, 69
location and, 45
London, 245–50
problem definition, 241–2
from regional policy, 240–1
structuralist attitude to, 64
Urban Programme, 248–9
urban-rural shift, 10–12, 14–15, 57, 58, 64
labour and capital, 72, 83
multiplants and, 161–2
Urry, J., 56, 59
USSR, 168, 183

Valente, J., 246
Value Added Tax, 129–30, 133, 138–9, 211, 216

Index

Van Duijin, J. J., 3
variable-cost model, generalized, 25–6
VAT *see* Value Added Tax
venture capital, 141, 144
Vickers, 172, 175, 176, 178, 179
Vipond, J., 71

Wachter, M. L., 83
wages, 70, 72, 83, 173
Wales
  capital, 79
  Development Agency, 87, 89, 164, 199
  employment, 11–13, 14, 15, 70, 83, 163, 227, 230, 238
  GKN in, 180–1
  innovation, 56, 99–102, 106
  motor vehicle industries, 115
  multinationals, 185–7
  multiplants, 153, 160, 163, 164
  public sector industries, 200–2, 204
  regional policy, 225–7, 230, 232-2, 236
  small firms, 138–43
  urban policy, 243, 244
Walker, R., 56, 59
Walker, S., 50, 71, 126
Ward, M. F., 6
Warren. K., 205
Watkins, D., 130
Watt, H. D., 6, 18, 45, 46, 47, 123, 126, 182, 190, 198 on multiplants, 149–66
Weber, A., 23–6, 29
welfare formulations of neoclassical location theory, 32–5
Wellbelove, D., 245
Wellcome Foundation, 172, 173, 174, 175, 177, 178
Wells, J. D., 220
West Midlands
  capital, 72
  core functions, 43, 47
  employment, 11–13, 14, 69, 70, 71, 72, 83, 227, 238
  GKN in, 180–1
  innovation, 99–102
  linkages, 110, 111
  locality studies, 56
  motor vehicles industries, 114–15
  multinationals, 185–6, 189, 190
  multiplants, 152, 153, 157, 159, 160, 162
  offices, 116
  public sector industries, 200–2
  regional policy, 227, 229, 231, 236
  small firms, 138–43
  urban policy, 241–3, 244
West Germany
  industry restructured, 172
  innovation, 3, 102
  investment in UK, 167, 183
  labour costs, 219
  motor vehicle industry, 7, 180
  multinationals, 173, 182, 183, 186–7, 188, 192–3
  small firms, 128
  trade, 218
Westaway, J., 47, 48, 58
Whitbread, 154, 157–8, 172, 178
White, R. R., 17
Whitelegg, J., 44
Whitting, G., 64
Whittington, R., 131
wholesale, small and new firms in, 126–7, 129, 130
Whyatt, A., 132
Williams, S., 215
Wilson, B., 189
Wilson, Sir H. (now Lord) and Committee, 125–8
Wilson, P., 134
women
  employment, 72–3, 153
  labour costs in electronics, 31
  unemployment, 71, 72–3
Wood, P. A., 21–2, 57, 71, 110, 111, 116, 234
  on behavioural approach to location studies, 38–55
Work Experience Programme, 217
working environment, small firms, 136
Wright, E. O., 58
Wright, M., 132

YOP (Youth Opportunities programme), 217
Yorkshire and Humberside
  employment, 11–13, 70, 227, 238
  GKN in, 180–1
  innovations, 99–102
  linkages, 113
  locality studies, 56, 65
  motor vehicles industries, 115
  multinationals, 185–6, 189
  multiplants, 150, 153, 157, 160
  offices, 116, 117–18
  public sector industries, 200–2, 203, 204
  regional policy, 225–7, 231–2, 236
  small firms, 138–43
  unemployment, 76, 77, 109
  urban policy, 242–3
  see also North
Young, S., 182, 231, 252, 253
Youth Opportunities Programme, 217
Youth Training Scheme (YTS), 217–18, 244, 254

Zegveld, W., 137, 141, 145
Zimbalist, A., 60